Indoor Air Quality Solutions
FOR
STATIONARY ENGINEERS

AMERICAN TECHNICAL PUBLISHERS, INC.
HOMEWOOD, ILLINOIS 60430-4600

American Technical Publishers, Inc., Editorial Staff

Editor in Chief:
    Jonathan F. Gosse
Vice President—Production:
    Peter A. Zurlis
Art Manager:
    James M. Clarke
Technical Editor:
    Russell G. Burris
Copy Editor:
    Diane J. Weidner
Cover Design:
    James M. Clarke

Illustration/Layout:
    Mark S. Maxwell
    Samuel T. Tucker
    Thomas E. Zabinski
Multimedia Coordinator:
    Carl R. Hansen
CD-ROM Development:
    Gretje Dahl
    Daniel Kundrat
    Nicole S. Polak

 This book is printed on 10% recycled paper.

ACKNOWLEDGMENTS

Indoor Air Quality Solutions for Stationary Engineers is a collaborative effort between the International Union of Operating Engineers –National Training Fund (IUOE-NTF) and American Technical Publishers, Incorporated (ATP). This book was developed by and for the stationary engineers responsible for indoor environments across the United States and Canada and is based on the 1995 publication, *Stationary Engineers: The Indoor Air Quality Solution*.

Since Congress enacted the Clean Air Act of 1990, the IUOE-NTF has been a leader in Indoor Air Quality (IAQ) training. IAQ-trained stationary engineers are widely recognized for making a positive impact on indoor air quality and currently maintain over two billion square feet of commercial building space. With a continued emphasis on advanced skill training, IUOE stationary engineers are ready to meet the challenges of increasing building control sophistication and the demand for energy efficiency.

This book would not have been possible without the contributions of the IUOE-NTF indoor air quality review committee, the IUOE headquarters staff, and independent reviewers. This joint effort will make this book the new premier resource for those responsible for keeping the building air we breathe safe and healthy.

The International Union of Operating Engineers–National Training Fund and the publisher are grateful to the following companies and organizations for providing photographs, information, and technical assistance:

Air Zone, Inc.	Fire-Lite Alarms	Midsun Group, Inc.
Carrier Corporation	FirstAidStore.com	Saftronics, Inc.
Cintas Corporation	Fluke Corporation	Saylor-Beall Manufacturing Company
Danfoss Global Group	GrayWolf Sensing Solutions, Ltd	Siemens
Datastream Systems, Inc	Honeywell International, Inc.	SKC Gulf Coast, Inc.
DPSI (DP Solutions, Inc.)	Jackson Systems, LLC	The Staplex Company
DuctSox	KMC Controls	TSI Incorporated
Family Safety Products, Inc.	Lab Safety Supply, Inc.	

Technical content development and review:
Ron Auvil

The International Union of Operating Engineers–National Training Fund would also like to thank the Environmental Protection Agency for assistance in technical review and support of the IUOE-NTF training initiative. This document does not necessarily reflect the views or policies of the U.S. government

TABLE OF CONTENTS

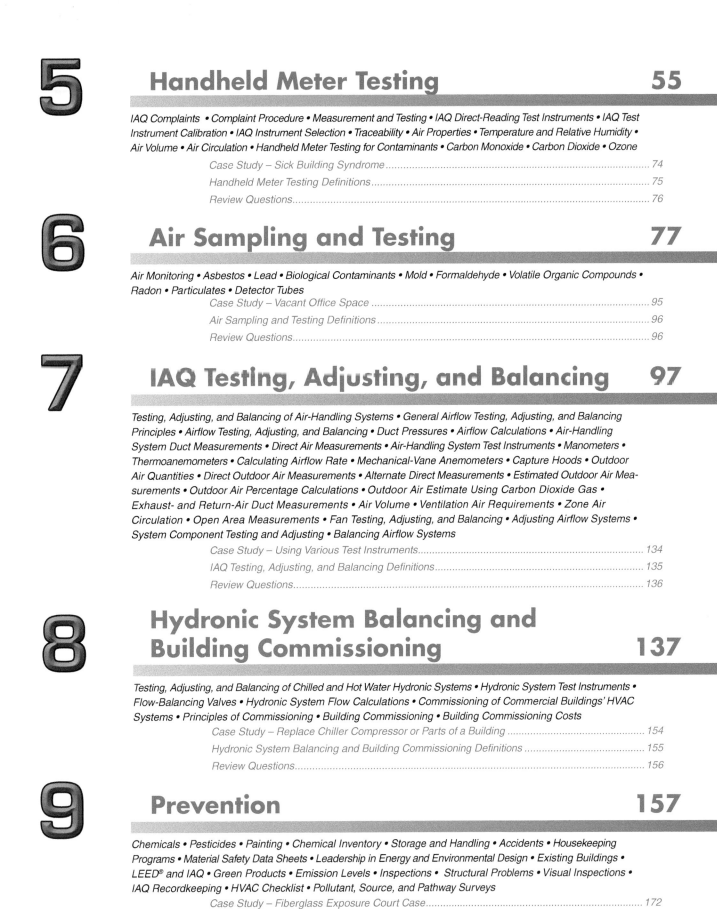

5 Handheld Meter Testing 55

IAQ Complaints • Complaint Procedure • Measurement and Testing • IAQ Direct-Reading Test Instruments • IAQ Test Instrument Calibration • IAQ Instrument Selection • Traceability • Air Properties • Temperature and Relative Humidity • Air Volume • Air Circulation • Handheld Meter Testing for Contaminants • Carbon Monoxide • Carbon Dioxide • Ozone

6 Air Sampling and Testing 77

Air Monitoring • Asbestos • Lead • Biological Contaminants • Mold • Formaldehyde • Volatile Organic Compounds • Radon • Particulates • Detector Tubes

7 IAQ Testing, Adjusting, and Balancing 97

Testing, Adjusting, and Balancing of Air-Handling Systems • General Airflow Testing, Adjusting, and Balancing Principles • Airflow Testing, Adjusting, and Balancing • Duct Pressures • Airflow Calculations • Air-Handling System Duct Measurements • Direct Air Measurements • Air-Handling System Test Instruments • Manometers • Thermoanemometers • Calculating Airflow Rate • Mechanical-Vane Anemometers • Capture Hoods • Outdoor Air Quantities • Direct Outdoor Air Measurements • Alternate Direct Measurements • Estimated Outdoor Air Measurements • Outdoor Air Percentage Calculations • Outdoor Air Estimate Using Carbon Dioxide Gas • Exhaust- and Return-Air Duct Measurements • Air Volume • Ventilation Air Requirements • Zone Air Circulation • Open Area Measurements • Fan Testing, Adjusting, and Balancing • Adjusting Airflow Systems • System Component Testing and Adjusting • Balancing Airflow Systems

8 Hydronic System Balancing and Building Commissioning 137

Testing, Adjusting, and Balancing of Chilled and Hot Water Hydronic Systems • Hydronic System Test Instruments • Flow-Balancing Valves • Hydronic System Flow Calculations • Commissioning of Commercial Buildings' HVAC Systems • Principles of Commissioning • Building Commissioning • Building Commissioning Costs

9 Prevention 157

Chemicals • Pesticides • Painting • Chemical Inventory • Storage and Handling • Accidents • Housekeeping Programs • Material Safety Data Sheets • Leadership in Energy and Environmental Design • Existing Buildings • LEED® and IAQ • Green Products • Emission Levels • Inspections • Structural Problems • Visual Inspections • IAQ Recordkeeping • HVAC Checklist • Pollutant, Source, and Pathway Surveys

10 Preventive Maintenance 175

Overview of Maintenance Types • Predictive Maintenance • Proactive Maintenance • Reliability-Centered Maintenance • Enterprise Asset Management • Run-To-Fail Maintenance • Preventive Maintenance • Fundamentals of Preventive Maintenance Programs • Economic Benefits • Resolving Problems • Preventive Maintenance Types • Scheduled Maintenance • Unscheduled Maintenance • Example of a PM Schedule • Establishing a New PM System • PM Organizational Tools • Do You Need a CMMS? • Preventive Maintenance Surveys • Equipment History Records • Operating Manuals and System Prints • PM Charts • PM Work Orders • Master Schedule • PM System Operation • Paperless Systems • Construction and Renovation • Safe Practices • Sink Effect • Chemical Emissions

11 HVAC Air Systems 199

HVAC Systems • HVAC System Characteristics • Common HVAC System Designs • Outside-Air Systems • Mixed-Air Systems • Economizer Systems • Air-Handling Unit Components • Fans • Mixing Plenum and Dampers • Example: Summer Operation • Example: Winter Operation • Example: CO_2 Operation • Air-Cleaning Methods • Electronic Air Cleaners • Ultraviolet Air-Cleaners • Carbon Filters • Filters • Ventilation Air • HVAC System Ductwork • Cooling Coils • Heating Coils • Common Air-Handling Systems • Single-Zone Systems • Multizone Systems • Dual-Duct Systems • Reheat Systems • Variable-Air-Volume Systems • Facility Pressures • Odors • Energy Conservation Techniques • All-Water Systems • Air and Water Systems • Electric Radiant Heating Systems

12 HVAC Controls 239

Pneumatic Control Systems • Air Compressor Stations • Air Compressors • Auxiliary Components • Transmitters and Controllers • Pneumatic Thermostats • Limit Thermostats • Pneumatic Humidistats • Pneumatic Pressurestats • Receiver Controllers • Auxiliary Components • Switching Relays • Minimum-Position Relays • Positioners • Electric/Pneumatic Switches • Pneumatic/Electric Switches • Controlled Devices • Actuators • Dampers • Valves

13 Building Automation Systems for IAQ 261

Building Automation Systems • Distributed Direct Digital Control Systems • Building Automation System Controllers • Application-Specific Controllers • Network Communication Modules • Operator Interface Methods • On-Site Operator Interface Methods • Off-Site Operator Interface Methods • Server Data • Building Automation System Input Devices • Analog Input Devices • Digital Input Devices • Building Automation System Output Components • Analog Output Components • Digital Output Components • Direct Digital Control Strategies • Closed-Loop Control • Open-Loop Control • Direct Digital Control Features • Setpoint Control • Reset Control • Low-Limit Control • High-Limit Control • High/Low Signal Select • Direct Digital Control Algorithms • Proportional Control Algorithms • Integration Control Algorithms • Derivative Control Algorithms • Adaptive Control Algorithms • IAQ Strategies and Energy Considerations • IAQ Strategies for Air-Handling Systems • Carbon-Dioxide-Controlled Ventilation • Air-Side Economizers • Night Precooling

14 Homeland Security 293

Homeland Security • Background • Building Layout • Specific CBR Attack Prevention Recommendations • CBR Preventive Activities to Avoid • Building Physical Security • Securing Outdoor-Air Intakes • Ventilation and Filtration • Maintenance, Administration, and Testing • Emergency Plans, Policies, and Procedures • Wi-Fi • Conclusions on Specific Recommendations

15 Troubleshooting and Mitigating IAQ Problems 317

Troubleshooting and Mitigating IAQ Problems • Step-by-Step Complaint Investigation Procedures • IAQ Complaint Solutions • Overview of Non-IAQ Problems • Work-Related Complaints • Indoor Air Quality Control Strategies • Contaminant Source Control • Ventilation • Modification of HVAC Systems • Air Cleaning Exposure Control • Mitigation Strategies • Remedies for Complaints Not Attributed to Poor Air Quality • Judging Proposed Mitigation Designs and Their Success • Permanence • Operating Principles • Degree to Which a Strategy Fits the Job • Ability to Institutionalize the Solution • Durability • Installation and Operating Costs • Conformity to Codes • Judging the Success of a Mitigation Effort • Sample Case Study Problems and Solutions • Hiring Professionals to Solve an IAQ Problem • Hiring Professionals with an Effective Approach • Selection Criteria

CD-ROM Contents

- Quick Quizzes®
- Illustrated Glossary
- Flash Cards

- Test Tool Procedures
- IAQ Survey Checklists
- Virtual Meters

- IAQ Equipment Forms
- Media Clips
- ATPeResources.com

Appendix 351

Glossary 373

Index 381

INTRODUCTION

Indoor Air Quality Solutions for Stationary Engineers is a unique learning tool that outlines typical ways to monitor and control indoor air quality (IAQ) in institutional and commercial facilities. This comprehensive textbook focuses on several key factors such as safety, air contaminants, common test tool applications, and the HVAC systems used for controlling indoor air quality. *Indoor Air Quality Solutions for Stationary Engineers* provides learners with the capabilities and knowledge needed to monitor, test, and mitigate IAQ problems.

The many full-color illustrations serve as visual aids that enhance concepts presented in the text. *Indoor Air Quality Solutions for Stationary Engineers* includes chapter review questions, IAQ scenarios based on chapter content, and procedures for the various test instruments used to monitor IAQ and to test for contaminants. *Indoor Air Quality Solutions for Stationary Engineers* provides a solid foundation of knowledge for maintaining a healthy indoor environment for facility occupants.

The interactive CD-ROM is a self-study aid that complements the content in the book and includes the following features:

- Quick Quizzes® that provide an interactive review of topics covered in the chapters, with 10 questions for each chapter
- An Illustrated Glossary that provides terms and definitions as well as visual references to selected terms
- Flash Cards that facilitate a review of glossary terms and test tools used for IAQ
- Test Tool Procedures that explain specific instrument usage
- An IAQ Survey Checklist that provides helpful forms used to assess IAQ conditions
- Virtual Meters for IAQ and related fields
- Indoor Air Quality Equipment Forms that provide a checklist for testing equipment
- A Print Library that provides three examples of HVAC systems prints
- Media Clips that illustrate principles discussed in the book using video clips and animations
- Access to ATPeResources.com, an on-line source providing a variety of instructional resources

To obtain information about related products from American Technical Publishers, visit our web site at www.go2atp.com or log onto www.ATPeResources.com.

The Publisher

FEATURES

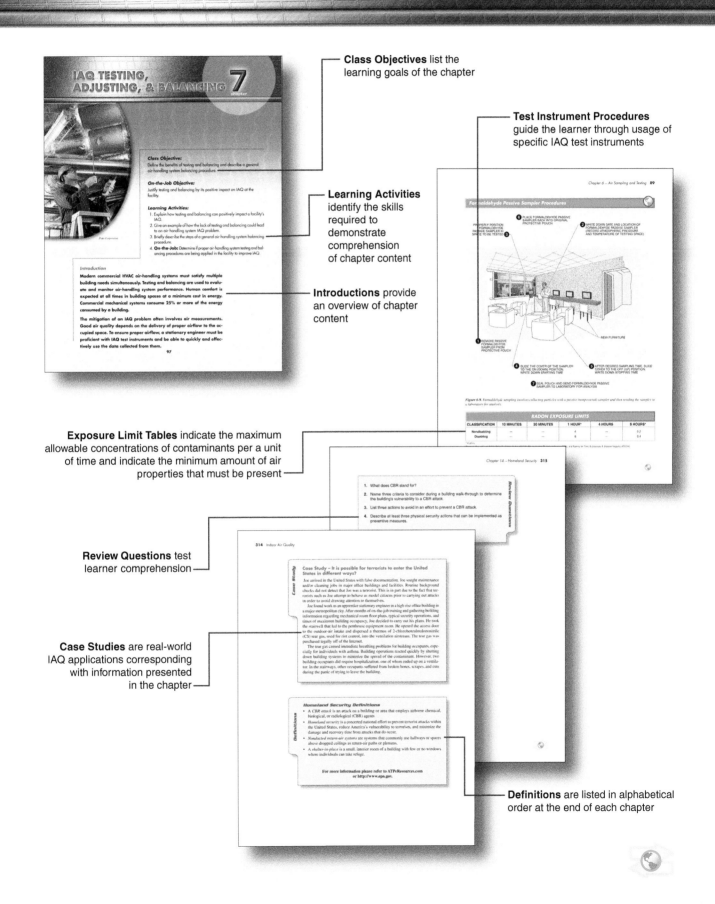

Class Objectives list the learning goals of the chapter

Test Instrument Procedures guide the learner through usage of specific IAQ test instruments

Learning Activities identify the skills required to demonstrate comprehension of chapter content

Introductions provide an overview of chapter content

Exposure Limit Tables indicate the maximum allowable concentrations of contaminants per a unit of time and indicate the minimum amount of air properties that must be present

Review Questions test learner comprehension

Case Studies are real-world IAQ applications corresponding with information presented in the chapter

Definitions are listed in alphabetical order at the end of each chapter

INDOOR AIR QUALITY

Fluke Corporation

Class Objective
Define indoor air quality (IAQ) and describe some of the effects of poor IAQ.

On-the-Job Objective
Establish a written procedure for handling IAQ complaints. Develop a strategy for handling communications between occupants and the staff of the facility. Only stationary engineers or trained IAQ management staff should handle this type of complaint.

Learning Activities:
1. Define indoor air quality (IAQ).
2. Describe why is it important to treat any IAQ complaint seriously.
3. Identify two effects of poor IAQ in schools and commercial buildings.
4. Identify three components of an effective IAQ communications program.
5. Explain the role of a stationary engineer regarding IAQ.
6. **On-the-Job:** Describe the procedures used in your building for communicating to building occupants during an IAQ investigation.

Introduction

Indoor air quality (IAQ) is determined by the heating, ventilating, and air conditioning (HVAC) systems used in buildings to provide comfort for the occupants of the buildings. A feeling of comfort results when temperature, humidity, circulation, filtration, and ventilation are all at their proper levels. IAQ problems can result in adverse health effects for building occupants, such as dizziness, scratchy throat, dry skin, and nausea, all of which create a lower level of comfort. Building occupants need to know that building stationary engineers will work diligently to resolve any and all IAQ problems.

INDOOR AIR QUALITY

Indoor air quality (IAQ) is the comfort level of indoor air based on a combination of factors, including temperature, humidity, airflow, and contaminant levels. *Good IAQ* is when the indoor air of a building is free of harmful amounts of chemicals and particles, and the temperature and humidity of the air is comfortable. *Poor IAQ* can occur when the indoor air temperature, humidity, chemicals, or contaminants rise to harmful or uncomfortable levels. Poor IAQ affects the health of building occupants and can lead to serious health problems as more time is spent in the undesirable air.

There have been documented instances where the air within a building was more contaminated than the outdoor air in the largest industrialized cities. Research also indicates that many people spend approximately 90% of their time indoors. This group is at greater risk from exposure to poor IAQ in comparison to exposure to outdoor air pollution. **See Figure 1-1.** In other words, people who are exposed to indoor air pollutants for the longest periods of time are most susceptible to the effects of indoor air pollution. Within this group, the most highly susceptible populations to poor IAQ include the young, the elderly, and the chronically ill, especially those suffering from respiratory or cardiovascular diseases.

HISTORY OF INDOOR AIR QUALITY

Early commercial building designs and construction allowed a large percentage of outdoor air to infiltrate buildings. Building openings, such as doors, windows, and openings through wall piping and sheet-metal work, were rarely tightly sealed. In addition, building designs included windows that were not permanently sealed and could be opened by building occupants. Non-sealed window air conditioning units were also being used. At that point in time, paying higher energy bills was less expensive than spending the required funds for improvements, such as sealing buildings against outdoor air infiltration.

During the 1970s, the cost of energy began to increase dramatically. In response, building owners were faced with the need to conserve energy to minimize costs. This affected the entire range of building design, construction, operation, and control. Building designs and the operation of heating, ventilation, and air conditioning (HVAC) systems were updated to conserve energy costs. Design changes included better insulation, permanently sealed windows, and tighter building seals to minimize air leakage. **See Figure 1-2.** Over time, building codes changed and energy-efficient buildings and operations became common practice.

TYPICAL STATE OUTDOOR AIR POLLUTION EPISODE LEVELS				
POLLUTANT	**ADVISORY**	**YELLOW ALERT**	**RED ALERT**	**EMERGENCY**
Particulate Matter*	2-hour 420	24-hour 350	24-hour 420	24-hour 500
Sulfur Dioxide[†]	2-hour 0.30	4-hour 0.30	4-hour 0.35	4-hour 0.40
Carbon Dioxide[†]	2-hour 30	8-hour 15	8-hour 30	8-hour 40
Nitrogen Dioxide[†]	2-hour 0.40	1-hour 0.60 24-hour 0.15	1-hour 1.20 24-hour 0.30	1-hour 1.60 24-hour 0.40
Ozone[†]	1-hour 0.12	1-hour 0.20	1-hour 0.30	1-hour 0.50

* measured in micrograms per cubic meter (ug/m²)
[†] measured in parts per million (ppm)

Source: Environmental Protection Agency (EPA)

Figure 1-1. *People are usually notified in some manner (by TV, radio, or internet) as to the quality of outdoor air, but the quality of indoor air is usually unknown by building occupants.*

Energy-Efficient Building Design

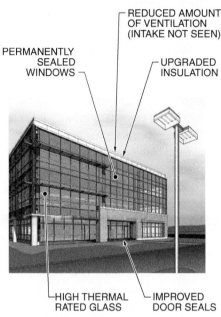

PERMANENTLY SEALED WINDOWS

REDUCED AMOUNT OF VENTILATION (INTAKE NOT SEEN)

UPGRADED INSULATION

HIGH THERMAL RATED GLASS

IMPROVED DOOR SEALS

Figure 1-2. Since the 1970s, buildings are more tightly sealed, reducing heating and air conditioning costs.

In many buildings, the HVAC system became the sole source of outdoor air. In addition, a lower percentage of outdoor air (ventilation) was being used by HVAC systems to conserve energy and to lower the operating costs. Reducing the amount of ventilation meant increasing the use of recirculated indoor air. The possible health consequences of these actions were given little thought because the primary emphasis was on lowering energy and operation costs.

Several years later, an IAQ case arose concerning American Legion members attending a convention at a Philadelphia hotel. As the American Legion members breathed the hotel's indoor air, they inhaled tiny bacteria. The bacteria lodged in their lungs. After the convention, 182 people became sick and 29 died. The illness became known as Legionnaires' disease. Legionnaires' disease is caused by bacteria known as Legionella. It was believed that the most probable cause for the outbreak was bacteria growing in contaminated water in the hotel's cooling system. **See Figure 1-3.** The bacteria could potentially enter the HVAC system through the outdoor-air intake and then be dispersed through the building. The Legionnaire's outbreak dramatically raised IAQ awareness among building operating personnel, building owners, and the general public.

Contaminated Cooling Water Systems

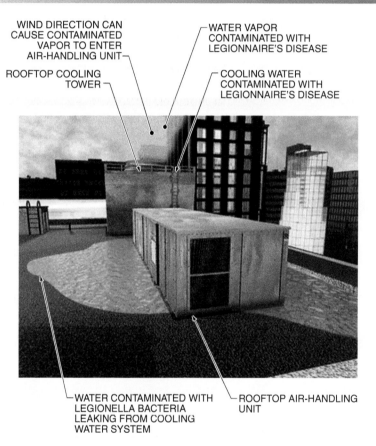

WIND DIRECTION CAN CAUSE CONTAMINATED VAPOR TO ENTER AIR-HANDLING UNIT

ROOFTOP COOLING TOWER

WATER VAPOR CONTAMINATED WITH LEGIONNAIRE'S DISEASE

COOLING WATER CONTAMINATED WITH LEGIONNAIRE'S DISEASE

WATER CONTAMINATED WITH LEGIONELLA BACTERIA LEAKING FROM COOLING WATER SYSTEM

ROOFTOP AIR-HANDLING UNIT

Figure 1-3. Cooling towers and the area around a cooling tower may be a source of IAQ problems as bacteria grows in the moist conditions.

IAQ Fact

According to the Department of Energy (DOE), 10 CFR Parts 434 and 435, Energy Code for New Federal, Commercial, and Multi-Family High-Rise Residential Buildings—Final Rule, Section 402.2.1, Air Barrier System, a barrier against leakage shall be installed to prevent the leakage of air through the building envelope.

IAQ FACTORS

Because it is less expensive for commercial buildings to heat or cool recirculated indoor air than outdoor air, adding outdoor air is an extra expense. If a building did not add outdoor air to return air, occupants would continuously breathe recirculated indoor air. Contaminant levels would rise as contaminants would be trapped indoors without the benefit of ventilation for dilution. The air in the building eventually becomes unhealthy.

A healthy indoor environment is one in which the surroundings contribute to productivity, comfort, and a sense of health and well being. The indoor air is free from significant levels of odors, dust, and contaminants. The air also circulates to prevent stuffiness without creating drafts. The temperature and humidity levels are appropriate to the season and to the clothing and activity of the building occupants. IAQ is a constantly changing interaction of complex factors that affect the types, levels, and severity of pollutants in indoor environments. Factors affecting IAQ include pollutant and odor sources, moisture and humidity, occupant perceptions and susceptibilities, and the design, maintenance, and operation of building ventilation systems.

The quality of indoor air can vary by building, date, time of day, and season. Indoor air could potentially contain more than 3000 different contaminants. Contaminants come from cleaning products, building materials, carpet, furniture, combustion sources, photocopiers, fax machines, and certain products like perfumes and deodorants. Indoor air may contain dust, dirt, pollen, mold, bacteria, and many other types of particles and pesticides brought in from outdoors or generated indoors. **See Figure 1-4.**

Indoor Air Contaminants

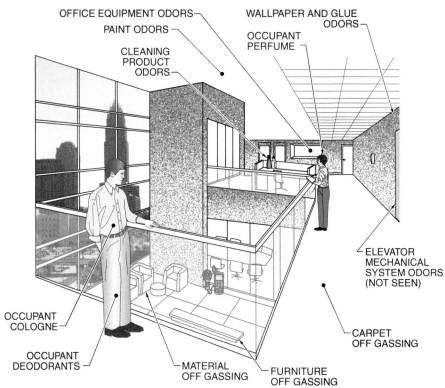

Figure 1-4. Indoor air can contain vapors from cleaning products, building materials, carpet, furniture, combustion sources, photocopiers, fax machines, and particulates like dust, dirt, pollen, mold, bacteria, and many other types of particles and pesticides brought in from outdoors or generated indoors.

Poor IAQ costs money in terms of lost productivity, direct medical and legal costs, and damage to building materials and equipment. Performance losses may be subtle and occur even when people do not complain about working conditions. Poor IAQ increases absentee rates and contributes to poorer occupant performance while at work. Approximately 150 million workdays are lost each year due to IAQ problems. The Environmental Protection Agency (EPA) estimates that productivity lost to IAQ problems costs more than $5 billion annually. Implementing IAQ recommendations from the EPA may save $75 billion to $150 billion per year for the United States economy.

Health Effects

The health effects of poor IAQ are serious. The EPA ranks poor IAQ among the top five environmental health risks faced in modern societies. In the United States, asthma leads to 2 million emergency room visits per year and 5000 deaths per year. Asthma accounted for more than 14 million missed school days in the year 2000. Poor IAQ is suspected to be one of the many causes of asthma. Rates of asthma have risen sharply over the past 30 years, particularly among children aged 5 to 14.

Schools must be especially concerned about IAQ. Twenty percent of the United States population spends their weekdays in schools. In the 1990s, studies showed that one in five of the nation's schools reported unsatisfactory indoor air quality, and one in four schools reported unsatisfactory ventilation. **See Figure 1-5.** Children are at great risk for IAQ-related illnesses because they are still growing and their immune systems are not as fully developed as those of adults.

IAQ Fact

The EPA has developed the **Indoor Air Quality (IAQ) Tools for Schools (TfS)** *program to reduce exposure to indoor environmental contaminants in schools by identifying, correcting, and preventing IAQ problems through sound indoor air quality management practices. http://www.epa.gov/iaq/schools*

SCHOOL SPACE	WINTER TEMP*	SUMMER TEMP*	VENTILATION RATE†	1ST STAGE AIR FILTER	2ND STAGE AIR FILTER
Classrooms, auditoriums, libraries, and administrative areas	72	78	15	20%-30%	60%-85%
Corridors	68	80	5	20%-30%	–
Laboratories	72	78	–	20%-30%	85%
Locker rooms and shower areas	75	–	5	20%-30%	–
Mechanical rooms	60	–	5	20%-30%	–
Industrial technologies	72	78	5	20%-30%	60%-85%
Storage	65	–	5	20%-30%	–
Restrooms	72	–	20	20%-30%	–

RECOMMENDATIONS FOR INDOOR AIR QUALITY OF EDUCATIONAL FACILITIES

* in degrees Fahrenheit
† flow rate in cubic feet per minute per person

NOTE: Corridors, mechanical rooms, industrial technology shops, and storage areas are not usually air conditioned. Shops, laboratories, and restrooms usually have additional exhaust system requirements.

Figure 1-5. Millions of Americans spend a majority of their week in schools. One in five schools report unsatisfactory indoor air quality, while one in four schools report unsatisfactory ventilation.

STATIONARY ENGINEER'S ROLE

When IAQ problems arise, the stationary engineer is often responsible for identifying the source and responding to IAQ complaints. As a result, the stationary engineer's role must involve incorporating the building's IAQ as an everyday concern.

Besides solving IAQ issues as they arise, operating an effective preventive maintenance program is another way for stationary engineers to reduce IAQ problems. HVAC systems and controls must be maintained and operated correctly to prevent the development of IAQ issues. Stationary engineers must work with facility or building occupants to solve IAQ issues before serious illnesses develop.

People Skills

There are many reasons why people are affected by IAQ problems. Indeed, often the causes of complaints are not completely understood by experts in the field. Stationary engineers work with chemicals all the time and may not understand that someone can be made sick by what they take for granted as part of their jobs.

The person complaining may be more sensitive to whatever is in the air than others working nearby. This person might serve as the first warning of problems that could affect others later. This person can be thought of as serving the same function as the canary that was often taken into coal mines as an oxygen check. When oxygen ran low, the small bird passed out before the miners were affected. This alerted them of the danger. Just as not everyone gets the measles, mumps, flu, or the cold that is making the rounds, the same is true of the effect of some indoor contaminants on individuals. **See Figure 1-6.**

Good Manners. It is important to be polite and patient. The anger and ill will the person complaining would like to direct to a boss or coworkers might be directed at the stationary engineer. It is vital to never respond rudely to a building occupant for any reason. The goodwill of building occupants will help the stationary engineer solve an IAQ problem. Sometimes just paying attention to someone's complaints and making a few changes will reduce or stop IAQ complaints.

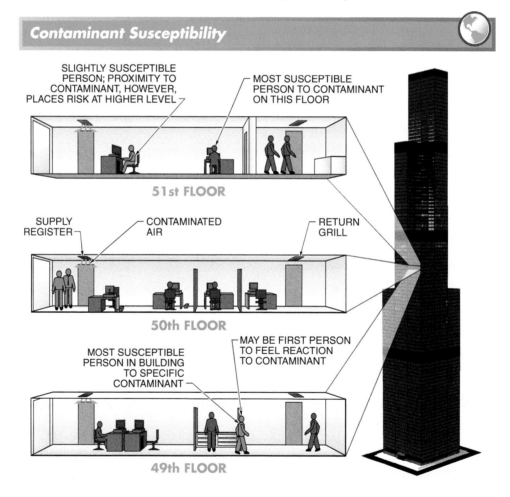

Figure 1-6. The location of building occupants, the amount of contaminants in the air, and occupants' varying susceptibility to a specific contaminant are all factors that determine which occupants are affected by the contaminant and when they feel the effects.

Communication

One of the tasks of the stationary engineer is to participate in the IAQ communication process. A good communication program encourages building occupants to improve their work environment through specific actions that help to reduce IAQ problems. **See Figure 1-7.** The following guidelines are important to consider when communicating with building occupants:

• Provide accurate information about factors that affect indoor air quality.

• Clarify the responsibilities of different parties (e.g., building management, staff, tenants, and contractors).

• Establish an effective system for logging and responding to IAQ complaints that occur.

One of the most important tasks in dealing with IAQ in a commercial building is ensuring adequate communication. Establishing and maintaining a communication system can prevent IAQ problems or resolve problems quickly when they do arise. Communication becomes more critical when there are delays in identifying and resolving IAQ problems and when serious health concerns are involved. The stationary engineer responding to an IAQ problem must convey that management is dedicated to providing a healthy and safe building, that good IAQ is an essential component of a healthful indoor environment, and that all complaints about IAQ are taken seriously.

IAQ Communication Documents from the EPA

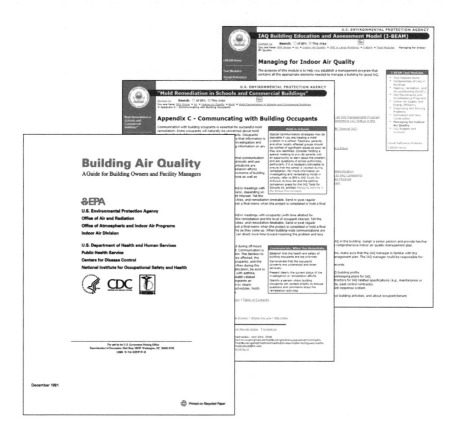

Figure 1-7. The Environmental Protection Agency (EPA) has a number of documents that address IAQ communication with building occupants.

Information that management must share in the communication process includes the proper use of building spaces, design occupancy rates, building modifications, and notification of planned activities such as construction projects. The responsibilities of building management, staff, and occupants regarding indoor air quality must be well defined. These responsibilities can be formalized by incorporating them into documents such as employee handbooks or tenant lease agreements.

Complaints

In many cases, occupants may be the first to know and alert building managers about potential indoor air quality problems. Occupant observations about patterns of symptoms or building conditions do provide valuable information. However, occupant complaints are often vague. For example, someone may complain of feeling sick or that they have noticed an unusual odor. Even when a complaint is vague, the person is usually reacting to a real problem, so complaints must be thoroughly investigated and accurately documented.

Many building owners have established procedures for responding to occupant complaints, including indoor air quality concerns. **See Figure 1-8.** Building occupants must know how to contact the responsible person(s) that can receive and respond to IAQ complaints. Ways to express complaints may include phone number contacts, posted information, and complaint forms. Tenants can also have an internal system for channeling complaints, such as through a building health and safety representative, supervisor, or facility doctor.

Complaints must be handled promptly and every incident given serious attention. Indoor air quality complaints that can be resolved quickly (e.g., harmless odors coming from an easily identified source) and that involve small numbers of building occupants can be handled with little confusion and disruption. Many commercial buildings and facilities have tracking and recordkeeping systems that can be helpful to identify specific patterns related to IAQ complaints.

Health and Safety Committee

Establishing a health and safety committee that works to promote good working conditions can prevent many IAQ problems. Health and safety committees are useful in disseminating information to building occupants and fostering a sense of cooperation. Committees that represent the different interests of the building owner, stationary engineers, facility personnel, health and safety officials, worker representatives, and tenants or other occupants have the greatest success. **See Figure 1-9.**

Water-stained ceiling tiles are usually noticed first by building tenants, not the building staff.

IAQ Fact

The National Fire Protection Agency (NFPA) has established the High-Rise Building Safety Advisory Committee. The committee has proposed changes to the NFPA 101®, Life Safety Code®, and the NFPA 5000™, Building Construction and Safety Code. New language states that the protection of high-rise buildings from terrorist attacks will require protection methods beyond those mandated by code.

Building Processing of IAQ Complaint

6 MANAGEMENT PROCESSES THE WORK ORDER TICKET ONCE IT IS GIVEN BACK. DECISIONS ON FUTURE WORK MAY BE REQUIRED.

2 MANAGEMENT OFFICE FORMALLY DOCUMENTS THE CALL AND DISPATCHES WORK ORDER TICKET TO THE ENGINEERING DEPARTMENT (WORK ORDER INCLUDES OCCUPANT NAME, LOCATION, AND DESCRIPTION OF IAQ PROBLEM).

1 BUILDING OCCUPANT REPORTS IAQ COMPLAINT TO THE BUILDING MANAGEMENT OFFICE (REPORTING MAY BE IN WRITTEN FORM OR BY PHONE CALL).

3 CHIEF ENGINEER ASSIGNS WORK ORDER TICKET (WITH ALL KNOWN INFORMATION) TO A STATIONARY ENGINEER.

4 STATIONARY ENGINEER RESPONDS (INVESTIGATES AND MITIGATES IAQ PROBLEM).

5 STATIONARY ENGINEER RECORDS THE DATES OF RESPONSE AND FINAL RESOLUTION OF THE COMPLAINT ON THE WORK ORDER TICKET, ALONG WITH THE DESCRIPTION OF WHAT WAS FOUND AND THE ACTIONS TAKEN.

7 ALL INFORMATION IS ENTERED INTO BUILDING'S DATABASE TO ALLOW MORE EFFICIENT TRACKING OF INCIDENTS AND ACTIONS, AND TO POSSIBLY BE EXAMINED AS PART OF PREVENTIVE MAINTENANCE.

Figure 1-8. Building occupants must know how to contact the responsible person(s) that can receive and respond to IAQ complaints. Complaints must be handled promptly and every incident given serious attention.

Building Health and Safety Committees

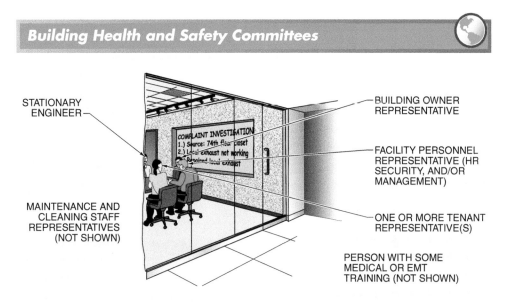

STATIONARY ENGINEER

BUILDING OWNER REPRESENTATIVE

FACILITY PERSONNEL REPRESENTATIVE (HR SECURITY, AND/OR MANAGEMENT)

MAINTENANCE AND CLEANING STAFF REPRESENTATIVES (NOT SHOWN)

ONE OR MORE TENANT REPRESENTATIVE(S)

PERSON WITH SOME MEDICAL OR EMT TRAINING (NOT SHOWN)

Figure 1-9. Health and safety committees are useful in disseminating information to building occupants and should be made up of building owner personnel, stationary engineers, facility personnel, health and safety officials, and tenant representatives.

Widespread or potentially serious problems should be discussed with the health and safety committee. When no health and safety committee exists, building management should consider forming one, or should consider establishing a joint management-tenant IAQ task force. Productive relations are enhanced when building occupants are given basic information during the investigation and remediation process. Potential critics of a proposed IAQ solution can become allies when they are a part of the problem-solving process and are better educated about IAQ and building operations. Building management and staff should be encouraged to talk directly with occupants at the time a complaint occurs, as well as during the complaint investigation.

Release of Information

Investigative activities should be explained with the known facts in order to dispel rumors and suspicions. Notices or memoranda can be delivered directly to selected occupants through mailings or posted in general-use areas. **See Figure 1-10.** Newsletter articles or other established communication channels could also be used to keep building occupants up to date.

Problems can arise from saying either too little or too much. A premature release of information before all data has been gathered can produce confusion, frustration, and mistrust. Similar problems can result when no information is released at all, leading to occupants assuming the worst-case scenario. Also, when progress reports are not given, people may assume that no progress is being made or that something terrible is happening in the building. The best practice is to obtain the consent of the facility manager, building owner, and/or legal counsel before releasing information. Management should always be factual and to the point when presenting information such as the following:

- definition of the areas of the building where the IAQ complaints are occurring based on the location and distribution of complaints, which may be revised as the investigation progresses

- progress of the complaint investigation, including the kind of information being gathered and ways that occupants can help

- factors that have been evaluated and found not to be causing or contributing to the IAQ problem

Complaint Investigation Postings

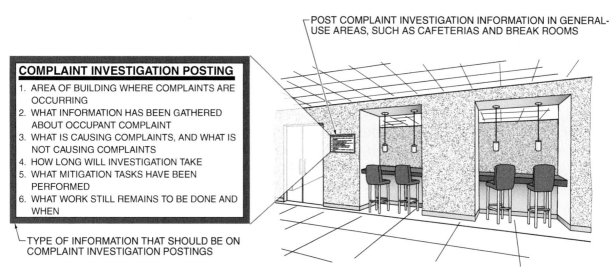

POST COMPLAINT INVESTIGATION INFORMATION IN GENERAL-USE AREAS, SUCH AS CAFETERIAS AND BREAK ROOMS

COMPLAINT INVESTIGATION POSTING
1. AREA OF BUILDING WHERE COMPLAINTS ARE OCCURRING
2. WHAT INFORMATION HAS BEEN GATHERED ABOUT OCCUPANT COMPLAINT
3. WHAT IS CAUSING COMPLAINTS, AND WHAT IS NOT CAUSING COMPLAINTS
4. HOW LONG WILL INVESTIGATION TAKE
5. WHAT MITIGATION TASKS HAVE BEEN PERFORMED
6. WHAT WORK STILL REMAINS TO BE DONE AND WHEN

TYPE OF INFORMATION THAT SHOULD BE ON COMPLAINT INVESTIGATION POSTINGS

Figure 1-10. Notices or memoranda of a complaint investigation can be delivered directly to selected occupants by mailings or posted in general-use work areas, cafeterias, or floor lobby areas.

- projected duration of the investigation
- attempts made to improve indoor air quality
- remaining work that is required and the schedule for its completion

IAQ Response

Implementing a quality preventive maintenance program, performing regularly scheduled random air sampling, and maintaining a well-trained team of stationary engineers who can identify and mitigate IAQ problems eliminate most IAQ problems before they occur. When problems do occur, these steps may demonstrate due diligence to limit any legal liability.

IAQ problems are often difficult to investigate because of simultaneous factors such as intermittent symptoms, vague discomfort, and the complex interactions of job stress with environmental factors. These factors may also hinder the effects of IAQ remediation efforts. Even after the proper remediation strategy is in place, it may take days or weeks for contaminants to dissipate and symptoms to disappear.

LEGAL LIABILITIES

A large variety of legal issues can arise from IAQ problems. Three of the most common issues include negligence (professional malpractice), constructive eviction (termination of the landlord-tenant relationship), and workers' compensation. Negligence is the most common issue cited for indoor air pollution injuries. *Negligence* is a legal issue that occurs when there is an injury because of a lack of performance by building personnel. In other words, building occupants might accuse building management of negligence when management fails to remedy an IAQ problem in a reasonable fashion or time frame. In addition, professionals such as architects and engineers may be sued for their alleged negligence in the performance of their professional

services. *Constructive eviction* is a legal issue that occurs when a tenant is unable to use a building space due to special circumstances, such as those related to an unresolved IAQ problem. Depending on the terms of the lease and on the various services that the landlord may have agreed to perform, the occurrence of indoor air quality problems may provoke a tenant's claim for constructive eviction.

In some cases, employees who become ill due to IAQ problems may be awarded worker's compensation. *Workers' compensation* is a legal issue referring to payments that are offered to employees who are temporarily unable to work because of a job-related injury. However, workers' compensation is in fact more than just income insurance because it may pay compensation for economic loss, reimbursement or payment of medical and medical-like expenses, and damages for pain and suffering. It also pays benefits to the dependents of workers killed during employment. While negligence and constructive eviction tend to be more tenant based, workers' compensation is concerned with the health effects on building employees who work in the affected building or building space.

Confidentiality of records is important to occupants who are concerned that IAQ complaints will lead to negative reactions from their employers. Investigators are more likely to obtain honest and complete information by reassuring occupants that their privacy is respected.

Many IAQ-related lawsuits have been filed, and several have resulted in large private settlements. Different types of people have been sued over IAQ issues including designers, owners, outside contractors, and equipment installers and manufacturers. At the same time, lawsuits may be hard to win because it is often hard to prove that IAQ is the sole source of a problem. In one case, a $1 million award was overturned because the plaintiff failed to prove the exact source of the IAQ problem.

Future IAQ Regulations

Concerns about IAQ have increased since energy conservation measures were started in commercial buildings in the 1970s. The energy conservation measures minimized the infiltration of outdoor air and allowed contaminants to build up in commercial buildings.

Due to the health problems associated with IAQ and the lawsuits that have already been filed in courts, federal and state officials are considering enacting laws to govern IAQ. Comprehensive indoor air quality bills have been introduced in both the United States House of Representatives and the Senate in recent years. However, none of these bills have ever been passed. Still, the Occupational Safety and Health Administration (OSHA) and the Federal Register do have standards and notices that apply to IAQ. **See Figure 1-11.** The various bills included measures such as:

- establishing specific ventilation standards
- establishing guidelines to identify and prevent indoor air hazards
- providing funds for research into IAQ contaminants
- requiring OSHA inspections of buildings and facilities with repeated IAQ complaints

Per Ricin-OSHA standards, twenty-four states, Puerto Rico, and the Virgin Islands have OSHA-approved state plans for government and private sector buildings. For the most part, these states adopt standards that are identical to those put forward by federal agencies. However, some states have adopted different standards applicable to this topic or may have different enforcement policies. Highlights of the rules used by some of the states include:

- Employers must develop and implement IAQ compliance programs and designate an individual as an IAQ coordinator to ensure compliance. The program must include written documents, including building specifications and functions, operation procedures, a preventive maintenance program, a visual inspection checklist for building systems, and a record of employee complaints.
- HVAC systems must be operated during all regular work shifts.
- HVAC systems must be maintained and operated within the original design specifications and provide at least the minimum outdoor ventilation rate based on actual occupancy and required by the code applicable at the time the facility was constructed, renovated, or remodeled, whichever is most recent.
- Smoking must either be banned or restricted to designated smoking lounges under negative pressure and exhausted to the outdoors.
- General or localized exhaust ventilation must be used when activities could reasonably be expected to result in hazardous chemical or particulate exposures to employees in other areas of the building or facility.
- The need for alterations to the compliance program must be evaluated in response to employee complaints of building-related illnesses, and necessary remedial measures must be taken.
- Employers must monitor carbon dioxide levels during routine maintenance and take action when the levels exceed 800 ppm.
- Requirements are specified for controlling the entry of outdoor air contaminants, microbial contamination in the building, chemical use and application, and air quality during renovation and remodeling.
- The employer must provide training for maintenance and facility engineers on maintaining adequate ventilation during building cleaning and maintenance, on the use of personal protective equipment, and on how to minimize adverse effects on indoor air during chemical (and other agent) use and disposal.

Other associations, such as the American National Standards Institute (ANSI) and the American Society of Heating, Refrigerating, and Air Conditioning Engineers

(ASHRAE), have also published their own standards to protect members and building occupants against possible IAQ problems. ANSI/ASHRAE Standard 62.1-2007, *Ventilation for Acceptable Indoor Air Quality,* specifies minimum rates of ventilation and IAQ considered acceptable for occupants to avoid health risks.

OSHA-Related IAQ Standards

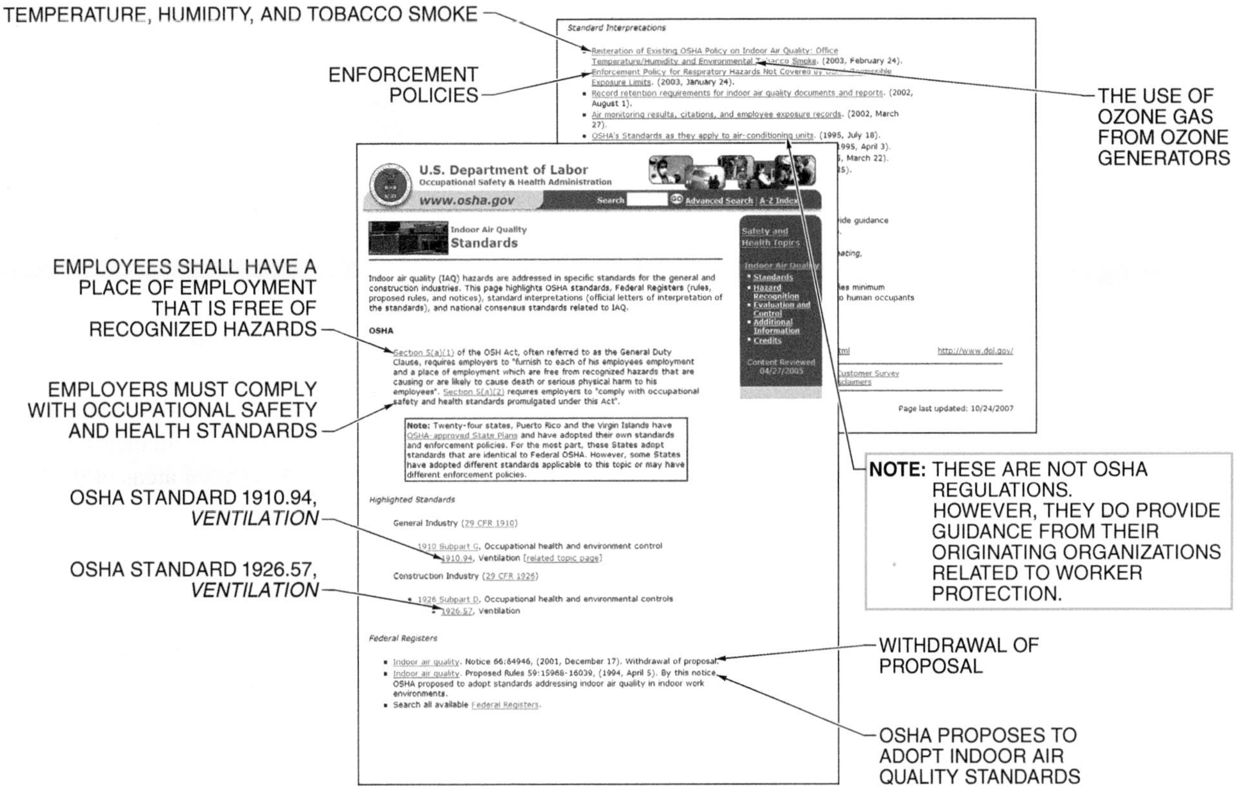

TEMPERATURE, HUMIDITY, AND TOBACCO SMOKE
ENFORCEMENT POLICIES
THE USE OF OZONE GAS FROM OZONE GENERATORS
EMPLOYEES SHALL HAVE A PLACE OF EMPLOYMENT THAT IS FREE OF RECOGNIZED HAZARDS
EMPLOYERS MUST COMPLY WITH OCCUPATIONAL SAFETY AND HEALTH STANDARDS
OSHA STANDARD 1910.94, VENTILATION
OSHA STANDARD 1926.57, VENTILATION
NOTE: THESE ARE NOT OSHA REGULATIONS. HOWEVER, THEY DO PROVIDE GUIDANCE FROM THEIR ORIGINATING ORGANIZATIONS RELATED TO WORKER PROTECTION.
WITHDRAWAL OF PROPOSAL
OSHA PROPOSES TO ADOPT INDOOR AIR QUALITY STANDARDS
U.S. Department of Labor
Occupational Safety & Health Administration
www.osha.gov
Indoor Air Quality
Standards

Figure 1-11. *Even though the federal government has not passed any bills into law regarding IAQ standards, the Occupational Safety and Health Administration (OSHA) and the Federal Register do have IAQ standards and notices that some individual states have passed into law.*

Case Study – Chain-of-Command

The EPA recommends establishing a formal process for handling IAQ problems. Proposed legislation mandates that a formal process be in place for receiving and handling IAQ problems. Possible legislation proposes that one person on staff (IAQ manager or coordinator) be responsible for overseeing IAQ management. That person may be a member of building management or a stationary engineer. The legislation proposed for the state of Washington stipulates that such a person be appointed.

There are general outlines being used that state the procedures for processing an occupant's complaint. These procedures are used by many well-run facilities. When there are no laws regulating IAQ in a community, the following steps can serve as minimum guidelines on how to process occupant complaints.

1. Building occupants report IAQ complaints to the management office (by phone or in writing).

2. The management office formally documents the call and dispatches it to the engineering department by completing a work order ticket that includes the occupant's name, location, and description of the problem.

3. The chief engineer assigns the work order ticket containing all known information to a stationary engineer.

4. The stationary engineer responds by taking all required actions.

5. The stationary engineer records the dates of the response and final resolution of the complaint on the work order ticket, along with a description of what was found, what actions were taken, and why. The engineer also notes whether the job is complete or if follow-up is needed. Full documentation of the actions taken and reasons for them are recorded daily in the appropriate engineer's watch/shift/ incident log and IAQ log (when air sampling is performed). This provides an accurate historical record of activities and keeps all engineers in the department informed.

6. Management processes the work order ticket. The ticket might include recommendations for parts or work that require money or time, which must be approved by management.

7. When a building automation system is in use, all information is keyed into a database to permit more efficient tracking of incidents and actions, as well as routine preventive maintenance.

Indoor Air Quality Definitions

- *Constructive eviction* is a legal issue that occurs when a tenant is unable to use a building space due to special circumstances, such as those related to an unresolved IAQ problem.
- *Good IAQ* is when the indoor air of a building is free of harmful amounts of chemicals and particles, and the temperature and humidity of the air is comfortable.
- *Indoor air quality (IAQ)* is the comfort level of indoor air based on a combination of factors, including temperature, humidity, airflow, and contaminant levels.
- *Negligence* is a legal issue that occurs when there is an injury because of a lack of performance by building personnel.
- *Poor IAQ* can occur when the indoor air temperature, humidity, chemicals, or contaminants rise to harmful or uncomfortable levels.
- *Workers' compensation* is a legal issue referring to payments that are offered to employees who are temporarily unable to work because of a job-related injury.

For more information please refer to ATPeResources.com or http://www.epa.gov.

1. What is IAQ?

2. How did building designs change during the energy crisis of the 1970s?

3. Name three important guidelines to consider when communicating with occupants regarding IAQ issues.

4. Explain the function and importance of a health and safety committee.

5. Why are lawsuits that attempt to recover IAQ-related damages difficult to win?

RUSH
NORTH SHORE
MEDICAL CENTER

← Emergency

← Hospital Main Entrance
Medical Office Building
Golfside Pavilion
Knox Building
Kenton Building

Class Objective

Describe different medical problems caused by poor IAQ.

On-the-Job Objective

Identify common IAQ health problems.

Learning Activities:

1. Identify the difference between acute and chronic health effects.
2. Compare and contrast four types of building-related illnesses.
3. Explain the difference between building-related illness and sick building syndrome.
4. Describe the signs of an outbreak of mass psychogenic illness.
5. **On-the-Job:** Review all previous IAQ complaints in the building. List all IAQ-related health effects currently experienced in the facility. Describe possible IAQ contaminants that may be the cause of these health issues.

Introduction

Spending a large amount of time indoors increases the possible number of indoor air quality health problems that a building occupant may encounter. Short-term exposure to contaminants can produce a variety of health symptoms. However, long-term exposure to contaminants tend to lead to more serious symptoms and illnesses. It may be difficult to link certain illnesses to a specific contaminant or contaminant group.

HEALTH EFFECTS

IAQ contaminants vary widely and symptoms of IAQ-related illness can vary widely as well. Health problems due to poor IAQ can be either dramatic or subtle. Symptoms of poor IAQ may include irritated eyes and throat, nasal congestion, coughing, chest pain, fatigue, headache, muscle ache, and irritability. **See Figure 2-1.**

These symptoms are also the general symptoms of the flu and stress, so diagnosing an IAQ-related illness can be difficult. *Note:* A professional should be consulted for any health problems that are severe or remain unresolved.

COMMON SYMPTOMS OF POOR IAQ	
allergies	irritability
chest pain	muscle aches
chills	nausea
coughing	shortness of breath
dizziness	sinus congestion
fatigue	throat and eye irritation
fever	weakness
headaches	

Figure 2-1. A wide array of symptoms may manifest in building occupants who are subjected to poor indoor air quality.

Indoor air contaminants may be unnoticed, annoying, or life threatening. Some contaminants have highly variable effects because people's susceptibility can differ. In many cases, the harm that each contaminant has on the body increases as its concentration in the air increases. Reactions to an indoor air contaminant are either acute or chronic.

IAQ Fact

Poor IAQ can cause a significant decrease in occupants' ability to concentrate or perform normal daily tasks. Several factors affect whether or not a person has a reaction to indoor contaminants, including their age and any preexisting medical conditions. Some individuals are also more sensitive than others to certain types of contaminants.

Acute Health Effects

Acute health effects are health problems in which symptoms develop rapidly and often subside after the exposure stops. Acute health effects usually develop within 24 hours after exposure to substances like formaldehyde, ammonia, cleaning agents, fiberglass, and pesticides. Some health effects at high exposures can lead to death. Even short-term exposure to high concentrations of substances such as carbon monoxide can be life threatening.

Health problems may show up after a single exposure and are usually short-term and treatable. **See Figure 2-2.** Acute health effects include headaches, dizziness, and fatigue. Sometimes the only treatment necessary is eliminating the person's exposure to the source of the problem when it can be identified.

Symptoms of some diseases, including asthma, hypersensitivity pneumonitis, and humidifier fever, may also show up soon after exposure to indoor air pollutants. The likelihood of acute reactions in humans to indoor air pollutants depends on several factors. Age and preexisting medical conditions often affect susceptibility to reactions. In other cases, a person's reactions to a pollutant depend on individual sensitivity, which varies from person to person.

Some people become sensitized to biological or chemical pollutants after repeated exposures. Certain immediate effects are similar to those from colds or other viral diseases, so it is often difficult to determine when symptoms are a result of exposure to indoor air contaminants. For this reason, it is important to keep records of the time and place symptoms occur.

Communication and documentation are important in diagnosing IAQ-related symptoms. For example, if the symptoms fade or go away when a person is away from work, an effort should be made to identify the contaminant sources that may be possible causes. Some effects may be made worse by an inadequate supply of outdoor air or from the heating, cooling, or humidity conditions prevalent in the building.

ACUTE HEALTH EFFECTS VS. CHRONIC HEALTH EFFECTS			
EFFECT	**TIME FRAME**	**RELATED SYMPTOMS**	**SOURCES**
Acute	Short-term exposure	Headaches, dizziness, and fatigue	VOCs and fiberglass
Chronic	Long-term exposure	Asthma, heart disease, and cancer	Environmental tobacco smoke and asbestos

Figure 2-2. Reactions to contaminants are either acute or chronic. Acute effects are seen immediately. Chronic effects are seen long term.

At low contaminant air concentrations, there may be no observable symptoms. For example, at low concentrations of carbon monoxide, such as 9 parts per million (ppm), there are no reported health effects. As the contaminant concentration increases, and thus the occupant's exposure, the severity of the symptoms also increases. Death usually occurs at extremely high levels of contaminant concentration.

Chronic Health Effects

Chronic health effects are health problems in which symptoms occur frequently or develop slowly over a long period of time. Chronic health effects, which include some respiratory diseases, heart disease, and cancer, can be severely debilitating or fatal. Indoor air quality should be improved even when symptoms are not noticeable. Contaminants that may cause chronic health effects include environmental tobacco smoke, radon gas, and asbestos fibers in the indoor environment. Some contaminants, such as formaldehyde, can have both acute and chronic health effects.

Contaminant Categories

Contaminants commonly found in indoor air are responsible for many harmful effects. Indoor air contaminants generally fall into one or more categories. Categories include irritants, asphyxiates, allergens, neurotoxins, pathogens, and carcinogens. **See Figure 2-3.**

An *irritant* is a substance that can cause irritation of the skin, eyes, or respiratory system. Examples of irritants include formaldehyde, cleaning agents, fiberglass, and some pesticides. An *asphyxiate* is a substance that causes a lack of oxygen.

Examples of asphyxiates include carbon monoxide and many modern refrigerants. Modern building codes may require refrigerant monitoring and ventilation systems in mechanical rooms with chillers using refrigerants. **See Figure 2-4.** Stationary engineers and technicians should be aware of the lack of oxygen and the possibility of asphyxiation when entering confined spaces.

An *allergen* is a substance that causes an allergic reaction in individuals sensitive to it. Allergic reactions range from mild irritation to life-threatening fevers, asthma, and debilitating illness. Examples of allergens include pollen, molds, excrement, and animal dander (particles from skin, hair, or feathers). **See Figure 2-5.**

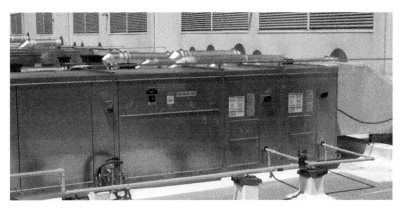

Scheduled monitoring and maintenance of HVAC systems, including rooftop units, can help to reduce the accumulation of contaminants.

CONTAMINANT CATEGORIES

CATEGORY	HEALTH EFFECTS	EXAMPLES
Irritants	Inflammation, redness, and discoloration	Fiberglass, pesticides, cleaning agents, and perfume
Asphyxiates	Drowsiness, headaches, and slowed breathing	Carbon monoxide and refrigerants
Allergens	Extreme fever and asthma	Pollen, molds, and animal dander
Neurotoxins	Cardiac problems, convulsions, and loss of memory	Lead and chemicals in paint and glue
Pathogens	Fever, chills, and chest congestion	Some molds and bacteria
Carcinogens	Cancer	Radon gas, asbestos fibers, and environmental tobacco smoke

Figure 2-3. Contaminants may be categorized according to the health effects they produce.

Asphyxiate Monitor

REFRIGERANT SENSOR

Figure 2-4. Refrigerant monitoring systems alert personnel if leaks occur.

A *neurotoxin* is a contaminant that damages the central nervous system. Neurotoxins can cause cardiac problems, drowsiness, headache, depressed respiration, convulsions, nerve tremors, restlessness, and loss of memory. Examples of neurotoxins include lead and certain chemicals in paints, pesticides, glues, and organic solvents.

A *pathogen* is a disease-causing biological agent. Pathogens may cause fever, chills, or chest congestion. Legionnaires' disease and hypersensitivity pneumonitis are two dangerous diseases caused by bacterial pathogens. Examples of pathogens include some molds and bacteria. A *carcinogen* is a substance that is capable of causing cancer. Cancer may develop years after exposure to carcinogens. Examples of carcinogens include environmental tobacco smoke, benzene, radon, and asbestos.

Allergens

OUTSIDE
AIR INTAKE

ANIMAL DANDER   POLLEN

Figure 2-5. Mold is the most common contaminant in industrial buildings. However, it is possible for pollen and animal dander to enter a building's air supply.

BUILDING-RELATED ILLNESSES

A *building-related illness* (BRI) is a diagnosable illness whose cause and symptoms can be directly attributed to a specific contaminant source within a building. Symptoms vary depending on the illness. Victims may require prolonged recovery time after leaving the building. A BRI is traceable to a specific cause or causes in a building. Building-related illnesses include Legionnaires' disease, hypersensitivity pneumonitis, humidifier fever, and asthma.

Legionnaires' Disease

Legionnaires' disease is an infection of the lungs caused by Legionella bacteria. Water over 140°F and/or a chlorine flush can be used to kill the bacteria. Legionella bacteria can be found anywhere untreated water is in contact with air or soil. Legionella bacteria has been found in cooling towers, humidifiers, and warm water sprays or mists, shower heads, fountains, and a variety of other sources. **See Figure 2-6.**

IAQ Fact

Several thousand people are hospitalized in the United States each year with Legionnaires' disease. Most cases occur in the summer and early fall, but the disease can occur at any time of the year.

Legionnaires' disease begins like a cold or the flu and gets worse, damaging the kidneys and lungs. Approximately 15% of those who contract Legionnaires' disease die. People most susceptible to the disease are individuals over age 50, especially those who smoke, abuse alcohol, or have immune system deficiencies.

Legionella Bacteria Location

Figure 2-6. Cooling towers may be a breeding area for Legionella bacteria.

Hypersensitivity Pneumonitis

Hypersensitivity pneumonitis is a disease that irritates a person's lungs after inhaling certain organic dusts. Hypersensitivity pneumonitis begins with flu-like symptoms that eventually result in lung problems. Outbreaks of hypersensitivity pneumonitis in office buildings have been traced to air conditioning and humidification systems contaminated with bacteria and molds.

Building-related illnesses may stem from short or long periods of time spent indoors in almost any type of building, such as a hotel, workplace, or school.

Hypersensitivity pneumonitis usually affects between 1% to 5% of the individuals exposed to these indoor air pollutants. Symptoms may occur within hours of exposure and then subside soon after the person leaves the contaminated building. Sometimes symptoms appear months or years after exposure occurs. Continued exposure may lead to end-stage pulmonary fibrosis requiring a lung transplant or resulting in death. Hypersensitivity pneumonitis symptoms are similar to symptoms of pneumonia, allowing this illness to be misdiagnosed.

Humidifier Fever

Humidifier fever is a respiratory illness caused by exposure to toxins from microorganisms found in wet or moist areas in humidifiers and air conditioners. Humidifier fever has symptoms similar to hypersensitivity pneumonitis, but it occurs more often and has a shorter recovery time. Symptoms of humidifier fever often occur a few hours after exposure and include a fever, headache, chills, and a general feeling of discomfort. Humidifier fever normally subsides within 24 hours without residual effects, and a physician is rarely needed.

Scientists believe humidifier fever is associated with biological agents (bacterial and fungal) that grow in humidifier reservoirs and air conditioners. All humidifiers using untreated water have the potential to emit these biological agents. Airborne concentrations of microorganisms have been noted during operation of this equipment and might be quite high for individuals using ultrasonic or cool mist units.

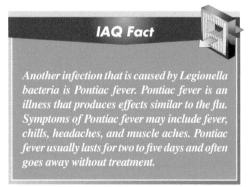

IAQ Fact

Another infection that is caused by Legionella bacteria is Pontiac fever. Pontiac fever is an illness that produces effects similar to the flu. Symptoms of Pontiac fever may include fever, chills, headaches, and muscle aches. Pontiac fever usually lasts for two to five days and often goes away without treatment.

Drying and chemically cleaning humidifiers can temporarily eliminate the problem, though manual cleaning of the reservoirs can expose humans to these contaminants. Drying and chemical disinfection are useful over a short period of time, but they cannot be considered to be as reliable as consistent maintenance. Only following a strict preventive maintenance program that includes humidifier water treatment has been shown to be effective.

Asthma

Asthma is a respiratory disease in which airway obstruction is triggered by various stimuli, such as allergens or a sudden change in air temperature. Asthma attacks often occur in response to one or more triggers. **See Figure 2-7.** Attacks may be triggered by exposure to an environmental stimulant or allergen (i.e., secondhand smoke and dust mites), mold, outdoor pollution, and pests.

Other triggers may include drastic changes in air temperature, moist air, exercise or exertion, or emotional stress. For children, the most common triggers are viral illnesses, such as those that cause the common cold. The narrowing of the airway causes symptoms such as wheezing, shortness of breath, chest tightness, and coughing. Indoor allergens and irritants play a significant role in triggering asthma attacks.

Between asthma attacks, most patients feel well but can have mild symptoms. Patients with asthma may also remain short of breath after exercise for longer periods of time than people without asthma. The symptoms of asthma, which can range from mild to life threatening, can usually be controlled with a combination of drugs and environmental changes. Many resources are available for additional ways to reduce asthma triggers. **See Appendix.**

SICK BUILDING SYNDROME

Sick building syndrome (SBS) is a situation in which building occupants experience acute health and comfort effects that appear to be linked to time spent in a building, though no specific illness may be diagnosed and the cause can be difficult to determine. Frequently, problems result when a building is operated or maintained in a manner that is inconsistent with its original design or prescribed operating procedures. However, indoor air problems are sometimes the result of poor building design or occupant activities. People suffering from SBS often feel worse as the day or week progresses and usually recover when they are away from the facility.

ASTHMA TRIGGERS		
TRIGGER	**COMMON SOURCES**	**SOLUTION**
Mold	Plumbing leaks; excessive humidity	Fix leaks; control humidity levels; and remediate mold
Secondhand smoke	Smoking lounges; smoking near entrances	Develop smoking ban and/or a designated smoking location away from entrance and air intakes
Dust mites	Carpeting and furniture	Vacuum and clean hard surfaces often
Outdoor air pollution	Outdoor-air intakes	Divert outdoor pollution sources; check air filter system
Pests	Storing food improperly	Remove trash often; keep food well sealed

Figure 2-7. Asthma may be triggered by many different environmental factors. To minimize risk, certain basic steps should be followed.

Complaints may involve one area of a building or may be building-wide. Generally, a broad spectrum of specific and nonspecific complaints are involved. The key factors of SBS are a commonality of symptoms among building occupants and an absence of symptoms among building occupants when the individuals are not in the building. Sick building syndrome should be suspected when occupants voice complaints of nonspecific symptoms or a variety of symptoms, occurring in one or more locations within a building.

It is important, however, to distinguish SBS from problems stemming from building-related illness. **See Figure 2-8.** Building-related illnesses are conditions in which the signs and symptoms of a diagnosable illness are identified and can be attributed directly to a specific airborne building contaminant or contaminants.

Possible causes of SBS may include poor design, maintenance, and/or operation of the structure's ventilation system. Interior redesign, such as the rearrangement of offices or installation of partitions, may also interfere with the efficient functioning of such systems. Humidity may be another factor. While high relative humidity may contribute to biological pollutant problems, an unusually low relative humidity (below 30%) may heighten the effects of mucosal irritants that also prove bothersome.

However, the extent of SBS problems is unknown. Various surveys of United States office workers show that the majority believes their office has air quality problems. Due to the difficulty of diagnosis, specific percentages of contaminants contributing to SBS are usually unavailable.

Multiple Chemical Sensitivity

Multiple chemical sensitivity (MCS) is a negative physical reaction to low levels of common chemicals. MCS is also known as chemical hypersensitivity or environmental illness. These substances may include recognizable contaminants and material that would normally be considered harmless.

The predominant medical opinion is that MCS is an unproven hypothesis. Some doctors believe that the condition has a purely psychological basis. Further research into causes and treatments is necessary before drawing definite conclusions. Regardless, these complaints should be taken seriously. A stationary engineer receiving a complaint from someone claiming to have MCS should display the same concern as for any other work-related health complaint. Steps to follow include:

- Listen to the complaint.
- Check suspected contaminant sources and pathways.
- Follow the original equipment manufacturer troubleshooting steps for all HVAC equipment and test instrument procedures.
- Adjust the occupant's work environment as much as possible.

SICK BUILDING SYNDROME (SBS) VS. BUILDING-RELATED ILLNESS			
ILLNESS	**SYMPTOMS**	**CAUSES**	**TIME PERIOD**
Sick building syndrome	Headache; eye, nose, and throat irritation; dry cough; dry or itchy skin; dizziness and nausea; difficulty concentrating; fatigue; sensitivity to odors	Unknown	Immediate recovery
Building-related illness	Cough; chest tightness; fever; chills; muscle aches	Identifiable	Prolonged recovery

Figure 2-8. Sick building syndrome and building-related illness both encompass diseases due to poor IAQ. Building-related illnesses have identifiable causes, whereas sick building syndrome conditions do not.

When complaints continue despite the stationary engineer's best efforts, management must be advised of the situation. Any remaining recommendations should be followed until a resolution has been found.

Mass Psychogenic Illness

Mass psychogenic illness is a condition that occurs when groups of people start feeling sick at the same time, even though there is no physical or environmental reason. Mass psychogenic illness has occurred all around the world in many different social settings. Outbreaks of mass psychogenic illness may start with an environmental trigger. The environmental trigger can be a bad smell, a suspicious-looking substance, or something else that makes people in a group believe they have been exposed to a germ or a poison.

When an environmental trigger incites a group of people to believe they might have been exposed to something dangerous, many of them may begin to experience signs of sickness at the same time. People may experience headaches, dizziness, faintness, weakness, and/or a choking feeling. In some cases, one person gets sick and then other people in the group also start feeling sick.

Signs that indicate that a group sickness may be caused by mass psychogenic illness include the following:

- A large group of people is sick simultaneously.
- Individual medical exams and tests show no physical abnormalities.
- Nothing is found in the group's environment, such as a poison, that would make people sick.

An outbreak of mass psychogenic illness causes anxiety and worry. During an outbreak, media coverage and the presence of ambulances or emergency workers can make people feel even more anxious and at risk. During an outbreak, if people hear about or see someone getting sick, it can be enough to make others feel sick as well.

This illness is not just psychological. People who are involved in these outbreaks have real signs of sickness that are not imagined. However, in cases of mass psychogenic illness, these symptoms are not caused by an IAQ problem.

Psychological and social factors, not indoor air pollution, may be responsible for these symptoms. Most of these outbreaks stop when people get away from the place where the illness started. The signs of illness tend to go away once people are examined and doctors find that they do not have a severe illness. It is important to keep the people who feel sick away from the commotion and stress related to the outbreak.

Mass psychogenic illness is an example of the way the human body can have a strong reaction to stressful situations. Another example of psychogenic illness is stage fright, which can cause reactions such as nausea, shortness of breath, headaches, dizziness, a racing heart, stomach pains, or diarrhea. Outbreaks of mass psychogenic illness show how stress, behavior, and people's feelings can affect the way a person's body responds.

ACCEPTABLE INDOOR ENVIRONMENTS

Acceptable indoor air quality is often defined as air that (1) does not contain any known contaminants or contains contaminants at concentrations that are not harmful and (2) does not provoke dissatisfaction among a substantial majority of the people exposed to it. The EPA applies several standards to declare an indoor environment as a healthy one, including the following:

- Surroundings contribute to human productivity, comfort, and a sense of health and well-being.
- Indoor air circulates to prevent stuffiness without creating drafts and is free from significant levels of odors, dust, and contaminants.
- The temperature and humidity are appropriate to the season, clothing, and activities of the building occupants.
- There is enough light to illuminate work surfaces without creating glare, and noise levels do not interfere with activities.

Case Study – Possible Mass Psychogenic Illness

Event

According to a report published by the National Institute for Occupational Safety and Health (NIOSH), in September 1987 NIOSH received a request for an evaluation of an illness outbreak at a company in Kentucky that manufactures small pressure gauges. At the time, nine company employees had been to the emergency room for various symptoms.

Employees first started complaining about a sewer-like odor. Following that, several employees experienced dizziness, burning of the nose and throat, and numbness of the lips, tongue, and other extremities.

Testing and monitoring was done in work areas where there were no employee complaints or symptoms, in the area where there were employee symptoms, and outside the building. The possible contaminants that were found included various lubricants, odorants, and solvents. However, concentration levels of the substances were found to be less than 1 ppm (parts per million). The concentration levels were lower than most environmental IAQ standards.

NIOSH also spoke with the physicians of the employees treated in the local hospital. In general, medical testing of the employees did not reveal any abnormalities common to the symptoms presented. A few employees did have mild cases of respiratory alkalosis. Respiratory alkalosis is a condition that results when the amount of carbon dioxide found in the blood drops to a level below the normal range.

Result

NIOSH was unable to definitively identify a contaminant that sparked the outbreak. It is possible that an unknown contaminant, related to the connection of a new sewer system, contributed to the employees' complaints and symptoms. This event caused facility-wide anxiety and stress, which appears to have lead to mass psychogenic illness among employees.

National Institute for Occupational Safety and Health

Health Concerns Definitions

- *Acute health effects* are health problems in which symptoms develop rapidly and often subside after the exposure stops.
- An *allergen* is a substance that causes an allergic reaction in individuals sensitive to it.
- An *asphyxiate* is a substance that causes a lack of oxygen.
- *Asthma* is a respiratory disease in which airway obstruction is triggered by various stimuli, such as allergens or a sudden change in air temperature.
- A *building-related illness* (BRI) is a diagnosable illness whose cause and symptoms can be directly attributed to a specific contaminant source within a building.
- A *carcinogen* is a substance that is capable of causing cancer.
- *Chronic health effects* are health problems in which symptoms occur frequently or develop slowly over a long period of time.
- *Humidifier fever* is a respiratory illness caused by exposure to toxins from micro-organisms found in wet or moist areas in humidifiers and air conditioners.
- *Hypersensitivity pneumonitis* is a disease that irritates a person's lungs after inhaling certain allergens.
- An *irritant* is a substance that can cause irritation of the skin, eyes, or respiratory system.
- *Legionnaires' disease* is an infection of the lungs caused by Legionella bacteria.
- *Mass psychogenic illness* is a condition that occurs when groups of people start feeling sick at the same time, even though there is no physical or environmental reason.
- *Multiple chemical sensitivity* (MCS) is a negative physical reaction to low levels of common chemicals.
- A *neurotoxin* is a contaminant that damages the central nervous system.
- A *pathogen* is a disease-causing biological agent.
- *Sick building syndrome* (SBS) is a situation in which building occupants experience acute health and comfort effects that appear to be linked to time spent in a building, though no specific illness may be diagnosed and the cause can be difficult to determine.

**For more information please refer to ATPeResources.com
or http://www.epa.gov.**

Review Questions

1. What are some of the most common symptoms of poor IAQ?

2. Explain the difference between acute and chronic effects.

3. List the contaminant categories and possible symptoms each can cause.

4. Describe two building-related illnesses.

5. Explain the difference between BRI and SBS.

6. List steps to be taken when encountering a complaint of MCS.

7. Explain why an outbreak of mass psychogenic illness may occur.

8. List at least three of the standards used in determining an acceptable indoor environment.

CONTAMINANTS

3
chapter

Class Objective
Describe outdoor and indoor air contaminants and their primary sources.

On-the-Job Objective
Identify possible indoor and outdoor contaminant sources for the facility.

Learning Activities:
1. Define particle, gas, and mixture and give two examples of each.
2. Describe how lead can become an indoor air contaminant.
3. Describe the symptoms and three sources of carbon monoxide exposure.
4. Name one method used to prevent environmental tobacco smoke (secondhand smoke) from becoming an IAQ problem.
5. Explain why volatile organic compounds (VOCs) are so common in an indoor environment.
6. **On-the-Job:** List all possible contaminant sources within the facility. Describe possible methods to limit the contaminants.

Introduction
Air contaminants directly affect indoor air quality and can cause a variety of undesirable health effects and illnesses. Preventing illnesses and improving IAQ requires a basic knowledge of the various contaminants and their sources. Once the different contaminants are identified, corrective actions can be implemented. Strategies for corrective action usually involve controlling chemicals, removing contaminants, and checking HVAC systems to ensure they are working according to design specifications.

29

POOR INDOOR AIR QUALITY

Poor indoor air quality (IAQ) can lead to maintenance problems, building damage, and health problems. Many buildings can contain IAQ problems. Contaminants may be produced from a variety of activities and/or environmental sources, which include the following:
- building design
- system and filter installation
- system operation
- building maintenance
- renovation
- occupant activities
- furniture
- off-gassing

Contaminants may originate outdoors and be drawn into the building or may originate inside the building itself. **See Figure 3-1.**

Outdoor Sources

Outdoor air can be contaminated by many different sources including vehicle exhaust, dry cleaning, trash and refuse areas, fertilizers, and smoking areas. Public awareness about the hazards of environmental tobacco smoke has increased. As a result, more local and state laws are in place to limit environmental tobacco smoke. However, the presence of tobacco smoke still presents some IAQ problems.

COMMON AIRFLOW PATHWAYS FOR CONTAMINANTS	
COMMON PATHWAY	**COMMENT**
Outdoors to Indoors	
Outdoor air intake	Polluted outdoor air or exhaust air enters building through air intake
Windows/doors Cracks and crevices	Negatively pressurized building draws air and outside pollutants into building through any available opening
Substructures and slab penetrations	Radon and other soil gases and moisture-laden air or microbial contaminated air travel through crawlspaces and other substructures into the building
Indoors	
Stairwell Elevator shaft Vertical electrical or plumbing shafts	Stack effect causes airflow by drawing air toward shafts on lower floors and away from shafts on higher floors, affecting flow of contaminants
Wall openings and penetrations	Contaminants enter and exit building cavities and move from space to space
Duct or plenum	Contaminants commonly carried by HVAC system throughout occupied spaces
Duct or plenum leakage	Duct leakage accounts for significant unplanned airflow and energy loss in buildings
Flue or exhaust leakage	Leaks from sanitary exhausts or combustion flues cause serious health problems
Room spaces	Air and contaminants move within room or through doors and corridors to adjoining spaces

Source: Environmental Protection Agency (EPA)

Figure 3-1. Contaminants may originate outdoors or indoors and often follow a certain airflow path.

Potential sources of contamination should be considered when examining outdoor air intakes. Outdoor air intakes located near contaminant sources may draw contaminants indoors and increase the chance of IAQ problems. Contaminated outdoor air must be cleaned for use indoors. This usually means improved filtration.

Indoor Sources

Indoor contaminants usually occur due to issues with building design or products and machines used within the building. Some of the sources of indoor contaminants include:

- cleaning supplies and storing chemicals
- glues and sealants used in building materials and furniture
- special-use and mixed-use areas like smoking lounges, laboratories, exercise rooms, beauty salons, dry cleaners, and restaurants
- redecorating, remodeling, and repair work
- locations that collect or produce dust or fibers such as carpeting, open shelving, and old or deteriorated furnishings
- fax and photocopy machines

As buildings and HVAC systems age, there are more opportunities for air infiltration and for contaminants to enter the building. Other indoor problems such as excessive moisture, standing water, and water-damaged furnishings may also promote the growth of indoor contaminants.

Effects of Poor Indoor Air Quality

Health and comfort complaints are common problems resulting from poor IAQ. When the concentration of contaminants increases to the point of causing health-related symptoms (even minor symptoms), the number of occupant complaints generally increases. The increased number of illnesses, symptoms, and complaints caused by increased concentrations of harmful substances or contaminants represent the serious consequences of poor IAQ.

When people inhale poor-quality air, the contaminants affect their respiratory system (nose, airways, lungs) and, in some cases, the contaminants may be absorbed into the bloodstream. The resulting illnesses can be mild such as general fatigue or life threatening such as Legionnaires' disease. The economic impact of poor IAQ is serious.

In office buildings with poor IAQ, employees become less productive, costing their employers time and money. When people become seriously ill, they miss work and require expensive medical care. People who believe that they have been made ill by conditions at work could potentially sue a variety of parties for damages. If the courts agree, the amounts of money paid for damages can be astronomical.

The most important factor influencing indoor air quality is the presence of contaminant sources in the building.

PARTICLES

A *particle* is any very small part of matter, such as a molecule, atom, or electron. Each particle may be one material, such as asbestos, or a cluster of many materials, such as dust. Particles can land on surfaces and become airborne again when disturbed. Large particles that are inhaled are captured in the upper respiratory tract.

Soluble particles, which can be dissolved and then absorbed through the mucus lining, can also irritate the upper respiratory tract. Insoluble particles (and some soluble materials) are swept out of the upper lungs and nasal passages by mucus and enter the stomach. Particle contaminants include asbestos, lead, mold, and fiberglass.

Asbestos

Asbestos is a mineral fiber that can pollute air or water and cause cancer or asbestosis when inhaled. Asbestos has been used in a variety of building construction materials for insulation and as a fire-retardant. The EPA and the Consumer Product Safety Commission (CPSC) have banned several asbestos products.

Asbestos is not an everyday IAQ complaint because its health effects take years to materialize. However, asbestos is still dangerous when broken into minute fibers, and steps must be taken to prevent it from becoming airborne in the indoor air environment. Asbestos is often found in mechanical rooms, pipe shafts, pipe insulation, fireproofing materials, acoustical materials, shingles, siding, and ceiling and floor tiles. **See Figure 3-2.**

When asbestos fibers are inhaled, they may pierce the lung lining and remain lodged there. While disease can develop from exposure to clothing and equipment brought home from job sites, most people develop asbestos-related diseases due to exposure of elevated concentrations of asbestos on the job. The effects of asbestos can cause lung cancer, mesothelioma (a cancer of the chest and abdominal linings), and asbestosis (irreversible lung scarring that can be fatal). Smokers are at higher risk of developing asbestos-induced lung cancer. Symptoms of these diseases often appear years after exposure began.

Asbestos Sources

PIPE INSULATION

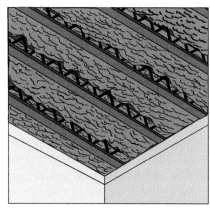

FIREPROOFING MATERIAL

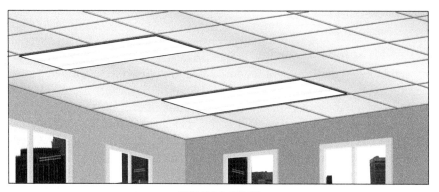

CEILING TILES

Figure 3-2. Asbestos is a mineral fiber that has been used commonly in a variety of building construction materials including pipe insulation, ceiling tiles, and building fireproofing.

Lead

Lead is a heavy metal that is a health hazard when breathed or swallowed. Lead has long been recognized as a harmful environmental pollutant. Humans are exposed to lead through air, drinking water, food, contaminated soil, deteriorating paint, and dust. **See Figure 3-3.**

Lead becomes an indoor air quality problem only when lead-based particles become airborne. Airborne lead enters the body when an individual breathes or swallows lead particles. Lead is a harmful substance that was used in many products including paint, gasoline, and water pipes. Airborne lead problems are caused by old paints, lead-based plumbing solder, electronics repair, stained-glass work, and firing ranges.

Lead affects the reproductive, circulatory, and central nervous systems in humans. Fetuses, infants, and children are more vulnerable to lead exposure than adults since lead is more easily absorbed into growing bodies. Similarly, the tissues of small children are more sensitive to the damaging effects of lead.

IAQ Fact

Though asbestos and lead particles are harmful contaminants, mold is a more common health risk. Maintaining a 30% to 60% relative humidity level indoors is recommended. This range ensures the optimal comfort range for humans and helps to prevent mold growth.

Lead Sources

Old Paint

Lead-Based Solder

MAJOR SOURCES

Drinking Water

Soil

MINOR SOURCES

Figure 3-3. Lead can be found in common sources such as drinking water, air, and soil. However, in sources such as old paint and lead pipes, there is a larger risk of lead becoming a harmful pollutant.

Mold

Molds are fungi that can be found both indoors and outdoors. No one knows how many species of fungi exist, but estimates range from the tens of thousands to the hundreds of thousands or more.

Molds grow best in warm, damp, and humid conditions. Molds spread and reproduce by making spores. Mold spores can survive harsh, dry environmental conditions, which are not normally considered conducive to mold growth. Water damage is not the only prerequisite to mold formation. High humidity levels are more common conditions for mold formation than the presence of water. Mold in buildings can form on any material with cellulose, such as ceiling tiles, paper, and cardboard. Mold is also found on wood, soap, and the skin of humans. **See Figure 3-4.**

Approximately 10% of the United States population is allergic to one or more types of mold. A certain level of mold must be in the air before a person suffers an allergic reaction. The higher the concentration of mold, the worse the reaction is likely to be. Under certain conditions, some types of mold produce toxins (poisons) or suspected carcinogens (cancer-causing chemicals). The toxins are usually concentrated in the spores. The effects of toxic molds may include liver damage and immune system impairment.

Fiberglass

Fiberglass is a material made from small fibers of glass twisted together. One form of fiberglass, known as glass wool, is made of fine (small-diameter) strands of glass and is used as an insulating material. Fiberglass is a common insulating material used in building walls, ceilings, and ventilation systems. Large quantities of fiberglass particles can be released into the air during remodeling.

Mold Growth Sources

WATER-DAMAGED CEILING TILE — CONCRETE WALL — DRYWALL

Environmental Protection Agency

Figure 3-4. *Mold spores grow in areas with excessive moisture or accumulated water. Mold can grow on a variety of materials including ceiling tiles and wallboard.*

Fiberglass contamination in buildings most often occurs through the ventilation systems. Fiberglass may be introduced to the air pathway through fans and blowers. Fiberglass can be an irritant to skin, eyes, and upper respiratory systems. Exposure to large quantities of fiberglass may result in rashes, sinus headaches, coughing, and conjunctivitis.

GASES

A *gas* is matter that has no definite volume or shape. Gases cannot be collected by ordinary particle filters typically used in a building's air management system. Certain gases, such as carbon monoxide, are odorless, whereas other gases, such as formaldehyde, produce odors. Some gases are emitted from liquids, such as solvents or paints.

Water-soluble gases (e.g., formaldehyde and nitrogen dioxide) are absorbed by the upper or lower respiratory system. Other inhaled gases (e.g., carbon monoxide) are dissolved in the bloodstream and are carried through the body. Gases include carbon dioxide, carbon monoxide, nitrogen dioxide, formaldehyde, ozone, and radon.

Carbon Dioxide

Carbon dioxide (CO_2) is an odorless, colorless gas that does not support combustion and is normally present in indoor air. Carbon dioxide is used in some industrial processes and in restaurants for carbonated beverages. CO_2 is the product of complete combustion. Another source of CO_2 is the exhaled breath of building occupants. An estimated 4% to 5% of exhaled breath is CO_2. Carbon dioxide is also a by-product of automobiles, so parking garages and heavy traffic may impact a facility as well. Environmental tobacco smoke is another source of CO_2.

Indoor CO_2 concentrations can, under some test conditions, provide a good indication of the adequacy of ventilation. The level of CO_2 in the air indicates how much outdoor air is entering the space. Carbon dioxide readings are useful in determining when a problem is being caused or aggravated by an inadequate supply of outdoor air. Peak CO_2 concentrations over 1000 parts per million (ppm) are a general indicator of inadequate ventilation. At higher concentrations, people may experience dizziness, headaches, and restlessness.

The most common source of carbon dioxide is the exhaled breath of humans. Carbon dioxide is also one of the many by-products of cars and trucks.

Carbon Monoxide

Carbon monoxide (CO) is a colorless, odorless, poisonous gas produced by incomplete fossil fuel combustion. Carbon monoxide is the product of incomplete combustion. Worn or poorly adjusted combustion devices such as those found in boilers and natural gas heating systems may become sources of CO emissions. Improperly vented heaters and gasoline engine-powered equipment also emit CO.

Carbon monoxide problems may also occur when the discharge flues of fossil-fueled equipment, such as boilers and gas heating equipment, are improperly sized, blocked, disconnected, or leaking. Cars, buses, and trucks are major sources of CO, so attached unventilated garages, parking areas, or heavy traffic patterns may affect a facility. Environmental tobacco smoke is another major source of CO. **See Figure 3-5.** Low levels of CO exposure can lead to drowsiness, chest pain, impaired vision, and nausea. High levels of CO exposure can result in a coma or even death.

Nitrogen Dioxide

Nitrogen dioxide (NO₂) is a toxic gas that is the result of nitric oxide combining with oxygen in the atmosphere. The average recommended NO_2 limit in outdoor air within a 24 hr period is 0.053 ppm. Sources of NO_2 include truck and bus emissions, kerosene heaters, unvented gas stoves and heaters, power plant emissions, gasoline engines operated indoors, and environmental tobacco smoke. **See Figure 3-6.** Diseases related to NO_2 exposure include lung damage, chronic bronchitis, and respiratory infections.

Formaldehyde

Formaldehyde is a colorless, pungent, and irritating gas used as a preservative in building materials. Formaldehyde is used in many types of glue, preservatives, and coatings for products such as particleboard, plywood, paper products, paints, cosmetics, permanent press fabrics, and certain insulations. **See Figure 3-7.** Formaldehyde is also used to preserve tissue specimens in medical and research facilities. Formaldehyde is released through auto emissions and tobacco smoke and by burning wood, kerosene, and natural gas.

IAQ Fact

Parking garages should be kept under negative pressure relative to adjoining areas. Otherwise, contaminants such as CO and CO₂ could be drawn into an adjoined building through doorways, cracks, and other available openings.

Carbon Monoxide Sources

TRAFFIC

UNVENTILATED GARAGES

HEATING EQUIPMENT

TOBACCO SMOKE

Figure 3-5. Carbon monoxide is an odorless, colorless, and toxic gas that can be produced from traffic, unventilated parking garages, heating equipment, and environmental tobacco smoke.

Nitrogen Dioxide Sources

CAR AND TRUCK EMISSIONS

GAS STOVES AND HEATERS

POWER PLANTS

TOBACCO SMOKE

Figure 3-6. *Sources of NO$_2$ include truck and bus emissions, kerosene heaters, unvented gas stoves and heaters, gasoline engines operated indoors, and environmental tobacco smoke.*

Formaldehyde becomes an IAQ issue due to off-gassing. *Off-gassing* is the release, over time, of a chemical compound from a material to the air. Off-gassing is higher when products are new. The rate of off-gassing is increased by higher temperatures and humidity. Exposure to formaldehyde can cause itchy and watery eyes, a sore or dry throat, a tingling nose, and itchy skin. Higher levels of exposure can cause coughing, chest tightness, and breathing difficulties.

Ozone

Ozone is a gas that is a variety of oxygen. Ozone is a molecule composed of three atoms of oxygen. The third oxygen atom can detach from the ozone molecule, and reattach to molecules of other substances, thereby altering their chemical composition.

Ozone is a powerful oxidizer of organic matter. Ozone combines readily with other contaminants in the air and changes their nature, which is why ozone can be used in air cleaning devices. However, these same properties can also make ozone hazardous and air cleaning devices that use ozone are not recommended, according to the EPA.

IAQ Fact

Sewer gas is another possible contaminant to be aware of. Sewer gas often enters through floor drain pipes with traps that have lost their prime. Sewer gas entry can be addressed through preventive measures, such as periodically priming the traps manually or installing automatic trap primers.

Formaldehyde Sources

PARTICLEBOARD/PLYWOOD

COSMETICS

BURNING WOOD

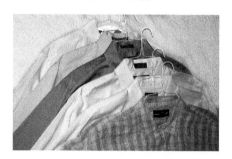

PERMANENT PRESS FABRICS

Figure 3-7. Formaldehyde is used in a wide variety of products including particleboard, plywood, cosmetics, and permanent press fabrics. Formaldehyde is also released during certain activities, such as burning wood.

Ozone is produced by high-voltage electrical equipment such as laser printers, photocopy and fax machines, welding equipment, ion generators, and some air cleaners. Ozone is the major component of smog and can be brought indoors by a heating, ventilation, and air conditioning system during high outdoor pollution episodes. Several United States government agencies have different standards regarding ozone concentration. **See Figure 3-8.** Low exposures for short time periods can cause symptoms such as headaches, lung congestion, aggravation of asthma, and shortness of breath. People with a history of lung problems are especially affected by ozone.

Radon

Radon is a colorless, naturally occurring, radioactive, inert gas formed by the radioactive decay of radium atoms in soil or rocks. Radon is a proven carcinogen and a form of ionizing radiation. Radon is found in outdoor air and in the indoor air of buildings of all kinds.

Radon is a gas produced by the radioactive decay of uranium in common materials, soil, and water. Radon may enter a facility from naturally occurring formations, underground garages, or any underground structure. Exposure to radon may lead to lung cancer. **See Figure 3-9.**

MIXTURES

A *mixture* is any combination of two or more chemical substances. Mixtures such as wood smoke may contain thousands of separate components. Mixtures include environmental tobacco smoke, wood smoke, pesticides, and volatile organic compounds.

OZONE HEALTH EFFECTS AND STANDARDS		
HEALTH EFFECTS	**RISK FACTORS**	**HEALTH STANDARDS**
Potential risk of experiencing:	Factors expected to increase risk and severity of health effects are:	Standards regulating ozone amounts are:
Decreases in lung function	Increase in ozone air concentration	The Food and Drug Administration (FDA) requires ozone output of indoor medical devices to be no more than 0.05 ppm
Aggravation of asthma	Greater duration of exposure for some health effects	The Occupational Safety and Health Administration (OSHA) requires that workers not be exposed to an average concentration of more than 0.10 ppm for 8 hr
Throat irritation and cough	Activities that raise the breathing rate (e.g., exercise)	
Chest pain and shortness of breath		The National Institute of Occupational Safety and Health (NIOSH) recommends an upper limit of 0.10 ppm, not to be exceeded at any time
Inflammation of lung tissue	Certain pre-existing lung diseases (e.g., asthma)	EPA's National Ambient Air Quality Standard for ozone is a maximum 8 hr average outdoor concentration of 0.08 ppm
Higher susceptibility to respiratory infection		

Source: Environmental Protection Agency (EPA)

Figure 3-8. High levels of exposure to ozone has certain health effects and risk factors. Several United States government agencies have developed standards relating to the allowable amount of ozone exposure for humans.

Environmental Tobacco Smoke

Environmental tobacco smoke (ETS) is a mixture of smoke from the burning end of a cigarette, pipe, or cigar and the smoke exhaled by a smoker. Environmental tobacco smoke, also referred to as secondhand smoke, contains a few thousand substances, most of which are harmful to humans and can even cause cancer. The substances in secondhand smoke may also cause eye irritation, headaches, nausea, heart problems, cancer, and can aggravate asthma. According to the EPA, secondhand smoke causes approximately 3000 lung cancer deaths in nonsmokers annually.

Wood Smoke

Wood smoke is a complex mixture of gases and fine particles produced when wood and other organic matter burn. Exposure to wood smoke can cause eye irritation, headaches, nausea, and dizziness. Two sources of wood smoke include wood-burning stoves and fireplaces. Cooking and food preparation should be restricted to areas that are negatively pressurized and vented to the outdoors.

Many public buildings have prohibited smoking indoors. Smoking areas are often designated outdoors with ashtray receptacles.

	EXPOSURE TO RADON	
Radon Level*	**If 1000 people who smoked were exposed to this level over a lifetime†...**	**The risk of cancer from radon exposure compares to‡...**
20	About 260 people could get lung cancer	250 times the risk of drowning
10	About 150 people could get lung cancer	200 times the risk of dying in a home fire
8	About 120 people could get lung cancer	30 times the risk of dying in a fall
4	About 62 people could get lung cancer	5 times the risk of dying in a car crash
2	About 32 people could get lung cancer	6 times the risk of dying from poison
1.3	About 20 people could get lung cancer	(Average indoor radon level)
0.4	About 3 people could get lung cancer	(Average outdoor radon level)

* in pCi/L
† Lifetime risk of lung cancer deaths from EPA Assessment of Risks from Radon in Homes (EPA 402-R-03-003)
‡ Comparison data calculated using the Centers for Disease Control and Prevention's 1999-2001
  National Center for Injury Prevention and Control Reports

Source: Environmental Protection Agency (EPA)

Figure 3-9. *Exposure to high levels of radon may lead to lung cancer. Radon levels are measured in picocuries per liter (pCi/L).*

Volatile Organic Compounds

Volatile organic compounds (VOCs) are organic, carbon-based compounds that evaporate at room temperature. VOCs are emitted as gases from certain solids or liquids. VOCs include a variety of chemicals, some of which may have short- and long-term adverse health effects. VOCs are emitted by a wide array of products.

Concentrations of many VOCs are consistently higher indoors (up to 10 times higher) than outdoors. The presence and concentration of VOCs changes as conditions and activities change. VOCs are used in solvents for many products that are designed to be applied while wet and then dry out.

VOCs include acetone, alcohols, ammonia, hydrocarbons, benzene, and phenols. Among the hundreds of sources are cleaners, deodorizers, pesticides, solvents, paints, waxes, glues, marking pens, copy machine fluids, correction fluid, cosmetics, perfumes, tobacco smoke, furnishings, draperies, carpets, and newly dry-cleaned fabric items. **See Figure 3-10.** VOCs are often released during and after renovations, as well as from certain businesses such as printing or photography shops.

Organic chemicals vary greatly from those that are highly toxic, to those that have no known health effects. Eye and respiratory tract irritation, headaches, dizziness, visual disorders, and memory impairment are among the immediate symptoms that some people have experienced soon after exposure to some organics. Exposure to some VOCs at levels higher than those typically found indoors may cause cancer and damage to the central nervous system, liver, and kidneys.

Pesticides

Pesticides are substances intended to repel, kill, or control any species that is designated a pest, including weeds, insects, rodents, fungi, bacteria, and other organisms. Pesticides, rodent poisons, or herbicides applied indoors or outdoors can lead to high indoor levels of harmful gases. Pesticides can cause upper respiratory tract irritation, endanger unborn children, and damage the nervous system and liver. Symptoms can range from a mild headache, fatigue, and nausea to severe visual disturbances, convulsions, and unconsciousness.

Volatile Organic Compound Sources

CLEANERS AND PESTICIDES

PAINTS, SOLVENTS, AND GLUE

COSMETICS AND PERFUMES

DRAPERIES AND CARPETS

Figure 3-10. *Volatile organic compounds are used in many products including cleaners, pesticides, paints and glue, cosmetics, perfumes, draperies, and carpets.*

BIOLOGICAL CONTAMINANTS

Biological contaminants are usually living organisms or derivates that can cause harmful health effects when inhaled, swallowed, or otherwise taken into the body. Per the EPA, biological contaminants include molds, mildew, viruses, mites, pollen, and animal dander. Biological contaminants come from many different sources. Contaminated central air handling systems can become breeding grounds for mold, mildew, and other sources of biological contaminants and can then distribute these contaminants through the building.

Other major sources of biological contaminants include wet or damp materials, standing water, cooling towers, contaminated outdoor air, sneezing, and coughing. Biological contaminants, including mold, can also accumulate on outdoor air screens and filters. Bird droppings or dead rodents can also become a problem within air screens and filters. Filters and air screens should be inspected periodically. Dirty filters and screens should be replaced or cleaned as necessary.

Many biological contaminants are small enough to be inhaled. Symptoms may range from eye, nose, and throat irritation to more serious problems like asthma, Legionnaires' disease, pneumonitis, and life-threatening fevers.

Microbial Contaminants

Microbial contaminants are contaminants invisible to the naked eye that multiply using living cells or that are living organisms. Per the EPA, examples of microbial contaminants include viruses, bacteria, fungi, yeast, and mammal and bird antigens. Microbial contaminants can be inhaled and can cause many types of health effects including allergic reactions, respiratory disorders, hypersensitivity diseases, and infectious diseases.

Excessive concentrations of bacteria, viruses, fungi (including molds), dust mite allergen, animal dander, and pollen may result from inadequate maintenance and housekeeping, water spills, inadequate humidity control, condensation, or may be brought into the building by occupants, infiltration, or ventilation air. Microbial contamination occurs anywhere accumulated carbon-containing material such as dirt, plastic, wood, or paper becomes moist or wet. Exposure to indoor biological pollutants causes symptoms in allergic individuals and also plays a key role in triggering asthma episodes for an estimated 15 million Americans.

Case Study

Case Study – Massachusetts School District

IAQ Team

A Massachusetts school district, consisting of 15 schools and several additional buildings, made IAQ improvements in 1997. An IAQ team was formed of different individuals who worked within the district. The team included a member of the United Food and Commercial Workers (UFCW) union, a teacher, a school nurse, a maintenance representative, and others. The IAQ team began having meetings and constructed a plan to do school walk-throughs, identify IAQ problems, and implement remediation procedures.

IAQ Issues

One of the main problems that the team encountered was a lack of communication between maintenance staff and teachers. The team issued deadlines for schools within the district to perform inspections and submit completed checklists. Some of the IAQ issues that were found in the schools included mold found in carpet and on countertops, leaking windows and pipes, and minimal ventilation in classrooms.

Resolution

After understanding the IAQ issues within the district, the team began the solution-making process by encouraging communication between maintenance staff and teachers. The school district learned that it might avoid future IAQ concerns by communicating potential problems early. Some of the short-term solutions to the problems identified included cleaning and/or replacing carpet, replacing or caulking leaky windows, and placing fans in rooms where ventilation was a problem. Long-term solutions included plans to replace or clean air filters quarterly, schedule annual cleanings of ventilators, and train staff in the proper use of pesticides.

Environmental Protection Agency

Contaminants Definitions

- *Asbestos* is a mineral fiber that can pollute air or water and cause cancer or asbestosis when inhaled.
- *Biological contaminants* are living organisms or derivates that can cause harmful health effects when inhaled, swallowed, or otherwise taken into the body.
- *Carbon dioxide (CO$_2$)* is an odorless, colorless gas that does not support combustion and is normally present in indoor air.
- *Carbon monoxide (CO)* is a colorless, odorless, poisonous gas produced by incomplete fossil fuel combustion.
- *Environmental tobacco smoke (ETS)* is a mixture of smoke from the burning end of a cigarette, pipe, or cigar and smoke exhaled by a smoker.
- *Fiberglass* is a material made from small fibers of glass twisted together.
- *Formaldehyde* is a colorless, pungent, and irritating gas used as a preservative in building materials.
- A *gas* is matter that has no definite volume or shape.
- *Lead* is a heavy metal that is a health hazard if breathed or swallowed.
- *Microbial contaminants* are contaminants invisible to the naked eye that are either derived from living organisms or that are living organisms.
- A *mixture* is any combination of two or more chemical substances.
- *Mold* is a type of fungus that can be found both indoors and outdoors.
- *Nitrogen dioxide (NO$_2$)* is a toxic gas that is the result of nitric oxide combining with oxygen in the atmosphere.
- *Off-gassing* is the release, over time, of a chemical compound from a material to the air.
- *Ozone* is a gas that is a variety of oxygen
- A *particle* is any very small part of matter, such as a molecule, atom, or electron.
- *Pesticides* are substances intended to repel, kill, or control any species that is designated a pest, including weeds, insects, rodents, fungi, bacteria, and other organisms.
- *Radon* is a colorless, naturally occurring, radioactive, inert gas formed by the radioactive decay of radium atoms in soil or rocks.
- *Volatile organic compounds (VOCs)* are organic, carbon-based compounds that evaporate at room temperature.
- *Wood smoke* is a complex mixture of gases and fine particles produced when wood and other organic matter burn.

For more information please refer to ATPeResources.com
or http://www.epa.gov.

Definitions

Review Questions

1. How do outdoor contaminants become indoor contaminants?

2. List two particle contaminants and their sources.

3. Why are gas contaminants so dangerous to humans?

4. How are VOCs produced and what symptoms could they trigger in humans?

5. Describe the difference between biological contaminants and microbial contaminants.

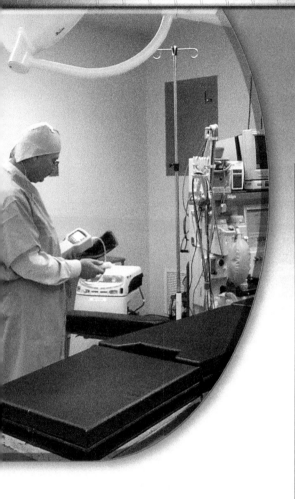

Class Objective
Learn about the various pieces of personal protective equipment (PPE) available for use when handling contaminants. CAUTION: There are comprehensive courses available for PPE that cover how to use the various types of PPE and what specific applications mandate the use of certain PPE.

On-the-Job Objective
Review current safety procedures and order more personal protective equipment if needed.

Learning Activities:
1. Compare and contrast minimum, limited, and full personal protective equipment.
2. Describe the different levels of protection for personal protective equipment.
3. Indicate what kinds of products require the use of gloves during handling.
4. List the different types of respiratory protection and explain when they are used.
5. **On-the-Job:** What personal protective equipment is already in the facility?

Introduction
When testing for contaminants, there are certain safety procedures and types of safety equipment that stationary engineers must be aware of. IAQ safety equipment is also known as personal protective equipment (PPE). The Occupational Safety and Health Administration (OSHA) sets regulations regarding the kind of protective equipment that is required for employees under different working conditions. The Environmental Protection Agency (EPA) recognizes certain levels of PPE based upon the kinds of contaminants (hazards) that may be encountered in an IAQ inspection or remediation.

PERSONAL PROTECTIVE EQUIPMENT (PPE)

Personal protective equipment (PPE) is clothing and equipment used to protect personnel from sickness, injury, or death by creating a barrier against IAQ hazards. **See Figure 4-1.** All IAQ personnel must assess the environment in which they are working in order to determine what types of PPE are required for the proper level of protection. Personal protective equipment should be used to provide safety to IAQ personnel at all times.

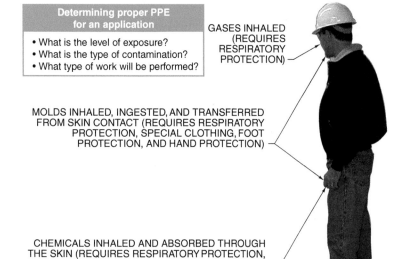

Saftronics Inc.

Figure 4-1. Personal protective equipment is used to protect personnel from possible hazards within a particular area.

The Occupational Safety and Health Administration (OSHA) 29 Code of Federal Regulations (CFR) 1910 Subpart I—*Personal Protective Equipment* requires employers to train employees on the use of PPE. The possible ways that gases, chemicals, or molds can enter the body of IAQ personnel are by skin contact, inhalation, and ingestion, and through the eyes, nose, or mouth. When determining the proper

PPE for an application, factors to be considered include the level of exposure, type of contamination, and type of work to be performed. The Environmental Protection Agency (EPA) recognizes levels of PPE protection as minimum PPE, limited PPE, and full PPE. **See Figure 4-2.**

Minimum PPE

Minimum PPE is the level of protection required to perform small-size IAQ tasks or to perform a task in a small area (under 10 sq ft). Minimum PPE includes hand protection, eye protection, and a particle respirator (usually N-95).

Limited PPE

Limited PPE is the level of protection required to perform medium-size tasks or to perform a task in a medium-size area (10 sq ft to 100 sq ft). Limited PPE also requires a half-face or full-face air-purifying respirator that includes a filter cartridge (HEPA). Air-purifying respirators have inhalation and exhalation valves that force air through the filter.

IAQ Fact

Per the Environmental Protection Agency (EPA) document entitled **Personal Protective Equipment,** *found on-line at www.epa. gov/emergencies/content/hazsubs/equip. htm, hazardous substance response activities for vapors, gases, and particulates require personal protective equipment. The EPA also publishes information on containment and PPE, found on-line at www.epa.gov/mold/moldcourse/chapter6/ index.html.*

The National Institute for Occupational Safety and Health (NIOSH) offers a document entitled **Emergency Response Resources—Personal Protective Equipment,** *found on-line at www.cdc.gov/niosh/topics/ emres/ppe.html.*

Many organizations such as the Environment, Health & Safety Division of Berkely Labs use OSHA standards 29 CFR 1910.132 through 29 CFR 1910.137 for PPE requirements.

Full PPE

Full PPE is the level of protection required to perform large-size tasks or to perform a task in a large-size area (over 100 sq ft). Full PPE also requires a full-face, powered air-purifying respirator that uses a blower to force air through a HEPA filter. Filtered air is sent to a full-face mask or hood that covers the entire head. The blower maintains a positive pressure in the mask or hood to prevent unfiltered air from entering through any openings. Training is required on powered respirators before IAQ personnel use them.

Indoor Air Quality – Personal Protective Equipment (PPE)

STAFF PERSON'S NAME

COMPANY OR BUILDING NAME

STAFF COLOR SHIRT

STAFF COLOR TROUSERS

SAFETY SHOES (NOT SHOWN)

Cintas Corporation

DAY-TO-DAY

EYE PROTECTION

N-95 RESPIRATOR (HALF FACE)

HAND PROTECTION

MINIMUM

EYE PROTECTION

PURIFYING RESPIRATOR WITH HEPA FILTER (HALF OR FULL FACE)

HAND PROTECTION

DISPOSABLE PAPER OVERALLS

Lab Safety Supply, Inc.

LIMITED

DISPOSABLE HEAD AND FOOT COVERINGS

POWERED AIR-PURIFYING RESPIRATOR (PAPR) OR SELF-CONTAINED BREATHING APPARATUS (SCBA) (FULL-FACE MASK OR HOOD)

DUCT TAPE TO SEAL GAPS AROUND ANKLES AND WRISTS

BODY SUIT AND GLOVES MADE OF BREATHABLE MATERIAL

Lab Safety Supply, Inc.

FULL

Figure 4-2. There are various levels of protective equipment, depending on the nature and level of exposure.

IAQ Personal Protective Equipment

IAQ personal protective equipment can include special clothing, foot protection, hand protection, head protection, eye protection, and respiratory protection. IAQ personnel using certain PPE equipment, such as special clothing and/or respirators, must be trained on its use, have medical clearance to use it, and be fit-tested by a qualified professional.

Special Clothing. *Special clothing* is a coverall or bodysuit that is treated to be impervious to specific fibers, gases, and/or chemicals. Many IAQ bodysuits also include a hood that is separate or is part of the one-piece bodysuit. A large percentage of clothing used in IAQ applications is considered to be disposable. **See Figure 4-3.**

Disposable clothing is usually found on medium and large IAQ projects. Disposable clothing is a limited type of protection used to prevent the movement of contaminants (mold) to clothing or to the skin of personnel. When full protection is required, a bodysuit made of a material such as Tyvek® is used that includes mold-impervious disposable coverings for the head and feet.

DISPOSABLE CLOTHING

HEPA FORCED-AIR HOOD

Lab Safety Supply, Inc.

Figure 4-3. *Special clothing is needed when encountering hazardous substances.*

Foot Protection. *Foot protection* is footwear such as shoes, boots, or covers that protect the foot and seal the bottom of the coverall or bodysuit leg. **See Figure 4-4.** IAQ footwear can include special-material boots or shoes. A standard IAQ practice is to use duct tape to seal the area between the coverall or bodysuit and the footwear being used.

Lab Safety Supply, Inc.

The level of appropriate personal protective equipment may vary according to the situation.

Hand Protection. *Hand protection* is gloves that protect the skin from contaminants and cleaning solutions. **See Figure 4-5.** Gloves should extend to at least the middle of the forearm. The material the gloves are made from should be specifically selected to protect workers from the type of contaminant that is being handled. When handling materials such as chlorine, bleach, or a strong cleaning solution, gloves made of neoprene, nitrile, or PVC are required. When handling soaps or detergents, gloves made of rubber are required.

Hand Protection

Lab Safety Supply, Inc.

SHORT GLOVES   LONG GLOVES

Foot Protection

DISPOSABLE BOOTS

DISPOSABLE SHOES

SHOE COVERS

Lab Safety Supply, Inc.

Figure 4-4. *Foot protection includes disposable boots, shoes, and covers. The protection chosen may vary depending on the situation.*

SIZING AND REMOVING GLOVES

TAPE MEASURE

INSULATION GLOVE SIZING

A proper insulating glove fit minimizes hand and finger fatigue. To measure a hand to determine insulating glove size, measure the distance around the palm of the hand. To remove an insulating glove:

1. Pull glove at fingertip on one hand and crumple glove to uncover thumb.
2. Use thumb under cuffs to remove glove from other hand.
3. Pull gloves inside out over the cover glove.

❶   ❷   ❸

INSULATION GLOVE AND COVER REMOVAL

Figure 4-5. *Gloves protect a person's hands against harmful substances. Glove materials vary depending on the type of substance that a worker is handling.*

Head Protection. *Head protection* may include a hood or bouffant cap that protects the head from contaminants or mold. **See Figure 4-6.**

Eye Protection. *Eye protection* is a properly fitting set of goggles or a full-face respirator mask that covers the eyes. If goggles are used for eye protection, they must prevent the entry of dust and small particles into the eyes. **See Figure 4-7.** Safety glasses or goggles with open vent holes are prohibited.

Eye Protection

**FITTED GOGGLES
(NO VENT HOLES)**

Head Protection

Lab Safety Supply, Inc.

HOOD

BOUFFANT CAP

Figure 4-6. *Head protection safeguards the head from contaminants.*

**FULL-FACE
RESPIRATOR MASK**

Lab Safety Supply, Inc.

Figure 4-7. *Eye protection must be close-fitting to prevent small particles from entering the eye. Full covering masks also work well for eye protection.*

Respiratory Protection

A *respirator* is a piece of equipment that protects IAQ personnel from inhaling dust, airborne mold, mold spores, and gases. Respiratory protection equipment is either an air-purifying type or an air-supplying type. **See Figure 4-8.** Air-purifying respirators include disposable dust masks, half-face respirators, full-face respirators, and full-face helmet or power air-purifying respirators (PAPRs). Air-supplying respirators include air-line respirators and self-contained breathing apparatuses (SCBAs).

Disposable Dust Mask. A *disposable dust mask* is a type of respiratory protection made of a filter material that is worn over the nose and mouth. Dust masks protect against solids such as dust and fiberglass. Dust masks are designed for a single use and are not considered to be adequate respiratory protection.

Half-Face Respirator. A *half-face respirator* is a type of respiratory protection that covers the nose and mouth and has filter cartridges that protect against dust, fumes, and particulates. Half-face respirators protect against contaminants up to a limited concentration, depending on the type of filter cartridge being used. Half-face respirators often operate under negative pressure and require fit testing prior to initial use. Special training is required prior to using a half-face respirator.

Full-Face Respirator. A *full-face respirator* is a type of respirator protection that covers the entire face including the eyes and has filter cartridges that protect against dusts, fumes, and particulates. Full-face respirators often operate under negative pressure, but have a much better face fit, which allows little leakage. Similar to a half-face respirator, special training is required prior to use.

RESPIRATORY PROTECTION		
AIR-PURIFYING		
Respirator		**Description and Suggested Use**
	Disposable dust mask	Low profile, lightweight, designed for limited use; low-cost protection against dusts, mists, and fumes (not for mists containing gases, vapors, or nonabsorbed contaminants); completely disposable; no cleaning or spare parts required
	Half-face respirator	Lightweight, easy to maintain, and very little restriction of movement or vision; uses replaceable cartridges and filters; limited number of parts; protects against chemical hazards such as dusts, fumes, mists, and vapors
	Full-face respirator	Offers greater eye and face protection than half-mask; uses replaceable cartridges and filters; easy to maintain (no intricate parts); protects against chemical hazards such as dusts, fumes, mists, and vapors
	Power air-purifying respirator (PAPR)	Cooler, less exhausting for worker; provides easier breathing for higher productivity; uses cartridges or filters; face- or belt-mounted with a battery for power; includes air blower that pulls air through the cartridges and filters into the face piece
AIR-SUPPLYING		
	Air-line respirator	Uses outside air source to keep worker cooler and offers greater protection than an air-purifying respirator; available in two styles: constant flow and pressure demand; uses grade D air supply from ambient air pump, plant compressor, or bottled air; not for use in immediately dangerous to life and health (IDLH) situations or where the oxygen content is less than 19.5%
	Self-contained breathing apparatus (SCBA)	Provides greatest protection available; pressurized bottle of air is carried on worker's back; for use in oxygen-deficient atmospheres, IDLH, and emergency situations; available in two different types of cylinders: aluminum and composite; provides good mobility with few restrictions because air source is carried on back

Lab Safety Supply, Inc.

Figure 4-8. *Respirators prevent personnel from inhaling contaminants. The two main types of respirators are air-purifying and air-supplying respirators.*

Full-Face Helmet. A *full-face helmet* is a type of respiratory protection that covers the entire head and has a filtered air blower to maintain positive pressure in the helmet or hood. Full-face helmets are also known as power air-purifying respirators (PAPRs). Due to positive pressure, no inward leakage occurs. Special training is required prior to using this equipment.

Air-Line Respirator. An *air-line respirator* is a type of respirator that uses a small-diameter hose from a compressed air cylinder or air compressor to supply air to the mask. Air-line respirators require that the wearer be connected to the air cylinder or air compressor by a hose at all times. Special training is required prior to using an air-line respirator.

Self-Contained Breathing Apparatus (SCBA). A *self-contained breathing apparatus (SCBA)* is a type of respiratory protection that uses air supplied by a compressed air tank on the back of the wearer. An SCBA allows the wearer to be very mobile. Due to the limits in size (weight) of the compressed air tank, the wearer usually has between 20 min to 60 min of usable air. Special training is required prior to using this equipment.

Controlling Bloodborne Pathogens

On April 18, 2001, OSHA revised the Bloodborne Pathogens Regulation, 29 CFR 1910.1030, which applies to all persons who may reasonably come in contact with blood or other potentially infectious materials while on the job. The regulation includes infectious materials contacting the skin, eyes, and mucous membranes, or entering the body through a scratch or cut. The focus of the regulation is the creation of a written Exposure Control Plan, which explains how

employers (building owners) will protect employees from infectious materials.

Building owners are required by regulation 29 CFR 1910.1030 to evaluate occupational exposure based on job descriptions and duties and to provide work practices and means of protection, including information on how to reduce potential exposure. As part of the Exposure Control Plan, buildings and facilities must have a bloodborne pathogen personal protection kit. When used properly, bloodborne pathogen personal protection kits allow building management to safely remediate a blood or bodily fluid incident. **See Figure 4-9.**

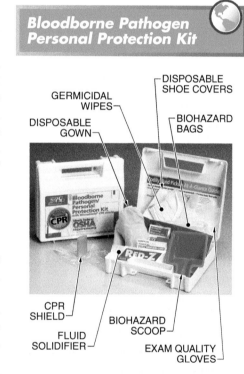

FirstAidStore.com

Figure 4-9. *Bloodborne pathogen personal protection kits allow building personnel to work safely when remediating a blood or bodily fluid incident.*

Case Study – Commercial Office Building

Complaint History

A commercial office building with 22 stories has commercial business tenants on all floors. The facility is 90% occupied and has been operating as a commercial facility for 10 years. Suddenly, the engineering department began receiving complaints from the third floor of the facility.

An investigation identified several items as potential problems. For example, a new tenant on the floor had recently had new furniture and carpet installed upon moving in. Complaints started almost immediately after the new tenant was in the building. Tenants on the third floor complained about a lack of airflow, high temperatures, and miscellaneous odors. Complaints were spread out intermittently across the floor. However, one corner of the building seemed to have universal temperature and air-flow problems.

Resolution

Results from temperature and relative humidity readings confirmed an inability to maintain setpoints for both properties. Airflow measurements showed an approximate 60% reduction from the original airflow design to the diffusers across the floor. The exception was one corner of the building, which showed zero airflow.

After inspecting all of the air-handler system-related controls and components, it was discovered that the floor damper linkage was loose and had slipped to a position allowing only 30% of the total airflow design. Correcting the damper linkage problem restored full airflow to the majority of the floor.

Further examination of the ventilation system revealed a broken fire damper link on a branch duct that served the corner of the building. A previous tenant had a fire damper installed in the branch duct to comply with the fire code for a special application room in that corner of the building. The repair of the fire damper resolved the remaining airflow problem.

Safety Definitions...

- An *air-line respirator* is a type of respirator that uses a small-diameter hose from a compressed air cylinder or air compressor to supply air to the mask.
- A *disposable dust mask* is a type of respiratory protection made of a filter material that is worn over the nose and mouth.
- *Eye protection* is a properly fitting set of goggles or a full-face respirator mask that covers the eyes.
- *Foot protection* is footwear such as shoes, boots, or covers that protect the foot and seal the bottom of the coverall or bodysuit leg.
- A *full-face helmet* is a type of respiratory protection that covers the entire head and has a filtered air blower to maintain positive pressure in the helmet or hood.
- A *full-face respirator* is a type of respirator protection that covers the entire face including the eyes and has filter cartridges that protect against dust, fumes, and particulates.

...Safety Definitions

Definitions

- *Full PPE* is the level of protection required to perform large-size tasks or to perform a task in a large-size area (over 100 sq ft).
- A *half-face respirator* is a type of respiratory protection that covers the nose and mouth and has filter cartridges that protect against dust, fumes, and particulates.
- *Hand protection* is gloves that protect the skin from contaminants and cleaning solutions.
- *Head protection* is a hood or bouffant cap that protects the head from contaminants or mold.
- *Limited PPE* is the level of protection required to perform medium-size tasks or to perform a task in a medium-size area (10 sq ft to 100 sq ft).
- *Minimum PPE* is the level of protection required to perform small-size IAQ tasks or to perform a task in a small area (under 10 sq ft).
- *Personal protective equipment (PPE)* is clothing and equipment used to protect personnel from sickness, injury, or death by creating a barrier against IAQ hazards.
- A *respirator* is a piece of equipment that protects IAQ personnel from inhaling dust, airborne mold, mold spores, and gases.
- A *self-contained breathing apparatus (SCBA)* is a type of respiratory protection that uses air supplied by a compressed air tank on the back of the wearer.
- *Special clothing* is a coverall or bodysuit that is treated to be impervious to specific fibers, gases and/or chemicals.

**For more information please refer to ATPeResources.com
or http://www.epa.gov.**

Review Questions

1. What are some of the possible dangers personnel may encounter when investigating or testing for contaminants?

2. When is full PPE used?

3. What is the purpose of special clothing?

4. Describe when a half-face respirator is used versus a full-face respirator.

5. How many minutes of usable air does a SCBA typically contain?

HANDHELD METER TESTING

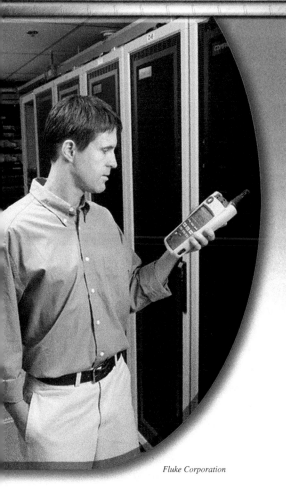

Fluke Corporation

Class Objective:
Describe different handheld measurement devices and testing procedures.

On-the-Job Objective:
Based on local facility conditions, create a list of handheld measurement devices that would be desirable in the facility.

Learning Activities:
1. Describe at least four steps for selecting a measurement device.
2. List the elements of a written quality assurance plan for measurement instruments.
3. Explain how wide variations in temperature and relative humidity measurements can indicate an IAQ issue.
4. Describe the purpose of a chemical smoke test.
5. **On-the-Job:** Use handheld measurement devices to evaluate indoor air quality in the building.

Introduction

A measurement and testing process should be in place in order to investigate IAQ complaints and resolve problems. While measurement processes are important, sampling for specific chemical contaminants is not always the first step to take when a problem is suspected. In many cases, a complaint can be resolved without any testing for contaminants. Basic air properties and indoor air quality must be understood before handling complaints and testing for contaminants. Handheld meters are used to test the levels of some contaminants.

IAQ COMPLAINTS

Complaints often vary from the vague to the precise, some of which can lead to legal problems. Legal liability can be a major concern for a company or building manager, and following a set procedure for handling IAQ complaints can help protect a company from legal problems. During the inspection and testing process following a complaint, all ideas and actions must be documented on a work order. A holding file should be established for these records. Notes from the work order should be recorded in an IAQ log, along with any HVAC system adjustments needed to meet acceptable air quality levels. An IAQ log, with detailed measurements and facility notes, may refute claims made in subsequent litigation.

Complaint Procedure

Once a complaint is received through the proper channels from a building or facility occupant, the response should be immediate. **See Figure 5-1.** Responding immediately increases the ability of stationary engineers to resolve IAQ problems before they become unmanageable. A quick response also saves money and helps to maintain good relations with building occupants.

The stationary engineer should speak with the occupant who filed the complaint. The person handling the complaint should be polite, concerned, and ask for further details on the nature of the complaint. All contributing factors that require investigation should be written down.

In addition, if possible, a time frame when the symptoms first occurred should be established. A time frame assists in evaluating possible reasons for the occupant's complaint.

An initial visual inspection of the building occupant's space is then conducted. An inspection provides a clear picture of the area and points out any obvious issues. During the inspection, various items should be checked around the complaint location, such as obstructed grills, occupancy level, occupant activity, equipment, odors, lighting, and noise.

The temperature and relative humidity in the area where the occupant feels the symptoms most strongly should be examined. Any thermostats or sensors that are operating improperly should be calibrated or repaired. High temperatures and relative humidity can cause a large emission of certain volatile organic compounds (VOCs), like formaldehyde, into the air from certain materials. Relative humidity that is too low can cause irritation in the eyes, nose, and throat.

The airflow from nearby diffusers and at the complaint source location should be checked. If airflow is not at the correct level, trained personnel should check the airflow at every diffuser on the heating, ventilation, and air conditioning (HVAC) loop and rebalance the entire loop.

The investigation should include an examination of space conditions for problems such as water damage, mold or mildew, odors, dust on surfaces, open chemical products, or damaged walls that allow gas or vapor penetration from other sources. Space activities and timing patterns should also be evaluated. Activities may include office and custodial tasks. Any activities that occurred before or occur during the time when symptoms are noticed should be recorded.

Records should also include any products used during these periods, such as cleaning products, glues, correction fluid, perfumes, and tabletop air fresheners. This list helps to identify possible contaminants generated by the activities or products. From the chemical inventory, the material safety data sheets (MSDSs) should be reviewed to determine what products are used in the facility space. This information should be saved for future reference.

Any changes made in the complaint area, such as the installation of new office partitions, the addition of laser printers, and the addition of people working in the space, should be noted. If no obvious contaminants are found in the complaint area, pathways leading out of the area should be inspected. Reviewing the contaminant pathway and accompanying prints may give insight into the problem. Spaces connected to the complaint area for contaminants are also evaluated. If these steps do not solve the problem, a more thorough inspection of the HVAC system should be considered.

When the problem is resolved, all forms used and a brief account of actions taken should be filed in the incident log for future reference. The facility's procedure for reporting all findings to the building occupants should be followed. Individual occupants should notify the appropriate authority concerning recurring problems.

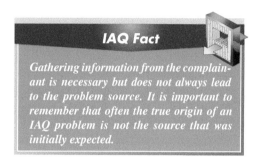

IAQ Fact

Gathering information from the complainant is necessary but does not always lead to the problem source. It is important to remember that often the true origin of an IAQ problem is not the source that was initially expected.

IAQ Compliant Investigation Procedures

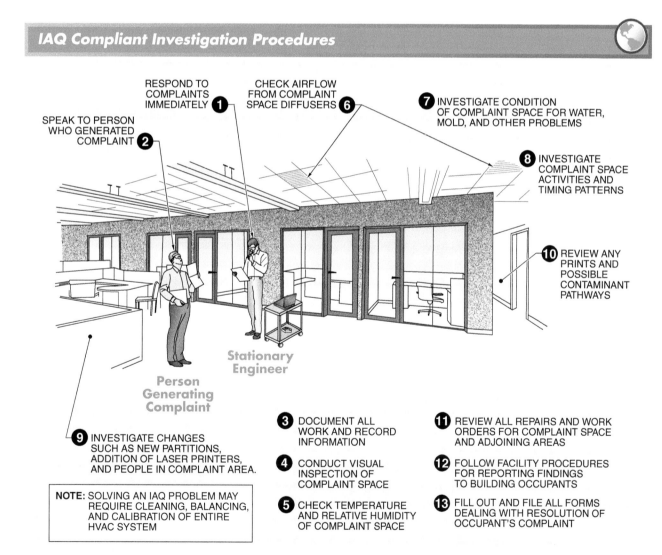

RESPOND TO COMPLAINTS IMMEDIATELY ❶

CHECK AIRFLOW FROM COMPLAINT SPACE DIFFUSERS ❻

❼ INVESTIGATE CONDITION OF COMPLAINT SPACE FOR WATER, MOLD, AND OTHER PROBLEMS

SPEAK TO PERSON WHO GENERATED COMPLAINT ❷

❽ INVESTIGATE COMPLAINT SPACE ACTIVITIES AND TIMING PATTERNS

❿ REVIEW ANY PRINTS AND POSSIBLE CONTAMINANT PATHWAYS

Stationary Engineer

Person Generating Complaint

❾ INVESTIGATE CHANGES SUCH AS NEW PARTITIONS, ADDITION OF LASER PRINTERS, AND PEOPLE IN COMPLAINT AREA.

NOTE: SOLVING AN IAQ PROBLEM MAY REQUIRE CLEANING, BALANCING, AND CALIBRATION OF ENTIRE HVAC SYSTEM

❸ DOCUMENT ALL WORK AND RECORD INFORMATION

❹ CONDUCT VISUAL INSPECTION OF COMPLAINT SPACE

❺ CHECK TEMPERATURE AND RELATIVE HUMIDITY OF COMPLAINT SPACE

⓫ REVIEW ALL REPAIRS AND WORK ORDERS FOR COMPLAINT SPACE AND ADJOINING AREAS

⓬ FOLLOW FACILITY PROCEDURES FOR REPORTING FINDINGS TO BUILDING OCCUPANTS

⓭ FILL OUT AND FILE ALL FORMS DEALING WITH RESOLUTION OF OCCUPANT'S COMPLAINT

Figure 5-1. A facility procedure should be followed when responding to possible IAQ complaints.

MEASUREMENT AND TESTING

Measurements should always be taken when a facility or building first opens in order to establish readings as part of a preventive maintenance (PM) program. Initial measurements provide baseline information about the facility, which helps when troubleshooting IAQ complaints. When establishing baseline information, it is necessary to identify sample testing areas that represent the different parts of the facility. Testing may need to be performed more frequently during the baseline measurement program to obtain a good profile.

Environmental conditions and the presence of chemicals and gases within a space can be easily measured with various test instruments (meters). **See Figure 5-2.** Test intervals vary depending upon the kind of PM program and visual inspection schedule that a facility

or building has in place. Whatever sampling program is established, the sampling program should also include testing the outdoor air.

Testing should be used to answer questions raised by specific symptoms or to confirm a hypothesis based upon an initial investigation. For example, testing is done to confirm that a formaldehyde problem is the cause of itchy skin and watery eyes in a building where new furniture was installed. Some testing may also be part of a regularly scheduled air-sampling strategy to log baseline measurements of the air quality in a facility.

Specialized tests are available, but the majority of IAQ issues can be resolved by testing for standard air conditions and common contaminants. If additional tests are required, facility management can consult with outside professionals.

IAQ Test Instrument Measurements

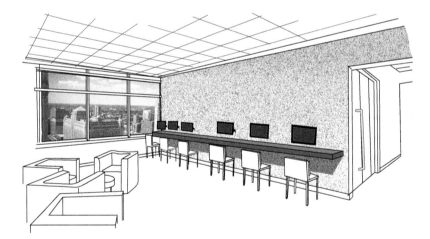

Facility or building space test instrument measurements	
Measurable Conditions	Measurable Contaminants
Temperature	Nitrogen dioxide
Pressure	Carbon monoxide
Relative humidity	Carbon dioxide
Air volume	Ozone
Air velocity	Formaldehyde

Figure 5-2. Certain conditions and contaminants can be measured easily with basic test instruments.

IAQ Direct-Reading Test Instruments

When conducting complaint investigations, personnel may use direct-reading instruments, chemical detector tubes, airflow measurement devices, and passive monitors. Direct-reading instruments, also known as handheld meters, are capable of measuring general building conditions such as air temperature, airflow, relative humidity, and carbon dioxide and carbon monoxide levels. Some handheld meters are capable of testing for multiple building conditions.

For complaint investigations, measurements are taken both in the complaint area and outside the complaint area. When checking ventilation adequacy, measurements should coincide with known peak occupant concentration periods. Periodic sampling helps determine when the pollutant peaks occur.

IAQ Test Instrument Calibration

In general, it is best to sample temperature, relative humidity, carbon dioxide, and carbon monoxide at periodic intervals. Each section of a facility should be part of the sampling. The frequency of the samplings is determined mainly by the type and size of the facility. The same instruments (for reading comparisons) are always used to take measurements when obtaining IAQ samples, and instruments must be kept properly calibrated. All handheld meters must be calibrated on a regular basis. Usually, carbon monoxide meters must be calibrated every month, while carbon dioxide meters require calibration once a year. **See Figure 5-3.**

When calibrating an instrument, facts to be noted include the frequency of calibration, the date calibrated, and the person who calibrated the instrument. Calibration notes must be recorded in an IAQ log, along with any HVAC system adjustments required to ensure proper air quality levels. An IAQ log, with detailed measurements and facility notes that include the exterior of the building, is helpful for future troubleshooting circumstances and also serves as support material in the defense of legal actions.

Extreme care must be used when taking IAQ measurements, including the use of the most accurate instrument available. Contaminant levels in commercial buildings are often low in comparison to industrial standards. Usually there is little difference between an acceptable and an unacceptable indoor environment in a commercial building. The error factor of inaccurate test instruments can imply that there is no problem, or that there is a problem where none exists. Errors of any kind can have serious health and legal consequences.

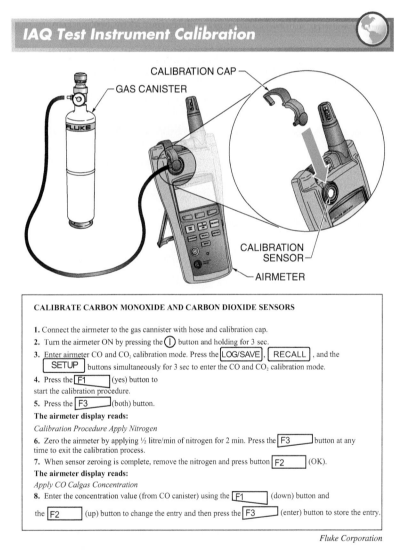

IAQ Test Instrument Calibration

CALIBRATE CARBON MONOXIDE AND CARBON DIOXIDE SENSORS

1. Connect the airmeter to the gas cannister with hose and calibration cap.
2. Turn the airmeter ON by pressing the ⏻ button and holding for 3 sec.
3. Enter airmeter CO and CO_2 calibration mode. Press the LOG/SAVE , RECALL , and the SETUP buttons simultaneously for 3 sec to enter the CO and CO_2 calibration mode.
4. Press the F1 (yes) button to start the calibration procedure.
5. Press the F3 (both) button.

The airmeter display reads:

Calibration Procedure Apply Nitrogen

6. Zero the airmeter by applying ½ litre/min of nitrogen for 2 min. Press the F3 button at any time to exit the calibration process.
7. When sensor zeroing is complete, remove the nitrogen and press button F2 (OK).

The airmeter display reads:

Apply CO Calgas Concentration

8. Enter the concentration value (from CO canister) using the F1 (down) button and the F2 (up) button to change the entry and then press the F3 (enter) button to store the entry.

Fluke Corporation

Figure 5-3. *Direct-reading test instruments are the instruments of choice for IAQ complaint investigations due to their mobility.*

Manufacturers' recommendations must be followed when operating air quality test instruments. It is important to read all instructions carefully and to pay special attention to operation warnings. Air quality test instruments can sometimes display faulty readings due to the presence of excessive electromagnetic radiation. Unexpected gases that are present in the space being tested can also cause inaccurate readings when detector tubes are used.

IAQ Instrument Selection

Selecting an IAQ test instrument is a process that consists of several steps. One of the main uses of an instrument selection process is to judge between acceptable and unacceptable conditions of air quality. IAQ test instruments must also have a level of precision high enough to provide acceptable statistical confidence. A test instrument that does not have a high level of accuracy for an application cannot be trusted. The detection limit of the test instrument should be above or below the precision level required for the application by a factor of at least 10.

The instrument range should be at least twice the level of concern in order to provide statistical confidence in the results. **See Figure 5-4.** Results from an IAQ measurement must be scientifically defensible. Any conclusions drawn from the data gathered must also adequately and truthfully represent the situation. Data correctly derived from test instruments can withstand scrutiny.

Legal claims and counterclaims are often based on the IAQ testing process. To select an IAQ test instrument, first a list of possible contaminants should be established. Separate reviews for each contaminant should then be set up. IAQ standards are used to develop the level of concern.

The next step is to define the data quality objectives (DQOs). Data quality objectives usually include instrument precision, the method of detection and detection limit, and the representative nature of the samples. The *method detection limit* is the minimum concentration of a contaminant that is reported as zero. The representative nature of the samples relates to the degree to which the measurements taken are characteristic of the whole area, medium, exposure, or dose for which the data is used to make decisions. When no regulations are in place for an application, the person doing the testing determines the DQOs.

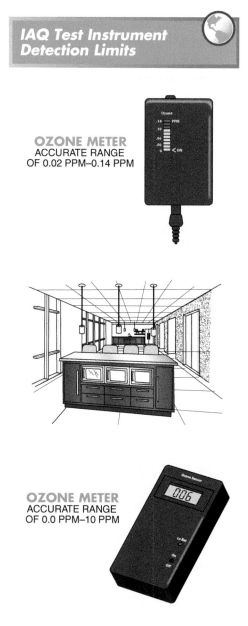

IAQ Test Instrument Detection Limits

OZONE METER
ACCURATE RANGE
OF 0.02 PPM–0.14 PPM

OZONE METER
ACCURATE RANGE
OF 0.0 PPM–10 PPM

Figure 5-4. Direct-reading IAQ test instruments are usually calibrated every month but some test instruments only require calibration once a year.

After the DQOs are established, it is necessary to evaluate the operational requirements, such as the proper frequency and duration for measurements being taken. Operational requirements are derived from the problem that exists and must be taken at the proper spot in the problem space or area. Other factors that may influence the instrument selection include the following:

- active or passive instrument operational mode
- output recording method, including continuous recording, single point in time, or weighted over time
- report generated, such as electronic signal, computerized data storage, or laboratory report
- portable or fixed-test instrument
- test instrument power requirements, such as batteries, AC, or mechanical
- instrument calibration, including frequency, voltage and current generators, particulate generators that follow laboratory procedures or factory recommendations
- cost of equipment
- personnel training required for use of instrument

After selecting the appropriate IAQ test instrument(s), the selection process, decisions, and rationale must be documented. The specific details that should be recorded include the objectives, levels of concern, data quality objectives, and actual instrument selection.

In addition to documenting the instrument selection process, a written quality assurance plan is desirable as well. A *written quality assurance (QA) plan* is a document designed to deliver defensible data acquired using test instruments about indoor air quality. **See Figure 5-5** and **Appendix.** A written QA plan should contain the following:

- data quality objectives
- assessment of data quality
- standard operating procedures
- internal quality control checks
- system performance and data audits

- document control
- corrective measures

The established protocol developed for each type of air sampling must be followed, especially when more than one person is taking measurements. Measurements taken with different test instruments or under varying methods and conditions can easily lead to false conclusions or interpretations. For example, IAQ readings that are outside of acceptable parameters must be double-checked by recalibrating the test instrument and retesting.

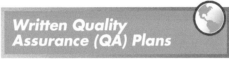

Environmental Protection Agency

Figure 5-5. *Written quality assurance plans contain the data acquired by measurements performed with IAQ test instruments.*

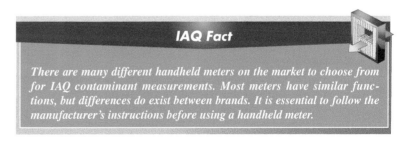

IAQ Fact

There are many different handheld meters on the market to choose from for IAQ contaminant measurements. Most meters have similar functions, but differences do exist between brands. It is essential to follow the manufacturer's instructions before using a handheld meter.

Traceability

Traceability is the unbroken chain of paperwork relating a test instrument's accuracy to a known standard. In the United States, national standards for weights and measures are maintained by the National Institute of Standards and Technology (NIST). Some IAQ measurement instruments come with a certificate of traceability that is used to prove that the instrument has been calibrated to a known standard. Not every instrument or situation requires traceability; however, traceability aids in providing defensible results.

AIR PROPERTIES

The basic thermal properties of air must be tested in addition to testing contaminants. Maintaining space temperature and relative humidity is critical to keeping occupants comfortable. **See Figure 5-6.** Complaints about excessively warm or cool temperatures should be corrected as soon as possible. Air comfort properties are also important for managing contaminants. Temperature and relative humidity play a vital role in controlling microbial contamination in building spaces.

Air Comfort Properties

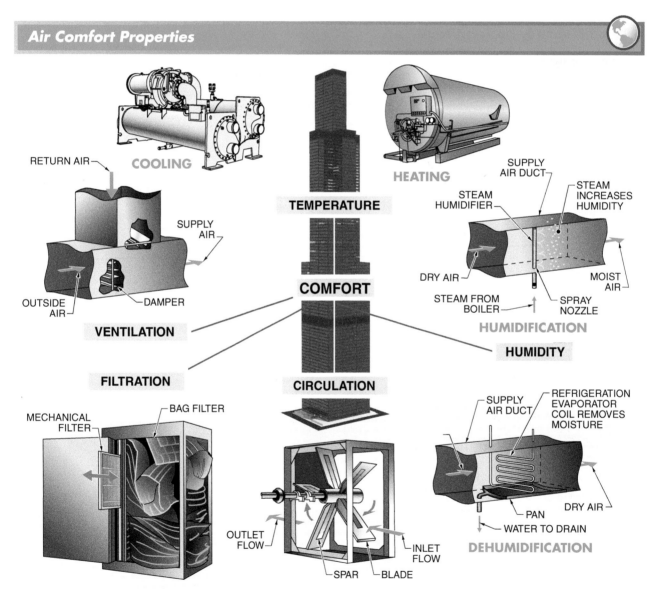

Figure 5-6. The measurement of basic air properties, in addition to contaminants, provides a baseline for the general condition of air quality within a building.

Temperature and Relative Humidity

Temperature measurements are made with a thermometer and relative humidity measurements are made with a sling or fan-operated psychrometer. Humidity measurements can also be performed by using a thermohygrometer. Today, thermometers, psychrometers, and hygrometers can be purchased with direct-digital readouts. Digital readout models minimize the need for calculations and using the psychometric chart.

Thermometers used for IAQ measurements should have a temperature accuracy of at least ±2°F (approximately ±1°C). A humidity-sensing instrument, such as a psychrometer or hygrometer, should have an accuracy of at least ±2% relative humidity. Laboratories and manufacturing facilities usually require a higher accuracy.

Digital humidity probes may be easiest to use because no calculations are needed to derive the relative humidity percentage. In many cases, multifunction test instruments provide carbon dioxide and carbon monoxide sensing as well as temperature and humidity measurements. **See Figure 5-7.**

Multifunction Air Test Meters

- TEMPERATURE MEASUREMENTS
- RELATIVE HUMIDITY MEASUREMENTS
- CARBON DIOXIDE MEASUREMENTS
- CARBON MONOXIDE MEASUREMENTS

Fluke Corporation

Figure 5-7. *Some IAQ test instruments can display several measurements simultaneously.*

Temperature and relative humidity measurements are often taken simultaneously. A good IAQ investigation includes indoor and outdoor measurements of the temperature and relative humidity for purposes of comparison. Taking indoor and outdoor readings provides a clearer picture of the conditions under which the system is operating and the adjustments that are feasible.

To establish baseline measurements, it is necessary to take daily temperature and relative humidity measurements in representative areas of HVAC zones. Occupancy levels, types of equipment in use, and the type and amount of lighting should also be documented. When testing temperature, personnel should ensure that room sensors are properly located. An HVAC system often responds only to the temperature at the room sensor, so the test instrument probe should be placed within a few inches of the room sensor.

When performing temperature or humidity measurements, personnel should not breathe on temperature or relative humidity sensor probes. Similarly, a stationary engineer should hold the probe so that his or her hands are far from the sensor or probe tip. The radiant heat from hands can cause sensors and probe tips to display incorrect temperature measurements.

Temperature and relative humidity readings should be taken at several locations and heights within an occupied space. Temperature measurements taken should be even throughout the space. Wide variations in temperature could indicate inadequate air circulation within the space or the presence of radiant heat from windows or equipment. The vertical temperature difference, from floor to ceiling, should not exceed 5°F within an occupied space.

IAQ Fact

*The occupied zone is defined by ASHRAE Standard 62.1-2004, **Ventilation for Acceptable Indoor Air Quality,** as "the region within an occupied space between planes 3" to 72" (75 mm to 1800 mm) above the floor and more than 2' (600 mm) from any walls or fixed air-conditioning equipment."*

Temperature measurements that vary widely suggest balance or control problems with the HVAC system.

Measurements are also taken in specific areas where mold or IAQ issues are more likely to be present. Cabinet doors and decorative furnishings can block airflow to small spaces, causing humidity to increase over the recommended value and contributing to mold growth.

Taking indoor and outdoor measurements provides a clearer picture of the conditions under which an HVAC system is operating and the adjustments that may be required. To establish baseline measurements, daily temperature and relative humidity measurements must be taken in representative areas of each HVAC zone.

Air Volume

An air volume measurement is taken to confirm that airflow out of each diffuser and register in a building meets design specifications. Comparisons can be made with a schematic duct system layout print or an outlet air balance report from the most recent written test and balance report.

Anemometers or air-capture hoods are used to measure the volume of air moving (flow rate) in cubic feet per minute (cfm). An *anemometer* is a device that measures average air velocity or average airflow rate. An anemometer is placed at the opening of a diffuser and is operated according to the manufacturer's instructions. **See Figure 5-8.** The manufacturer's specifications indicate the recommended number of readings and the location where measurements should be taken.

Velocity measurements are averaged and then multiplied by the K factor for the specific type of diffuser in use. The *K factor* is a number that represents the actual open area and design characteristics of a specific diffuser. The result of the K factor equation is the flow rate, in cubic feet per minute (cfm), out of the diffuser.

A *flow hood* is a test instrument that is placed over a diffuser to measure the volume of air moving in cubic feet per minute. **See Figure 5-9.** Flow hoods are easier to use than anemometers because flow hoods provide a direct reading without the need for calculations. Flow hoods also save a considerable amount of time when performing field measurements.

Testing Air Volume with Anemometer

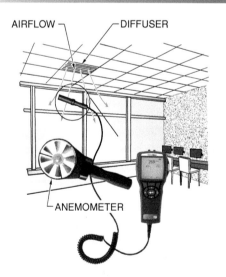

Figure 5-8. An anemometer is used to measure the airflow rate in cubic feet per minute (cfm) to grills and from registers.

Flow Hood

TSI Incorporated

Figure 5-9. A flow hood is a device that can be used to measure the air volume in cubic feet per minute from diffusers.

Air Circulation

Air must be allowed to circulate completely through an occupied zone. Air is needed to flush contaminants from occupied and unoccupied spaces, which cannot happen when air is blocked from reaching certain locations. The American Society of Heating, Refrigeration and Air-Conditioning Engineers (ASHRAE) sets the maximum air velocity in a building at 50 feet per minute (fpm) in the summer and 30 fpm in the winter. Higher air velocities may cause drafts. Anemometers or velometers with a range of 5 fpm to 100 fpm are appropriate tools for taking air velocity measurements. A *velometer* is a device that measures the velocity of air flowing out of a register.

A *chemical smoke test* is an IAQ test that shows the airflow inside a space or from one space into another space. Chemical smoke flows with the smallest of air currents because chemical smoke is heatless. **See Figure 5-10.**

Chemical smoke flows from high-pressure spaces to low-pressure spaces. Smoke flowing from a space, such as under a door or through window seals, shows that the space is under positive pressure. Smoke flowing back into a space, such as away from a door or window, shows that the space is under negative pressure. Smoke released at the shell of a building (by doors or windows) indicates that HVAC systems are maintaining interior spaces under positive pressure relative to the outdoors.

Chemical Smoke Test

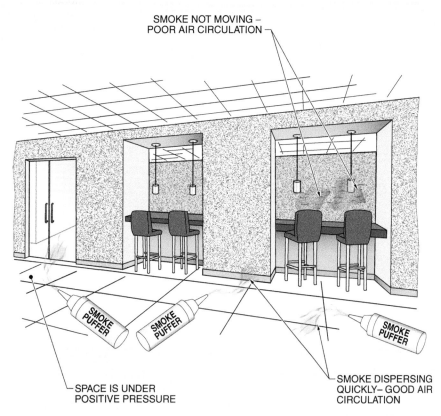

SMOKE NOT MOVING – POOR AIR CIRCULATION

SPACE IS UNDER POSITIVE PRESSURE

SMOKE DISPERSING QUICKLY – GOOD AIR CIRCULATION

SMOKE PUFFER

Figure 5-10. Chemical smoke tests help determine how airflow is moving inside a space or between spaces.

Smoke is initially released when an HVAC system is operating normally to acquire base airflow information. For the best smoke test results, testing personnel should release the smoke in short puffs. Smoke that disperses in a few seconds is evidence of good air circulation. Smoke that does not move shows that there is poor air circulation within the testing space.

Smoke can move away from a diffuser quickly or slowly, giving an idea of air velocity. Readings must be conducted throughout the day and under varied conditions. **WARNING:** Avoid inhaling the smoke and do not release smoke near building occupants.

HANDHELD METER TESTING FOR CONTAMINANTS

The presence of contaminants as part of the inside air of a building does not mean that occupants will experience harmful health effects. The duration, amount of exposure, and the type of contaminant determine the health effects felt by building occupants.

Some contaminants may be easily tested using handheld meters. Handheld meters are often used to test nitrogen dioxide, carbon monoxide, carbon dioxide, and ozone.

Nitrogen Dioxide

In the past, outside experts conducted real-time sampling for nitrogen dioxide. When elevated nitrogen dioxide (NO_2) concentrations are present, there are also elevated carbon monoxide (CO) concentrations. Therefore, stationary engineers would perform CO samplings to get a sense of NO_2 levels. Solving a CO problem generally solved any NO_2 problem. Today, portable nitrogen dioxide monitors are available with the required level of accuracy.

No standards have been agreed upon for permissible levels of nitrogen dioxide in indoor air. ASHRAE and the U.S. Environmental Protection Agency National Ambient Air Quality Standards (NAAQS) list 0.25 parts per million (ppm) to 1.0 ppm for an hourly average limit and 0.053 ppm as the average 24 hr limit for NO_2 in indoor air. **See Figure 5-11.**

NITROGEN DIOXIDE EXPOSURE LIMITS					
CLASSIFICATION	**10 MINUTES***	**30 MINUTES***	**1 HOUR***	**4 HOURS***	**8 HOURS***
Nondisabling	0.50	0.50	0.50	0.50	0.50
Disabling	20	15	12	8.2	6.7

EXPOSURE	**SYMPTOMS**	**PREVENTION**	**FIRST AID TREATMENT**
Inhalation	Burning sensation, cough, headache, dizziness, sore throat, nausea, and vomiting	Ventilation–area exhaust Respiratory protection–supplied air	Half-upright position with rest and fresh air; artificial respiration may be required.
Skin	Burns, redness, and pain	Protective clothing and neoprene gloves	Rinse with water for 15 min; remove clothes (discard); rinse again for 15 min
Eyes	Severe burns, redness, and pain	Nonvented eye protection or face shield	Rinse eyes with water for 15 min
Ingestion	—	Do not drink or eat in nitrogen dioxide environment	Rinse mouth for 15 min

* in ppm

Sources: Occupational Safety & Health Administration (OSHA), U.S. Department of Health and Human Services–Food and Drug Administration (FDA), and Praxair, Inc.

Figure 5-11. No official federal standards have been set for permissible levels of nitrogen dioxide; however, some basic exposure limits have been set by various agencies.

Nitrogen Dioxide Monitor Usage. A *nitrogen dioxide monitor* is a meter that measures the amount (in parts per million) of nitrogen dioxide in an air sample. Measurement of the concentration of NO_2 in an air sample is based on a chemical reaction between ozone (O_3) and nitrogen monoxide (NO), which produces light at a certain wavelength. The intensity of the light energy is proportional to the concentration of nitrogen monoxide and is measured by a photodetector.

Inside a nitrogen dioxide monitor, sampled air is drawn through two separate paths. In one path, the air is mixed directly with ozone and the resulting light energy is measured to determine the concentration of nitrogen monoxide (NO) in the air. In a separate path, sampled air is sent first through a catalytic reduction surface, which converts the NO_2 to NO. The sample is then mixed with ozone to create light energy, which is measured with the photodetector. This measurement represents the combined concentrations of NO and NO_2 in the original air sample. The difference between the two measurements is the level of NO_2 alone in the original air sample.

Nitrogen Dioxide Measurement Procedures. Some nitrogen dioxide monitors are connected to a personal computer (PC) to store and display continuous measurements. **See Figure 5-12.** Testing procedures first involve setting up the nitrogen dioxide monitor and synchronizing it with the PC. Software is typically required on the PC for communicating with the nitrogen dioxide monitor. This allows continuous nitrogen dioxide measurements to be displayed on both the handheld monitor and the PC.

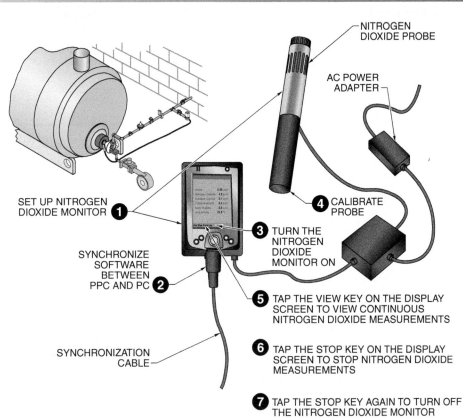

Nitrogen Dioxide Measurement Procedures

NITROGEN DIOXIDE PROBE

AC POWER ADAPTER

SET UP NITROGEN DIOXIDE MONITOR **1**

SYNCHRONIZE SOFTWARE BETWEEN PPC AND PC **2**

SYNCHRONIZATION CABLE

4 CALIBRATE PROBE

3 TURN THE NITROGEN DIOXIDE MONITOR ON

5 TAP THE VIEW KEY ON THE DISPLAY SCREEN TO VIEW CONTINUOUS NITROGEN DIOXIDE MEASUREMENTS

6 TAP THE STOP KEY ON THE DISPLAY SCREEN TO STOP NITROGEN DIOXIDE MEASUREMENTS

7 TAP THE STOP KEY AGAIN TO TURN OFF THE NITROGEN DIOXIDE MONITOR

Figure 5-12. Nitrogen dioxide monitors measure the amount of nitrogen dioxide in an air sample in parts per million (ppm).

Additional initialization procedures are indicated on the screen of the nitrogen dioxide monitor. These are used to view, start, and stop the continuous nitrogen dioxide measurements, in addition to accessing other device functions.

The probe sensor for nitrogen dioxide measurements reacts slowly to changes. After the nitrogen dioxide monitor is powered on, the probe sensor should be allowed to stabilize before measurements are taken. After short power downs, the probe will stabilize within 10 minutes to 1 hour. After long power downs, up to 24 hours should be allowed for the probe sensor to stabilize.

The probe should be calibrated every 6 months. This involves selecting the calibration function in the monitor for the nitrogen dioxide probe and placing the probe in the appropriate calibration environment. The monitor measures the nitrogen dioxide concentration and stores the information in its calibration parameters as a known value.

IAQ Fact

Many types of contaminants can be measured by a handheld meter designed for that specific contaminant only. However, some handheld meters can measure several contaminants simultaneously.

Carbon Monoxide

Carbon monoxide (CO) can be measured by direct-reading instruments or detector tubes. Measuring more than one sample per location improves the accuracy of the results. Direct-reading instruments or detector tube measurements should be in the range of 1 ppm to 50 ppm. **See Figure 5-13.**

The accuracy of detector tubes is ± 25%, at a minimum. Detector tubes, which take spot samples, are sufficiently accurate for screening samples. It is important to follow the manufacturer's directions when using detector tubes. Detector tubes are relatively inexpensive, and replacement tubes can be purchased as necessary.

CARBON MONOXIDE EXPOSURE LIMITS					
CLASSIFICATION	**10 MINUTES***	**30 MINUTES***	**1 HOUR***	**4 HOURS***	**8 HOURS***
Nondisabling	2500	1500	600	150	35–50
Disabling	6000	3000	800	400	80

EXPOSURE	**SYMPTOMS**	**PREVENTION**	**FIRST AID TREATMENT**
Inhalation	Asphyxiation	Ventilation–area exhaust Respiratory protection–supplied air	Artificial respiration and cardiopulmonary resuscitation
Skin	Possible frostbite	Appropriate gloves–not rubber or neoprene	Rinse with warm water
Eyes	Mechanical injury	Nonvented eye protection or face shield	Force eyes open and rinse with lukewarm water
Ingestion	—	—	—

* in ppm
Sources: Occupational Safety & Health Administration (OSHA), Spectra Gases, Inc., Washington State Department of Labor and Industries, and International Program on Chemical Safety (IPCS)

Figure 5-13. Several severe health symptoms may result when exposure limits of carbon monoxide are exceeded.

However, incorrect readings do occur because common chemical products such as cleaning fluids, perfumes, and hairspray are often present in the air samples.

In addition to regular annual readings, carbon monoxide measurements are also taken when a combustion source is suspected of having a problem. Measurements should be taken at high-peak use and potential exposure times. For example, measurements should be taken for automotive exhaust during morning and evening rush hour times.

Carbon Monoxide Meter Usage. A *carbon monoxide meter* is a meter that measures the amount (in parts per million) of carbon monoxide in an air sample. A carbon monoxide meter indicates that CO is present by displaying the amount on an LCD display and by using an alarm. The alarm is usually activated at a preprogrammed value of 35 ppm, but can be reprogrammed for any alarm value. Many CO meters indicate any amount displayed above 1000 ppm as OL (over measured limit).

CO is measured based on how it absorbs infrared radiation. An air sample is drawn into a chamber and exposed to a beam of infrared radiation. A decrease in the intensity of infrared radiation is proportional to the concentration of CO in the air sample. This is measured by a special detector. A duplicate beam passes through a reference chamber with no CO present. The difference between the two measurements corresponds to the measurement of CO concentration in the air sample.

Carbon Monoxide Measurement Procedures. When turned ON, a carbon monoxide meter performs a self-test. **See Figure 5-14.** Before testing, the meter establishes a baseline for comparing the CO concentration of the testing environment. It does this automatically after the self-test. If the meter senses more than 3 ppm, the meter then immediately starts taking CO measurements. However, for very low concentrations, such as less than 3 ppm, the meter establishes a new baseline before beginning continuous measurements.

Carbon Monoxide Measurement Procedures

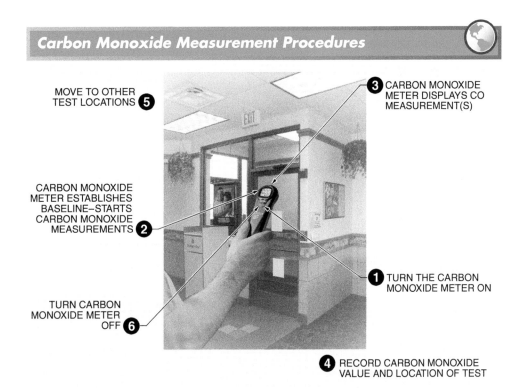

MOVE TO OTHER TEST LOCATIONS **5**

CARBON MONOXIDE METER ESTABLISHES BASELINE—STARTS CARBON MONOXIDE MEASUREMENTS **2**

TURN CARBON MONOXIDE METER OFF **6**

3 CARBON MONOXIDE METER DISPLAYS CO MEASUREMENT(S)

1 TURN THE CARBON MONOXIDE METER ON

4 RECORD CARBON MONOXIDE VALUE AND LOCATION OF TEST

Figure 5-14. Carbon monoxide meters measure the amount of carbon monoxide in an air sample in parts per million (ppm).

The carbon monoxide meter indicates CO measurements on its display. It may also beep to indicate a new maximum level. This feature is useful when moving the meter around to different test locations to determine the area with the highest concentration.

Carbon Dioxide

Normal outdoor air concentrations of carbon dioxide (CO_2) range from 300 ppm to 350 ppm. Indoor carbon dioxide concentrations are slightly higher because people exhale carbon dioxide into indoor air space. During the workday, CO_2 levels peak before lunch and again late in the afternoon before people leave for home.

Measurements of CO_2 peak levels are an indirect measurement of the amount of outdoor air being brought into a work area. According to ASHRAE Standard 62-2007, the upper indoor limit for CO_2 is 1000 ppm. **See Figure 5-15.** CO_2 levels that exceed 1000 ppm indicate inadequate supply or distribution of outdoor air for the level of occupancy in a space. Elevated CO_2 levels are associated with occupant complaints. Occupant complaints can occur at levels below 1000 ppm, particularly if there is a contaminant source that has not been controlled or the occupant space does not have local exhaust ventilation.

To prevent problems, corrective action must be taken anytime levels of CO_2 reach or exceed 800 ppm. Outdoor concentrations above 400 ppm may indicate an outdoor contaminant problem, such as vehicle exhaust or combustion sources. Monitoring outdoor CO_2 levels provides an indication of the quality of air being brought into a facility.

Concentrations of CO_2 are also a good indicator of the concentrations of other contaminants. If CO_2 levels are high, it is likely that there are other IAQ problems. This makes it possible to appropriately suspect other contaminants even if the stationary engineer does not have meters to test for them.

IAQ Fact

Elevated levels of carbon dioxide are remediated by introducing more outdoor air into the building. If there are any other contaminants in the indoor air, increasing ventilation will likely also help lower their concentrations.

CARBON DIOXIDE EXPOSURE LIMITS					
CLASSIFICATION	10 MINUTES*	30 MINUTES*	1 HOUR*	4 HOURS*	8 HOURS*
Nondisabling	15,000	10,000	7500	6000	5000
Disabling	30,000	20,000	15,000	10,000	10,000

EXPOSURE	SYMPTOMS	PREVENTION	FIRST AID TREATMENT
Inhalation	Asphyxiation–headaches and drowsiness	Ventilation–area exhaust Respiratory protection–supplied air	Artificial respiration using oxygen
Skin	Possible frostbite	Cuffless trousers–no rubber or neoprene gloves	Do not remove clothing–rinse with warm water
Eyes	Stinging	Face shield	Force eyes open and rinse with lukewarm water
Ingestion	—	—	—

* in ppm
Sources: Occupational Safety & Health Administration (OSHA), Praxair, Inc., and Valero Marketing & Supply Company

Figure 5-15. Some carbon dioxide exposure limits have been set by the American Society of Heating, Refrigerating and Air-Conditioning Engineers (ASHRAE).

The preferred measuring devices for CO_2 are direct-reading instruments that use infrared or electrochemical detection. The allowable instrument measurement range should be from 200 ppm to 2000 ppm, with an accuracy of ± 10% at 1000 ppm. Most direct-reading detector manufacturers bundle the CO_2 detector along with temperature and humidity measurements. Multifunction measuring devices often come with a data logger and software to store measurements and have the capability to retrieve the measurements later from a computer file.

Detector tubes can also be used, although detector tubes are less accurate. Detector tubes are also affected by hot or cold weather, which makes them less useful for outdoor measurement comparisons.

Baseline measurements are taken from multiple samples in representative spaces of each HVAC zone. Outdoor samples must also be taken, specifically near the outdoor air intake. During each sampling period, personnel should record the air damper settings, relative occupancy, and weather conditions. Initially, it is recommended to take monthly measurements in representative areas of each HVAC zone.

Carbon Dioxide Meter Usage. A *carbon dioxide meter* is a meter that measures the amount (in parts per million) of carbon dioxide in an air sample. A carbon dioxide meter indicates that carbon dioxide is present by displaying the amount of carbon dioxide on an LCD display. Carbon dioxide measurement is a common feature of multifunction meters, often combined with temperature, relative humidity, and carbon monoxide measurement functions.

Similar to carbon monoxide, carbon dioxide concentration is measured by its absorption of infrared radiation. It absorbs radiation at a slightly different wavelength, however, so with the proper filters and electronics, the concentrations of the two gases can be distinguished from each other. The similar methods of detection explain why concentrations of both gases can be measured by many multifunction meters.

Carbon Dioxide Measurement Procedures. When turned ON, a carbon dioxide meter performs a self-test and displays the battery charge, date, last calibration date, and absolute barometric pressure. **See Figure 5-16.** When taking measurements, the sensor should be held perpendicular to the airstream being measured. The measured CO_2 concentration appears on the meter's display. If the meter has a MIN/MAX mode, the display can be toggled to show the maximum, minimum, and/or average CO_2 values.

Stationary engineers should test and record carbon dioxide for multiple room locations by moving around the area with the meter. Each measurement should be carefully recorded in the IAQ testing log. Carbon dioxide should also be tested in various points in an HVAC system, including outside-air intakes, return-air grills, and mixed-air plenums. Some meters allow the user to store these measurements and to use them to automatically calculate the percentage of outdoor air used to ventilate the system.

Fluke Corporation

Handheld meters can be used throughout a building to measure contaminant levels in different spaces.

Ozone

According to the EPA, the permissible ozone (O_3) level for interior building spaces is 0.08 ppm. **See Figure 5-17.** Testing for ozone levels is conducted by outside experts in many facilities because the precision analyzers can be expensive and require extensive calibration. These analyzers are based on the measurement of ultraviolet absorption to determine ozone concentrations.

However, manufacturers have developed less expensive, portable ozone meters that can be used by stationary engineers. These meters use a measurement principle based on semiconductors and is not as accurate as the precision analyzers, but this measurement principle provides a useful estimate of the ozone levels. If these ozone meters detect an elevated concentration of ozone, further investigation can be conducted by outside experts with more sophisticated instruments.

Also, dry colormetric ozone sensors (detector cards) are readily available and affordable for the constant monitoring of personnel in potential ozone-contaminated environments. These cards are worn as badges and change color to indicate approximate ozone exposure.

Ozone Meter Usage. An *ozone meter* is a meter that measures the amount (in parts per million) of ozone in an air sample. An ozone meter indicates the ozone concentration on a multicolored bar graph or LCD display. Occupied space ozone meters usually have an accuracy range of 0.02 ppm to 0.14 ppm, or 0.0 ppm to 10 ppm.

Ozone Measurement Procedures. The ozone meter should be held in the appropriate air sampling space of a building or facility. **See Figure 5-18.** Measurements should be taken in multiple locations as necessary and recorded in the IAQ testing log.

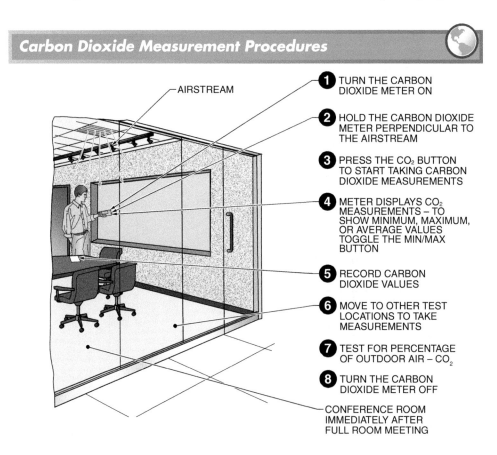

Carbon Dioxide Measurement Procedures

AIRSTREAM

1. TURN THE CARBON DIOXIDE METER ON
2. HOLD THE CARBON DIOXIDE METER PERPENDICULAR TO THE AIRSTREAM
3. PRESS THE CO_2 BUTTON TO START TAKING CARBON DIOXIDE MEASUREMENTS
4. METER DISPLAYS CO_2 MEASUREMENTS – TO SHOW MINIMUM, MAXIMUM, OR AVERAGE VALUES TOGGLE THE MIN/MAX BUTTON
5. RECORD CARBON DIOXIDE VALUES
6. MOVE TO OTHER TEST LOCATIONS TO TAKE MEASUREMENTS
7. TEST FOR PERCENTAGE OF OUTDOOR AIR – CO_2
8. TURN THE CARBON DIOXIDE METER OFF

CONFERENCE ROOM IMMEDIATELY AFTER FULL ROOM MEETING

Figure 5-16. Carbon dioxide measurement test instruments and procedures are similar to carbon monoxide measurement procedures.

OZONE EXPOSURE LIMITS

CLASSIFICATION	10 MINUTES*	30 MINUTES*	1 HOUR*	4 HOURS*	8 HOURS*
Nondisabling	0.3	0.2	0.12	0.11	0.1
Disabling	25	10	2	1	0.5

EXPOSURE	SYMPTOMS	PREVENTION	FIRST AID TREATMENT
Inhalation	Nose and throat irritation, pain in chest, nausea, drowsiness, and death	Full-face supplied air respirator or self-contained breathing apparatus	Artificial respiration using oxygen–do not breathe air from victim
Skin	Irritation and possible chemical burns	Plastic clothes and gloves	Remove clothing and shoes–rinse with warm water
Eyes	Stinging	Face shield	Force open eyes and rinse with lukewarm water
Ingestion	—	—	—

* in ppm
Sources: Environmental Protection Agency (EPA), Occupational Safety & Health Administration (OSHA), Illinois Department of Public Health (IDPH), and Praxair, Inc.

Figure 5-17. *The EPA has set permissible ozone limits for indoor building spaces.*

Ozone Measurement Procedures

PERFORM 2ND TEST IN SAME LOCATION ④

❺ PERFORM OZONE TESTING IN ALL SPACES TO BE OCCUPIED

OZONE GENERATOR USED OVERNIGHT TO KILL AIRBORNE BACTERIA

BACTERIA

TURN THE OZONE METER ON ❷

RECORD THE OZONE LEVEL MEASURED ❸

❶ HOLD THE OZONE METER IN THE APPROPRIATE TESTING SPACE

❻ TURN THE OZONE METER OFF

Figure 5-18. *Ozone meters measure the amount of ozone in an air sample in parts per million (ppm).*

Case Study – Sick Building Syndrome

History

A building owner hired a full-time stationary engineer for operation and maintenance duties. One of the first duties of the engineer was to address the high sickness rate among the staff.

The facility was a 60,000 sq ft office building with two main air handlers for each of the six floors. The air handlers were installed in small rooms. These air-handling rooms had louvers on one wall to let outdoor air in. The air handlers had a damper in the ductwork to draw the outdoor air in from the room instead of using ducted outdoor air.

Investigation

The stationary engineer first evaluated the activities of the building occupants and the building's mechanical design and operation. During the investigation, the engineer noted that there were several conditions contributing to poor IAQ.

The first issue was that one section of the building had a spray paint booth that was not ducted to the outside. The booth was located in a populated work area. All of the paint fumes were circulating in the facility. The booth also had several inches of built-up paint that were showing signs of biological growth.

The second problem was that the air-handling rooms were being used for storage. In some cases, the rooms were so packed with storage materials that the arrangement restricted outside airflow. The storage situation also made it hard to perform maintenance on the mechanical systems of the rooms.

The last issue encountered was the discovery that all of the air handlers on the east side of the building were covered with bird nests, and there were bird droppings on the floor of the air-handling rooms. In some cases, dead birds were found in the thick mass of nesting material. Birds and squirrels were able to enter the building through the louvered wall, which was designed to allow outdoor air to be drawn in.

Resolution

To remedy the first issue, a professional spray paint booth with a ducted exhaust fan was purchased. The new booth was located in an isolated area and in a location where the exhaust duct would not impact any outside air intakes. Secondly, within the air-handling rooms, all of the stored items were relocated to storage rooms or discarded. In regard to the bird issue, the engineer arranged for bird and rodent netting to be placed over all the outside air louvers to prevent further access by wildlife. The rooms were professionally cleaned and disinfected during off-hours with the air handlers off. Cleaning personnel were all required to wear appropriate personal protective equipment (PPE) because bird nesting and droppings can be dangerous to inhale.

Handheld Meter Testing Definitions

- An *anemometer* is a device that measures the average air velocity or average airflow rate.
- A *carbon dioxide meter* is a meter that measures the amount (in parts per million) of carbon dioxide in an air sample.
- A *carbon monoxide meter* is a meter that measures the amount (in parts per million) of carbon monoxide in an air sample.
- A *chemical smoke test* is an IAQ test that shows the airflow inside a space, or from one space into another space.
- A *flow hood* is a test instrument that is placed over a diffuser to measure the volume of air moving in cubic feet per minute.
- The *K factor* is a number that represents the actual open area and design characteristics of a specific diffuser.
- The *method detection limit* is the minimum concentration of a contaminant that is reported as zero.
- A *nitrogen dioxide monitor* is a meter that measures the amount (in parts per million) of nitrogen dioxide in an air sample.
- An *ozone meter* is a meter that measures the amount (in parts per million) of ozone in an air sample.
- *Traceability* is the unbroken chain of paperwork relating a test instrument's accuracy to a known standard.
- A *velometer* is a device that measures the velocity of air flowing out of a register.
- A *written quality assurance (QA) plan* is a document designed to deliver defensible data acquired using test instruments about indoor air quality.

For more information please refer to ATPeResources.com
or http://www.epa.gov.

1. Explain the first three steps in a typical IAQ complaint investigation.

2. List four factors that may affect the instrument selection process.

3. How do air volume and air circulation affect IAQ?

4. What organization sets an indoor limit for carbon dioxide?

5. List four different contaminants that can be tested using a handheld meter.

AIR SAMPLING AND TESTING

GrayWolf Sensing Solutions

Class Objective
Describe different measurement devices and testing approaches.

On-the-Job Objective
Based on local facility conditions, create a list of measurement devices that would be desirable in the facility.

Learning Activities:
1. Describe the difference between personal air-sampling pumps and area air-sampling pumps.
2. Describe the use of passive sampling devices.
3. Evaluate the calibration documentation for IAQ test instruments.
4. List the types of mold testing available.
5. **On-the-Job:** Inventory and evaluate the current measurement devices for indoor air quality found in the building.

Introduction

Air sampling helps identify contaminants in a facility. Stationary engineers should perform IAQ testing on a regular basis as a preventive measure in order to limit the occupants' exposure to contaminants. Specialized, personal air-sampling and area air-sampling devices are used to collect samples of contaminants, which are then analyzed by a laboratory.

AIR MONITORING

A wide variety of contaminants are found in air. Some of them are potentially hazardous and exposure to hazardous substances is regulated by federal, state, or local enforcement agencies. Airborne contaminants occur in solid, liquid, or gaseous forms, including dust, smoke, and fumes. Because of the potential risks involved from the exposure to airborne contaminants such as asbestos and lead, as well as biological contaminants such as mold and formaldehyde, airborne contaminant monitoring and testing should be performed.

Asbestos

When material containing asbestos is damaged or disintegrates with age, microscopic asbestos fibers are dispersed into the air, which can be harmful to occupants' health. **See Figure 6-1.** When asbestos contamination is suspected, the area must be evacuated and sealed, with access restricted. The asbestos findings must be reported in accordance with OSHA and EPA regulations. Asbestos abatement is regulated by federal, state, and local governments and requires certification, training, and the proper equipment. Many stationary engineers are certified to perform this work.

Asbestos air sampling is often performed to provide IAQ clearance before a building is occupied. Asbestos sampling is performed for schools any time renovation or construction work disturbs old walls, piping, ceiling tiles, or floors. Neither OSHA nor the EPA requires special training or accreditation for persons who collect air samples. However, some states require certification for the technicians or engineers who conduct air sampling for asbestos-abatement projects. A *personal air-sampling pump* is an instrument carried by a person and used to collect air samples in a special cassette for testing of the air sample by a laboratory. Samples are drawn into a cassette (canister) that is then sealed and sent to a laboratory for microscopic analysis. The sample rate is measured in fibers per cubic centimeter (f/cc).

> **IAQ Fact**
>
> *Chrysotile asbestos is the only curled fiber form of asbestos. The curled fiber makes chrysotile asbestos the least likely to be inhaled.*

ASBESTOS EXPOSURE LIMITS					
CLASSIFICATION	**10 MINUTES**	**30 MINUTES**	**1 HOUR**	**4 HOURS**	**8 HOURS**
Chrysotile	0.9*	0.7	0.5	0.3	0.2
Non-chrysotile	0.6	0.5	0.4	0.2	0.1

EXPOSURE	**SYMPTOMS**	**PREVENTION**	**FIRST AID TREATMENT**
Inhalation	May cause asbestosis, pulmonary fibrosis, and/or other lung disorders	Exhaust ventilation and full-face self-contained breathing apparatus	Rest in well ventilated area– oxygen possibly needed
Skin	Irritation	Special full body suit, hood, gloves, and boots	Wash with plenty of water and disinfectant soap
Eyes	Scratching of eyes	Splash goggles–do not wear contacts	Remove contacts; do not use ointments; flush eyes with water
Ingestion	May cause gastrointestinal irritation or cancer	Handling of materials, food, or drink prior to washing is prohibited	Do not induce vomiting–rinse mouth with water

* in fiber/cc

Sources: Occupational Safety & Health Administration (OSHA), National Institute of Occupational Safety and Health (NIOSH), National Institute of Standards and Technology (NIST), and Structure Probe, Inc. (SPI)

Figure 6-1. Asbestos exposure limits vary according to the classification of the fibers.

Personal Air-Sampling Pump for Asbestos. A *personal air-sampling pump for asbestos* is an instrument used to collect air samples in a special cassette, which is then sent to a laboratory for testing. The pump and the sampling cassette are separate units, each with an inlet and outlet. The inlet of the cassette is open to the indoor air environment. The cassette outlet is connected to the pump inlet by tubing. As the pump draws air in through the tubing, asbestos fibers in the air enter the cassette's inlet and become trapped in its filter. The sample air is then exhausted through the pump's outlet.

When sampling is complete, the cassette is disconnected from the tubing and sealed. The cassette must be sealed before and after sampling to avoid asbestos exposure beyond the sampling criteria and contamination from other particles. This would increase the amount of collected asbestos fibers, which would falsely indicate a higher contamination level. The cassette comes with inlet and outlet plugs that can be removed for sampling and then reinserted later to seal the sample.

The volume of airflow into the sampling cassette affects the number of asbestos fibers collected because greater airflow will draw more fibers into the cassette. For the sampling results to have meaningful correlation to the actual contamination of the indoor air, the airflow rate must be carefully calibrated. The airflow must also be constant throughout the sampling period. This provides consistently reliable samples that are acceptable for measuring and recording asbestos fiber concentrations.

Airflow calibrators (calibrating rotameters) are required to keep personal air-sampling pump flow rates in line with the EPA's and OSHA's air-sampling methods and with National Institute for Occupational Safety and Health (NIOSH) Method 7400.

Personal Air-Sampling Pump Procedures for Asbestos. The airflow rate of the personal air-sampling pump for asbestos must first be calibrated. **See Figure 6-2.** A calibrating rotameter is connected to the filter cassette inlet and the personal air-sampling pump is activated. The quick-check calibrating rotameter indicates the flow rate through the cassette. If this is not within sampling specifications, the flow rate must be adjusted. The pump includes a control dial to increase or decrease the flow rate until the rotameter indicates the desired value. (The flow rate should be between 0.018 scfm to 0.088 scfm or 0.5 Lpm and 2.5 Lpm.) When the flow rate is calibrated, the rotameter is removed from the filter cassette inlet.

If required, the appropriate PPE for protection from asbestos must be worn. When taking asbestos air samples, the personal air-sampling pump is attached to the stationary engineer's belt or waistband and the asbestos filter cassette is attached to the lapel or shirt collar. The inlet of the asbestos filter cassette must face down and be within breathing zone. This ensures that the collected sample accurately represents the indoor air that is inhaled by persons in that area.

Fluke Corporation

The outdoor air being brought into an HVAC system must be periodically tested for asbestos, lead, mold, and other contaminants.

The plug sealing the sampling cassette inlet is removed immediately prior to sample taking. When the pump is turned on, the airflow rate should be monitored to ensure that it is within specifications. When sampling is finished, the airflow rate should be verified by briefly connecting the quick-check calibrating rotameter to the cassette.

The cassette is then removed from the pump and tubing, sealed with the included inlet and outlet plugs, and sent to an approved laboratory for analysis. A record of the sampling is included with the sampling cassette and on the IAQ log. This includes the asbestos sampling date, sampling starting time, sampling stopping time, the flow rate that the personal air-sampling pump is set for (during calibration and actual flow at end of sampling), and a brief explanation of the conditions at the air-sampling site.

Personal Air-Sampling Pump Procedures for Asbestos

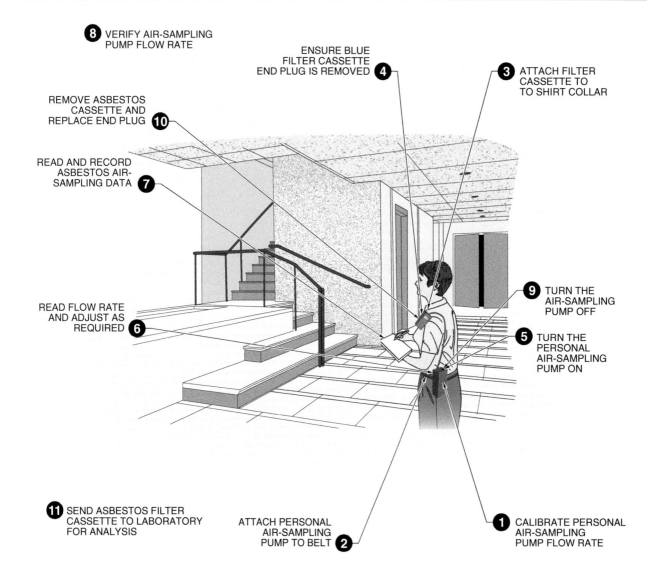

Figure 6-2. Asbestos fibers are collected with an asbestos sample pump that must maintain a constant airflow during sampling.

Lead

Any paint that is lead-based or contains lead is a source of lead exposure in the United States. Exposure to lead can become harmful when lead-based paint is improperly removed from surfaces by dry scraping, sanding, or open-flame burning. **See Figure 6-3.** Swabs are used for convenient screening of lead contamination in solder, paint, ceramics (dishes), soil, and dust. The test is a simple method to determine when corrective measures are required or if more definitive testing is necessary.

Results can be obtained in a few seconds by simply squeezing the swab shaft to release test chemicals, rubbing the swab on the surface to be tested, and waiting for a color change to confirm when lead is present. Advanced sampling for lead-based materials, or a sampling of indoor air space for contamination, should be performed by specialists. A report is then generated by the laboratory and returned to facility representatives.

According to EPA standards, if certain conditions are exceeded, lead is considered a hazard. For instance, lead is considered a hazard under the following circumstances:

- 40 micrograms of lead in dust per square foot exist on floors
- 250 micrograms of lead in dust per square foot exist on interior window sills
- 400 ppm of lead in bare soil exist in children's play areas
- 1200 ppm of lead (average) for bare soil exist in the rest of the yard

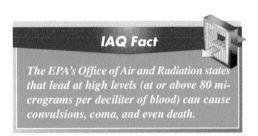

IAQ Fact

The EPA's Office of Air and Radiation states that lead at high levels (at or above 80 micrograms per deciliter of blood) can cause convulsions, coma, and even death.

LEAD EXPOSURE LIMITS					
CLASSIFICATION	**10 MINUTES**	**30 MINUTES**	**1 HOUR**	**4 HOURS**	**8 HOURS***
Nondisabling	—	—	—	—	0.03
Disabling	—	—	—	—	0.05

EXPOSURE	**SYMPTOMS**	**PREVENTION**	**FIRST AID TREATMENT**
Inhalation	Irritation of upper respiratory tract and lungs; possible chest and abdominal pain	Half-face high-efficiency particulate respirator or full-face positive-pressure air supplied respirator	Move to fresh air and provide oxygen; if not breathing, give artificial respiration
Skin	Short exposure may cause local irritation, redness, and pain; prolonged exposure absorbed through skin; lead poisoning	Use impervious protective clothing that includes boot, gloves, lab coat, apron, or coveralls	Flush skin with soapy water for 15 min; remove contaminated clothing and shoes
Eyes	Local irritation, blurred vision, and possibly severe tissue burns	Ventless safety goggles or full-face shield; eye wash station	Flush eyes with water for at least 15 min; move lower and upper eyelids while flushing
Ingestion	Poisoning that includes abdominal pain, spasms, nausea, vomiting, and headache; acute poisoning causes muscle weakness, insomnia, dizziness, loss of appetite, coma, and/or death	—	Induce vomiting immediately

* in mg/m^3

Sources: Occupational Safety & Health Administration (OSHA), National Institute of Occupational Safety and Health (NIOSH), American Conference of Governmental Industrial Hygienists (ACGIH), and Mallinckrodt Baker, Inc.

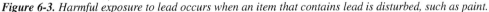

Figure 6-3. *Harmful exposure to lead occurs when an item that contains lead is disturbed, such as paint.*

Personal Air-Sampling Pump for Lead. Air sample testing for lead contamination is virtually identical to air sample testing for asbestos contamination. It requires a personal air-sampling pump and a special filter container, which are connected by a short length of tubing. A personal air-sampling pump for lead is an instrument used to collect air samples in a special canister for lead testing. Activating the pump draws an air sample through the canister, where lead contaminants are captured by the filter media inside. The amount of lead collected during the specified testing procedure is representative of the level of lead contamination in the indoor air.

Airflow calibrators (calibrating rotameters) are required to keep sampling-pump flow rates in line with EPA and OSHA air-sampling methods and with NIOSH Method 7400. Personal air-sampling pumps for lead must maintain a constant airflow during sampling to attain a canister that is acceptable for measuring lead.

Personal Air-Sampling Pump Procedures for Lead. For the stationary engineer conducting the air-sampling test, the procedure is identical to the procedure for using an air-sampling pump for testing for asbestos. **See Figure 6-4.** The difference is that a different type of sampling container is used to collect the contaminants, and the laboratory performs tests on the container specifically design to detect the concentration of lead.

As with other air-sampling pump tests, the airflow rate of the pump must be within certain specifications to ensure that the results are accurately representative of the indoor air. A calibrating rotameter is used at the inlet of the sampling canister to indicate the airflow rate while the pump is adjusted as necessary to achieve the desired rate.

The personal air-sampling pump is attached to the stationary engineer's belt or waistband, and the lead filter canister is attached to the lapel or shirt collar. If required, the appropriate PPE for protection from lead must be worn. The sample is taken by removing the canister's inlet plug and operating the pump for a certain amount of time. At the end of the test, the air-sampling pump flow rate is verified with the calibrating rotameter.

When finished, the canister is removed from the pump, sealed, and sent to an approved laboratory for analysis. The sampling record should include the lead-sampling date, sampling starting time, sampling stopping time, the flow rate that the personal air-sampling pump is set for (during calibration and actual flow at end of sampling), and a brief explanation of the conditions at the air-sampling site.

BIOLOGICAL CONTAMINANTS

Aeroallergens, such as mold and pollen, are a growing environmental concern. An indoor air quality sampler kit makes it easy to collect airborne contaminants including mold, pollens, fungal spores, fibers, dander, and insect residue within a few minutes. An area air-sampling pump is used to bring air into a collection canister (cassette) and the sample collection is then sent to a laboratory for analysis.

Mold

Two types of mold detection methods include tape sampling and vacuum samples. With both methods, mold samples are sent to a laboratory, where an attempt is made to grow the samples in a controlled environment. However, there are many different types of mold, and some types of mold may not grow easily or grow at all. In this case, additional testing is used to detect and identify mold spores. Most types of building mold can be identified by a microscopic examination of tape samples.

Tape sampling uses strips of self-adhesive tape to collect mold samples. The tape is applied to surfaces suspected of mold contamination. When the tape is removed, most of the mold spores remain stuck to the tape. Using the tape-sampling method has disadvantages. When using tape, investigators tend to collect samples that are easy to see, while missing light-colored mold that may be a greater health risk. Investigators may also miss highly airborne types of mold, which are more difficult to see.

Personal Air-Sampling Pump Procedures for Lead

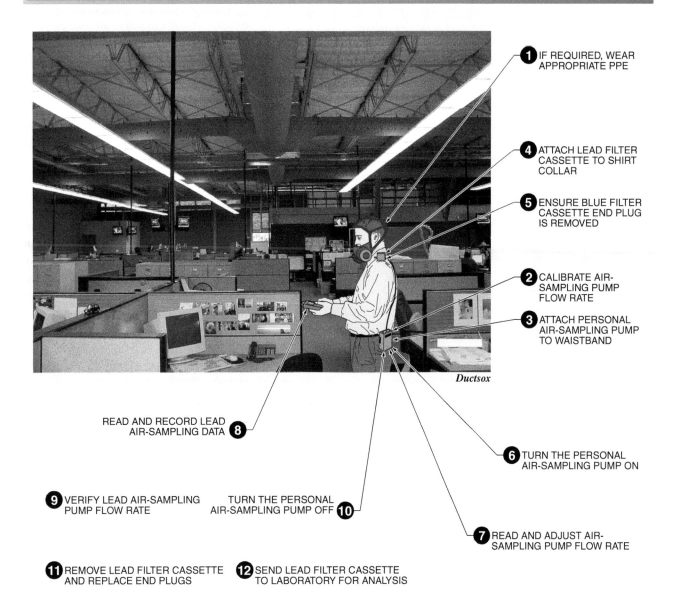

1 IF REQUIRED, WEAR APPROPRIATE PPE

4 ATTACH LEAD FILTER CASSETTE TO SHIRT COLLAR

5 ENSURE BLUE FILTER CASSETTE END PLUG IS REMOVED

2 CALIBRATE AIR-SAMPLING PUMP FLOW RATE

3 ATTACH PERSONAL AIR-SAMPLING PUMP TO WAISTBAND

Ductsox

READ AND RECORD LEAD AIR-SAMPLING DATA **8**

6 TURN THE PERSONAL AIR-SAMPLING PUMP ON

9 VERIFY LEAD AIR-SAMPLING PUMP FLOW RATE

TURN THE PERSONAL AIR-SAMPLING PUMP OFF **10**

7 READ AND ADJUST AIR-SAMPLING PUMP FLOW RATE

11 REMOVE LEAD FILTER CASSETTE AND REPLACE END PLUGS

12 SEND LEAD FILTER CASSETTE TO LABORATORY FOR ANALYSIS

Figure 6-4. Lead air samples are usually collected in a special canister that is then sent to a laboratory for testing.

Some small airborne mold spores do not settle out of the air rapidly, which can make the spores difficult to sample with tape. Also, the results of tape sampling are not specific. For instance, most laboratories offer a description of the density of fungal material found in the sample using generic terminology, such as levels one through four. **See Figure 6-5.**

IAQ Fact

Molds grow on almost any organic substance, as long as moisture and oxygen are present. Molds grow on food, paper, wood, carpet, and insulation. When moisture accumulates, mold growth will occur. It is impossible to eliminate all mold and mold spores in an indoor environment, but mold growth can be controlled by controlling the moisture indoors.

MOLD DENSITY LABORATORY TERMINOLOGY

DENSITY OF GROWING MOLD SPORES

Level <1	Very few mold spores and mold parts observed. Spores cover less than 10% of surface. Up to 20,000 spores/cm²
Level 1	Few spores and mold parts observed. Spores cover 10% to 25% of surface. Up to 200,000 spores/cm²
Level 2	A number of spores and mold parts observed. Spores cover 25% to 50% of surface. Up to 2 million spores/cm²
Level 3	Many spores and mold parts observed. Spores cover 50% to 75% of surface. Up to 4 million spores/cm²
Level 4	Many spores and mold parts observed. Spores cover more than 75% of surface. Up to and more than 4 million spores/cm²

DENSITY OF NON-GROWING MOLD SPORES

Very Few	Very few mold spores detected. Densities range up to 10 spores/cm²
Few	Few mold spores detected. Densities range up to 25 spores/cm²
Moderate	A number of mold spores detected. Densities range up to 50 spores/cm²
Many	Many mold spores detected. Densities range up to or more than 50 spores/cm²

Figure 6-5. Most laboratories offer a description of the density of fungal material found in a mold sample using generic terminology, such as levels one through four.

Roof puddles are a breeding ground for biological contaminants and are especially harmful when the puddles are located near air intakes that allow the contaminants to be brought into the HVAC system.

In addition to tape sampling, testing for mold contamination can include vacuum sampling. Vacuum sampling is very similar to the air sampling used to test for asbestos and lead contamination. A pump draws indoor air through a collection canister, which collects the airborne contaminants for later analysis at a laboratory.

However, vacuum sampling for mold uses an area air-sampling pump instead of the personal air-sampling pump. An *area air-sampling pump* is an instrument used to collect air samples from a relatively large area into a special filter container. Since this type of pump samples air from a much larger area, the results represent the contamination level of an entire room or open space instead of just the small area near where a person lives or works.

Area Air-Sampling Pump for Mold. Area air-sampling pumps are used to collect samples of mold present in an area. An *area air-sampling pump for mold* is an instrument used to collect mold from air samples by vacuuming the mold into a special canister, which is then sent to a laboratory for analysis. The laboratory transfers the mold particles from the filter onto a microscope slide or a medium for growing and preparing a culture for later analysis. **See Figure 6-6.**

While it is more precise than tape sampling, mold vacuuming has disadvantages as well. For instance, vacuuming does not collect identifiable structural components of mold as well as tape does. Vacuuming usually damages or destroys the structures of mold spores collected, which hinders the process of precisely identifying the mold type.

IAQ Fact

By controlling the relative humidity level in a building at 30% to 50%, the growth of some molds, mildews, bacteria, and insects can be minimized. Dust mites, the source of one of the most powerful biological allergens, grow in damp, warm environments.

Biological Airborne Contaminants

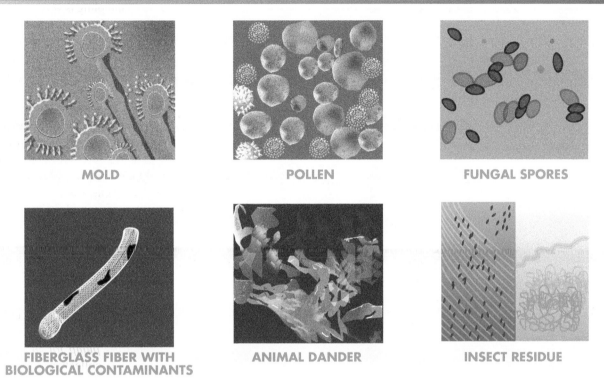

MOLD

POLLEN

FUNGAL SPORES

FIBERGLASS FIBER WITH
BIOLOGICAL CONTAMINANTS

ANIMAL DANDER

INSECT RESIDUE

Figure 6-6. Under a microscope, biological contaminants look quite different from one another.

In addition, vacuuming does not collect material on a hard surface as well as tape sampling. Particles that are easily lifted by the airflow are overrepresented in collection samples in comparison to particles that are stuck to the test surface. Mold vacuuming is even less effective when low airflow rates are used. A higher airflow rate is needed to draw the air from a large area into and through the filter canister.

Area Air-Sampling Pump Procedures for Mold. The airflow rate of the air-sampling pump must be calibrated before conducting an air-sampling test. This is accomplished by connecting the outlet of the filter canister to the inlet of the area air-sampling pump with tubing. With all plugs removed from the filter canister (and saved for sealing the canister later), the quick-check calibrating rotameter is connected to the inlet of the filter canister and the pump is turned ON. While the pump is operating, the airflow rate is adjusted on the pump until the rotameter indicates the desired flow rate. (The flow rate should be between 0.070 scfm to 0.565 scfm or 2 Lpm and 16 Lpm.)

For collecting mold samples with an area air-sampling pump, the procedures are virtually identical to those for a personal air-sampling pump, but the setup is slightly different. **See Figure 6-7.** While the location of the pump is not critical, the location of the filter canister is important because it must sample the air of an entire room or space. Therefore, the filter canister should be placed relatively high so that furniture and partitions do not significantly affect the movement of air from throughout the area. Likewise, the filter canister should be placed in the approximate center of the room, so that the sample reflects the average conditions in the room. A tripod stand is typically used to mount the canister in the best possible location. The canister inlet is pointed downward at a 45° angle.

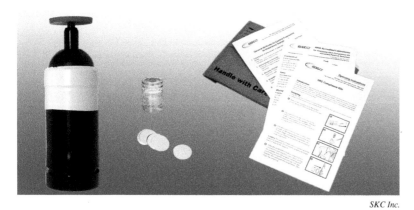

Mold filter canisters must be sent to a laboratory for analysis along with the paperwork that records all the information on the air-sampling process.

After the mold air sampling is complete, the area air-sampling pump flow rate is verified and the canister sealed with the included inlet and outlet plugs. The mold filter canister is then sent to an approved laboratory for analysis. Included with the canister and on the IAQ log is a record of the sampling test. This should include the mold-sampling date, sampling starting time, sampling stopping time, flow rate that the air sampler is set for (during calibration and actual flow at the end of sampling), and a brief explanation of conditions at the air-sampling site.

Area Air-Sampling Pump Procedures for Mold

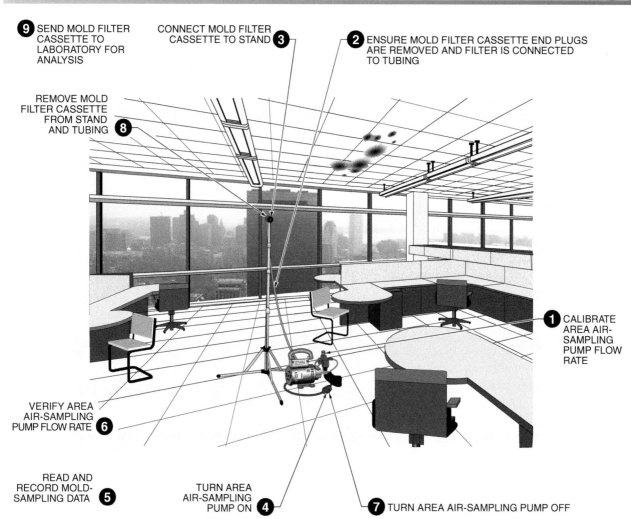

9 SEND MOLD FILTER CASSETTE TO LABORATORY FOR ANALYSIS

CONNECT MOLD FILTER CASSETTE TO STAND **3**

2 ENSURE MOLD FILTER CASSETTE END PLUGS ARE REMOVED AND FILTER IS CONNECTED TO TUBING

REMOVE MOLD FILTER CASSETTE FROM STAND AND TUBING **8**

1 CALIBRATE AREA AIR-SAMPLING PUMP FLOW RATE

VERIFY AREA AIR-SAMPLING PUMP FLOW RATE **6**

READ AND RECORD MOLD-SAMPLING DATA **5**

TURN AREA AIR-SAMPLING PUMP ON **4**

7 TURN AREA AIR-SAMPLING PUMP OFF

Figure 6-7. Mold sampling procedures vary when using the tape or vacuum air-sampling methods.

Formaldehyde

OSHA recognizes formaldehyde as a potential occupational carcinogen. OSHA regulations determine that the permissible exposure limit to formaldehyde is 0.75 ppm for an 8 hr, time-weighted average (TWA). **See Figure 6-8.** Also, OSHA has established a short-term exposure limit (STEL) of 2 ppm for a 15 min time period.

Concentrations of formaldehyde greater than 0.1 ppm irritate the eye, nose, throat, or skin in approximately 20% of the population. Several organizations, including the National Institute for Occupational Safety and Health (NIOSH), have adopted 0.1 ppm as a level that is acceptable, safe exposure for most people. Some people may be more sensitive and may have reactions at lower exposure levels.

One of the more practical methods to test for formaldehyde is to use a passive dosimeter. A *passive dosimeter* is a container in which air passes through a chemically treated badge or solution. Most passive dosimeters are not sensitive enough to measure formaldehyde below an exposure level of 0.1 ppm. Dosimeters should be placed approximately 4′ to 5′ above floor level in areas with adequate ventilation and be at least 12″ away from an inside wall. It is essential to always follow the manufacturer's instructions on using dosimeter badges.

Digital readout sensors are available that use a chemically treated element to change color in the presence of formaldehyde. Before purchasing equipment for this purpose, the range of a sensor must be carefully checked. Usually, formaldehyde samples are sent to a laboratory for analysis.

Baseline sampling for formaldehyde should be done after certain activities, such as after completing or receiving large quantities of new products containing formaldehyde. Product specification sheets are used to identify products containing formaldehyde. A sampling should be conducted during a 4 hour to 8 hour time period when the area is not occupied and air-handling units are not operating. The number of samples taken depends on the size and shape of the space. A minimum of two samples must be taken. When taking measurements, facts to be documented include the temperature, relative humidity, and occupancy level of the space.

FORMALDEHYDE EXPOSURE LIMITS					
CLASSIFICATION	**10 MINUTES***	**30 MINUTES***	**1 HOUR***	**4 HOURS***	**8 HOURS***
Nondisabling	2	1.5	1.25	1	0.5
Disabling	10	5	4	2	1
EXPOSURE	**SYMPTOMS**		**PREVENTION**		**FIRST AID TREATMENT**
Inhalation	Respiratory tract irritant; short exposure causes headache, vomiting, dizziness, and drowsiness; prolonged exposure causes sensory impairment		Wear full-face self-contained breathing apparatus		Remove to fresh air; if not breathing, provide artificial respiration; personnel may give oxygen
Skin	Short exposure may cause burns; prolonged exposure allows harmful amounts to be absorbed into body		Wear plastic or rubber gloves; wear coated protective clothing and footwear		Remove clothing, gloves, and shoes; flush skin with large amounts of warm water
Eyes	Severe irritation; vapors may cause tearing; liquid causes burns and permanent eye injury		For short exposure, wear ventless safety glasses and a full-face shield		Flush eyes with water for at least 15 min; move eyelids from eyeballs to ensure all surfaces are flushed
Ingestion	Causes severe irritation and inflammation of mouth, throat, and stomach; may cause nausea, blindness, and death		—		Do not induce vomiting; if conscious, provide water or milk

* in ppm
Sources: Occupational Safety & Health Administration (OSHA), Office of Environmental Health & Safety (EH&S), and Praxair, Inc.

Figure 6-8. OSHA has set formaldehyde exposure limits for indoor air quality.

When laboratory analysis indicates that the formaldehyde exposure is greater than 0.1 ppm, but there have been no complaints, the sampling process must be repeated. If high levels persist, the source must be isolated. When sampling is initiated by occupant complaints and high levels of formaldehyde have been identified through testing, it is necessary to take corrective actions, such as purging and flushing with outdoor air. In the presence of continued complaints, even with acceptable formaldehyde levels, outside assistance may be considered.

Formaldehyde Passive Sampler. A *formaldehyde passive sampler* is an instrument used to collect air samples on a special reactive tape that is sent to a laboratory for testing. The sampler is based on a simple chemical process that is reliable for detecting concentrations of 5 ppb to 5 ppm. The primary part of the sampler is a piece of chemically impregnated paper. As the reactive surface is exposed to air, molecules of formaldehyde that are present in the air react with the chemical to form a new compound that is very stable and remains fixed on the paper. This reaction requires no other chemicals to initiate or maintain the compound.

Formaldehyde Passive Sampler Procedures. The chemically reactive surface is sealed before use to avoid contamination from other environments. It must also be carefully resealed after the test. **See Figure 6-9.** The amount of formaldehyde reacting with the sampler depends in part on the length of the sampling time. Therefore, the stationary engineer conducting the test must be careful to expose the sampler to air for only the specified length of time and carefully seal the sampler after the test to avoid continued exposure, which would inaccurately affect the results.

Sampling times range from 15 min. to several days, depending on the specific sampling kit instructions. The dates and times of the start and end of the test are recorded on the sampler package, along with the location, atmospheric pressure, and temperature of the testing area.

Volatile Organic Compounds

A *volatile organic compound (VOC)* is a carbon-containing compound that easily evaporates at room temperatures. There are hundreds of VOCs, including formaldehyde, cetone, alcohols, ammonia, hydrocarbons, benzene, phenols, and radon. It would be impossible to identify all of them in a building. Their presence and concentration changes as conditions and activities change from moment-to-moment and day-to-day.

Radon

Radon in schools is an issue of major concern, especially for buildings with large footprints. A *footprint* is the amount of building area in contact with the ground. Schools tend to be low, sprawling structures with a larger footprint. Radon is generally less of a problem in commercial buildings because commercial buildings usually have smaller footprints in proportion to their overall size. Commercial buildings are also generally supported by properly functioning HVAC units, and radon gas tends to be removed by dilution. Areas of a building without proper ventilation contain the highest-expected risk for radon exposure. Basements and building areas next to the ground, particularly in unventilated spaces, are also areas of high-expected radon exposure.

The EPA has not finalized a measurement protocol for radon in large commercial buildings. However, the EPA does have a protocol for schools and large buildings during design and construction. According to the EPA, exposures exceeding 4.0 picocurie per liter of air (pCi/L) do increase the risk of lung cancer. **See Figure 6-10.** A *picocurie (pCi)* is a unit for measuring radioactivity and equals the decay of approximately two radioactive atoms per minute.

Formaldehyde Passive Sampler Procedures

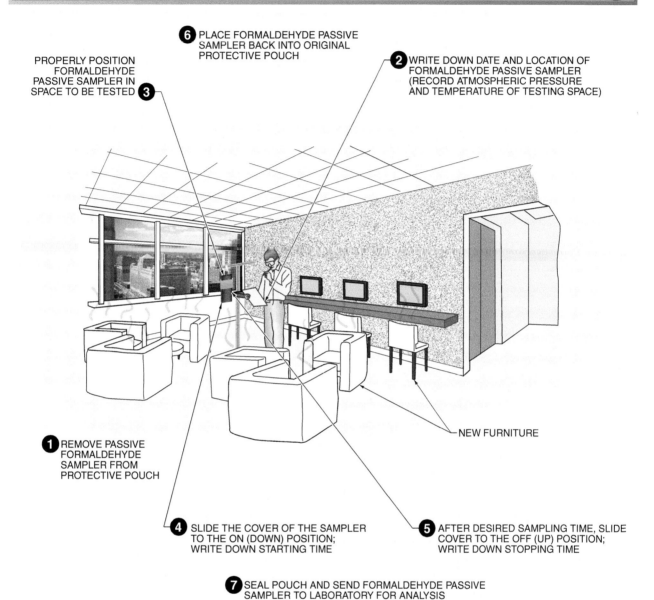

6 PLACE FORMALDEHYDE PASSIVE SAMPLER BACK INTO ORIGINAL PROTECTIVE POUCH

PROPERLY POSITION FORMALDEHYDE PASSIVE SAMPLER IN SPACE TO BE TESTED **3**

2 WRITE DOWN DATE AND LOCATION OF FORMALDEHYDE PASSIVE SAMPLER (RECORD ATMOSPHERIC PRESSURE AND TEMPERATURE OF TESTING SPACE)

1 REMOVE PASSIVE FORMALDEHYDE SAMPLER FROM PROTECTIVE POUCH

NEW FURNITURE

4 SLIDE THE COVER OF THE SAMPLER TO THE ON (DOWN) POSITION; WRITE DOWN STARTING TIME

5 AFTER DESIRED SAMPLING TIME, SLIDE COVER TO THE OFF (UP) POSITION; WRITE DOWN STOPPING TIME

7 SEAL POUCH AND SEND FORMALDEHYDE PASSIVE SAMPLER TO LABORATORY FOR ANALYSIS

Figure 6-9. Formaldehyde sampling involves collecting particles with a passive (nonpowered) sampler and then sending the sampler to a laboratory for analysis.

RADON EXPOSURE LIMITS					
CLASSIFICATION	**10 MINUTES**	**30 MINUTES**	**1 HOUR***	**4 HOURS**	**8 HOURS***
Nondisabling	—	—	4	—	0.2
Disabling	—	—	8	—	0.4

* in pCi/L

Sources: Occupational Safety & Health Administration (OSHA), Environmental Protection Agency (EPA), and Agency for Toxic Substances & Disease Registry (ATSDR)

Figure 6-10. The EPA has set some limits for radon exposure.

An *alpha track detector (ATD)* is a passive monitoring device used for radon testing. It is the preferred measurement device for radon because it can provide a stable estimate of average exposure. Professional radon gas test kits are also available. Professional radon gas test kits use the most advanced liquid-scintillation-type detectors that contain silica gel desiccants to remove all moisture for accurate and reliable radon test results.

A baseline radon measurement taken over a one-year period with an ATD provides the most reliable information about radon levels within a building. Measurements taken over shorter time periods such as three months are less reliable because of the day-to-day variation in naturally occurring radon gas levels.

The EPA recommends deploying one ATD for each 1000 sq ft of space in rooms with an earth-contact surface. The exception would be large halls, greater than 10,000 sq ft, where one detector per 5000 sq ft is acceptable. Both occupied and unoccupied ground-contact surface rooms should be measured, especially when a building has floors above the ground-contact surface floor. Unoccupied rooms may act as radon sinks. A radon sink is a location in a building where radon gases collect. The EPA has found that the leakage of radon-containing gas from unoccupied rooms may be an important factor in radon transport, especially to adjacent floors. **See Figure 6-11.**

The EPA further suggests not placing ATDs on nonground-contact floors until high radon levels are found on lower floors. Elevated levels of radon on upper floors are dependent upon the radon levels on lower floors. When the measurement of upper floors is warranted, it is important to include areas prone to high radon movement from lower floors such as elevator shafts, stairwells, and major pipe shafts.

Passive devices to detect radon, such as ATDs, are relatively inexpensive and readily available but require a laboratory to process the results. A special material is exposed to the space suspected of radon contamination

and then is sealed in a shipping package and sent to a laboratory. Radon is detected by analyzing the changes its radiation made to the special materials or by counting the tracks left on a material surface by the radioactive alpha particles.

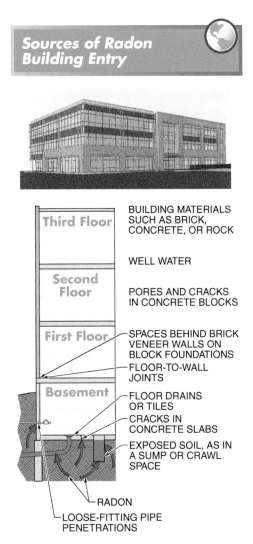

Figure 6-11. Radon leaks into various parts of a building from the basement or ground floor, making occupants susceptible to illness.

Radon Detector. Alternatively, active radon detectors are generally more expensive but provide continuous results. A *radon detector* is a test instrument used to collect and analyze air samples and indicates the amount of radon present on a numeric LED display. The display of a radon detector has a range of 0.1 pCi/L to 999 pCi/L. The radon detector continuously samples the air and updates the display every hour. The radon detector also displays the average radon level from the last seven days as a short-term measurement and displays the average radon level from the last startup, or five years as long-term measurement.

When the area is occupied, the approximate hours of occupancy should be noted. During the three-month measurement period, each detector must be rechecked monthly for placement and condition. When the results exceed EPA limits, it is necessary to seek outside assistance. The EPA always recommends resampling the necessary areas prior to taking corrective action.

Radon Detector Sampling Procedures. Radon detectors must be placed (mounted) in the appropriate position. For a basement or first floor, the radon detector is placed a minimum of 2′ above the floor, 1′ from exterior walls, and 3′ from any objects that interfere with airflow through the space. Detectors should not be placed near outside walls of rooms with operable windows, nor near drafts and heat sources. **See Figure 6-12.** The testing location should be recorded for each radon detector. This may require sketching a simple layout of the room with the distances to walls and obstructions indicated. Once a radon detector is placed, it should not be disturbed throughout its operation.

A radon detector may include modes for short-term (days) or long-term (months or years) measurements. Most also include self-test functions and settings for saving and clearing measurements in its memory. Measurements should also be recorded on the IAQ log, including multiple periodic measurements, if called for by the test procedure.

IAQ Fact

The EPA's web site can be used to find specific state information concerning radon testing. Two private organizations offer information on radon testing: the National Environmental Health Association (NEHA) and the National Radon Safety Board (NRSB).

Radon Detector Sampling Procedures

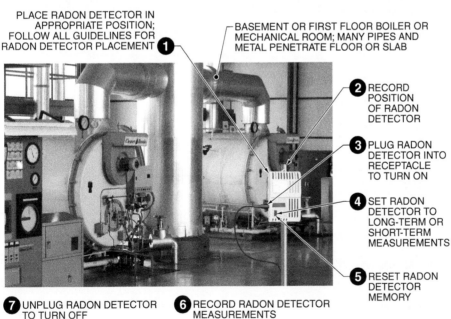

PLACE RADON DETECTOR IN APPROPRIATE POSITION; FOLLOW ALL GUIDELINES FOR RADON DETECTOR PLACEMENT **1**

BASEMENT OR FIRST FLOOR BOILER OR MECHANICAL ROOM; MANY PIPES AND METAL PENETRATE FLOOR OR SLAB

2 RECORD POSITION OF RADON DETECTOR

3 PLUG RADON DETECTOR INTO RECEPTACLE TO TURN ON

4 SET RADON DETECTOR TO LONG-TERM OR SHORT-TERM MEASUREMENTS

5 RESET RADON DETECTOR MEMORY

7 UNPLUG RADON DETECTOR TO TURN OFF

6 RECORD RADON DETECTOR MEASUREMENTS

Figure 6-12. When sampling for radon levels, the detectors must be placed 4′ to 5′ above the floor, no less than 4′ from objects in unobstructed areas, and must not be physically disturbed while operating.

Particulates

Particulates in the air are broken up into categories depending on type. For small particles, such as those from cooking, candles, and combustion, the concentration levels are listed as microgram per cubic meter. The acceptable limit for small particles is 15 micrograms per cubic meter. **See Figure 6-13.** The limit for other particle contaminants such as dust, smoke, and deteriorating materials is 50 milligrams per cubic meter. For contaminants with higher limits, handheld instruments with direct readouts can be used. Otherwise, a sample is drawn into a container and analysis by an off-site laboratory is required.

Particulate Detector and Counter. A *particulate detector and counter* is a test instrument used to collect and analyze air samples for the presence and number of particles in the air. In addition to the number of particulates, a particulate detector and counter also measures the temperature, relative humidity, and volume of the air sample. Moreover, the particulate meter records the date and time of each sample. All data can be downloaded to a personal computer for storing IAQ records, comparing with past measurements, and preparing IAQ reports.

Particulate detector and counter instruments consist of a handheld display unit with an attached probe that collects the particulates. The probe may be attached directly to the display unit or connected through a short length of tubing. Probes of various shapes and materials are available for testing different environments. An isokinetic probe is recommended for general use. This type of probe samples the air uniformly with low turbulence, which results in a proportional representation of particles of all sizes.

Particulate Detector and Counter Sampling Procedures. The necessary first step in conducting this type of sampling is to purge the particulate detector and counter. **See Figure 6-14.** This ensures that the probe is free of particulates that would interfere with accurate test results. A "zero count" filter is used to cover the inlet of the probe while the instrument is operated for 5 min. The filter allows no more than one particle greater than 0.3 microns into the probe during the test. When finished with the purge procedure, the filter is removed from the probe.

Particulate detectors and counters typically have three particulate counting modes. The concentration mode samples a small volume of air and calculates an extrapolated value for a larger volume. The totalize mode displays the particulate count until the end of the sample. The audio mode causes the particulate counter to beep when the alarm limit is reached.

PARTICULATE EXPOSURE LIMITS					
CLASSIFICATION	**10 MINUTES**	**30 MINUTES**	**1 HOUR**	**4 HOURS**	**8 HOURS***
Nondisabling	—	—	—	—	0.03
Disabling	—	—	—	—	0.05
EXPOSURE	**SYMPTOMS**		**PREVENTION**		**FIRST AID TREATMENT**
Inhalation	Symptoms, prevention measures, and first aid vary depending on the type of matter the particulate originated from				
Skin					
Eyes					
Ingestion					

* in mg/m^3

Sources: Occupational Safety & Health Administration (OSHA) and the National Institute of Occupational Safety and Health (NIOSH)

Figure 6-13. Some particulate exposure limits are in place for general particulates that may not be definable.

The desired volume of the air sample can be selected on the display unit, which determines the necessary sample time. Tests are typically conducted in a period between about 20 sec and 10 min. There is typically also an option for setting the method for counting the data. The cumulative method counts all particles greater than or equal to the selected particle sizes. The differential method counts all particles greater than or equal to and less than the selected particle sizes.

Once all the desired settings are entered into the display unit, the particulate detector and counter conducts the air sampling automatically. Measurements should be recorded on an IAQ log.

Detector Tubes

A *detector tube* is a device that indicates the average contaminant concentrations in a work area or other space for personal exposure monitoring. They are small glass tubes that are lined on the inside with a special chemical that is designed to react to the molecules of a specific gas by changing color. The tubes are initially sealed at both ends. The ends of the glass tubes are broken off to expose the reactive chemical to the indoor air. An approximate concentration of the contaminant gas is indicated directly from the amount of color change on the tube.

Some detector tubes are designed to react to the tested gas by diffusion. Exposure times greater than 1 hour require that the indicated dosage be divided by the number of exposure hours. **See Figure 6-15.** Other types of detector tubes are used with pumps that draw sampling air through the tube. This accelerates the exposure, which may require a conversion chart to determine the equivalent time-weighted average.

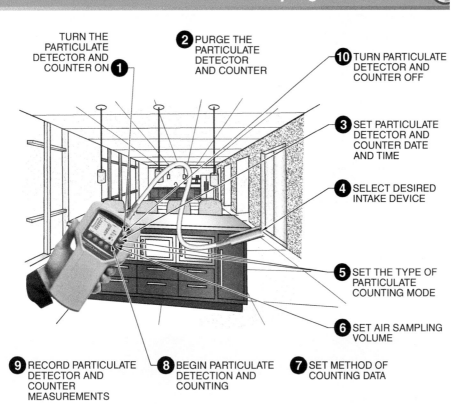

Particulate Detector and Counter Sampling Procedures

TURN THE PARTICULATE DETECTOR AND COUNTER ON **1**

2 PURGE THE PARTICULATE DETECTOR AND COUNTER

10 TURN PARTICULATE DETECTOR AND COUNTER OFF

3 SET PARTICULATE DETECTOR AND COUNTER DATE AND TIME

4 SELECT DESIRED INTAKE DEVICE

5 SET THE TYPE OF PARTICULATE COUNTING MODE

6 SET AIR SAMPLING VOLUME

7 SET METHOD OF COUNTING DATA

8 BEGIN PARTICULATE DETECTION AND COUNTING

9 RECORD PARTICULATE DETECTOR AND COUNTER MEASUREMENTS

Figure 6-14. Particulate detectors and counters sample a precise amount of air and then perform testing to identify the number of particles in the air sample.

Detector Tubes

LAPEL CLIP

DIFFUSION
DETECTOR
TUBE

500
400
300
200
150
100
50
25
10

DETECTOR
TUBE
HOLDER

COLORIZED
CALIBRATION

Figure 6-15. Detector tubes use time-weighted exposure (1 hr, 2 hr, 3 hr, etc.) to determine the average amount of contaminants in the air of a space.

Detector Tube Sampling Procedures. Detector tubes require no special training or equipment other than the tubes themselves. They are relatively inexpensive and readily available for detecting approximate concentrations of more than 160 different gases.

Detector tubes are very simple to use. One or both ends of the tube are removed and the tube is attached to a person's lapel with a special holder. **See Figure 6-16.** Exposure should be limited to full hour increments (1 hour, 2 hours, or 8 hours maximum). The concentration measurement is read from the tube and divided by the number of exposure hours (to determine the time-weighted average).

Detector Tube Sampling Procedures

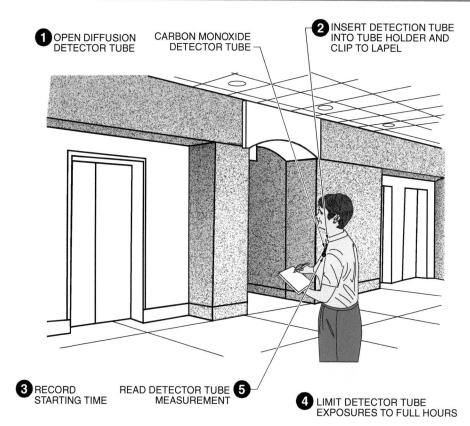

❶ OPEN DIFFUSION DETECTOR TUBE

CARBON MONOXIDE DETECTOR TUBE

❷ INSERT DETECTION TUBE INTO TUBE HOLDER AND CLIP TO LAPEL

❸ RECORD STARTING TIME

READ DETECTOR TUBE MEASUREMENT ❺

❹ LIMIT DETECTOR TUBE EXPOSURES TO FULL HOURS

Figure 6-16. A detector tube measurement is read by dividing the indicated detector tube measurement by the number of exposure hours.

Case Study – Vacant Office Space

Complaint History

Last year, the stationary engineer and building management of a 60-story office building began receiving IAQ complaints from occupants who reported having flu-like symptoms. Almost all of the calls came from occupants from the 40th, 42nd, and 43rd floors. The 41st floor had been vacant for about 5 months (November 2006 to April 2007).

Investigation

Upon entering the 40th floor, it was evident that water had been leaking from the bathroom pipework on the 41st floor. Apparently, a hot water pipe located above the drop ceiling on the 40th floor had been leaking for at least a few months. The prolonged presence of the moisture and the low-level heating due to vacancy created ideal conditions for fungal contamination, such as Stachybotrys. In addition, mushrooms were found growing through the carpeting in that section of the vacant space, and the walls and drop ceiling in that area were contaminated with mold.

Resolution

After the hot water leak is repaired, an outside company using proper mold mitigation procedures removed and disposed of the carpeting, drywall, soundproofing materials, and other contaminated materials such as ceiling tiles. The mitigation crew cleaned and disinfected the concrete floors, wall studding, and drop ceiling components.

Air and surface samples indicated that the abatement process was effective. Contractors installed new drywall, ceiling tiles, and carpeting. Building occupants ceased complaining about having IAQ-related flu-like symptoms, and the 40th floor was ready to be reoccupied.

Case Study

Definitions

Air-Sampling Testing Definitions

- *An alpha track detector (ATD)* is a passive monitoring device used for radon testing.
- An *area air-sampling pump* is an instrument, usually mounted on a tripod, used to collect air samples in a special cassette for testing of the air sample by a laboratory.
- A *detector tube* is a device that allows the determination of the average contaminant concentrations in a work area or other space for personal exposure monitoring.
- A *footprint* is the amount of building area in contact with the ground.
- A *formaldehyde passive sampler* is an instrument used to collect air samples on a special reactive tape for formaldehyde testing in a laboratory.
- A *particulate detector and counter* is a test instrument used to collect and analyze air samples for the presence and number of particles in the air.
- A *personal air-sampling pump* is an instrument carried by a person and used to collect air samples in a special cassette for testing of the air sample by a laboratory.
- A *picocurie (pCi)* is a unit for measuring radioactivity and equals the decay of approximately two radioactive atoms per minute.
- A *radon detector* is a test instrument used to collect and analyze radon air samples and indicates the amount of radon present on a numeric LED display.

**For more information please refer to ATPeResources.com
or http://www.epa.gov.**

Review Questions

1. How are airflow calibrators used?

2. When is IAQ clearance for asbestos sampling performed?

3. What type of measuring device is not usually used to measure formaldehyde and why?

4. Describe some of the possible factors to consider when measuring VOCs.

5. What is the minimum baseline measurement time for radon?

IAQ TESTING, ADJUSTING, AND BALANCING

7
chapter

Fluke Corporation

Class Objective:
Define the benefits of testing and balancing and describe a general air-handling system balancing procedure.

On-the-Job Objective:
Justify testing and balancing by its positive impact on IAQ at the facility.

Learning Activities:
1. Explain how testing and balancing can positively impact a facility's IAQ.
2. Give an example of how the lack of testing and balancing could lead to an air-handling system IAQ problem.
3. Briefly describe the steps of a general air-handling system balancing procedure.
4. **On-the-Job:** Determine if proper air-handling system testing and balancing procedures are being applied in the facility to improve IAQ.

Introduction

Modern commercial HVAC air-handling systems must satisfy multiple building needs simultaneously. Testing and balancing are used to evaluate and monitor air-handling system performance. Human comfort is expected at all times in building spaces at a minimum cost in energy. Commercial mechanical systems consume 25% or more of the energy consumed by a building.

The mitigation of an IAQ problem often involves air measurements. Good air quality depends on the delivery of proper airflow to the occupied space. To ensure proper airflow, a stationary engineer must be proficient with IAQ test instruments and be able to quickly and effectively use the data collected from them.

TESTING, ADJUSTING, AND BALANCING OF AIR-HANDLING SYSTEMS

Testing, adjusting, and balancing (TAB) of a commercial HVAC system is vital to ensure that an air-handling system has the desired operating characteristics. It is imperative for stationary engineers to have a comprehensive knowledge of air balancing and be able to apply that knowledge to facility systems. Without test instruments and procedures, proper verification of air-handling system operation is not possible. **See Figure 7-1.** This situation may lead to poor indoor air quality, lack of comfort, occupant dissatisfaction, high maintenance and operational costs, and possibly unhealthy or dangerous work environments.

Testing, balancing, and maintaining an HVAC air-handling system is of primary concern because these functions have a major impact on the indoor air quality of a building or facility. A preventive maintenance (PM) program cannot achieve peak performance unless an HVAC system is properly tested and balanced first.

HVAC system testing is the process of using specific IAQ test instruments to measure variables within an air-handling system in order to evaluate mechanical equipment and system performance. Variables such as rotational speed, air temperature and humidity, air pressure, noise level, air velocity, and volumetric airflow rates can all be tested. **See Figure 7-2.** *HVAC system adjustment* is the process of changing the settings of operating devices to obtain the final setpoint of control or air-handling system balancing. HVAC system adjustment is done with electric motor drives, dampers, thermostats, pressure controllers, blowers, fans, and controls.

Pollutant Causes

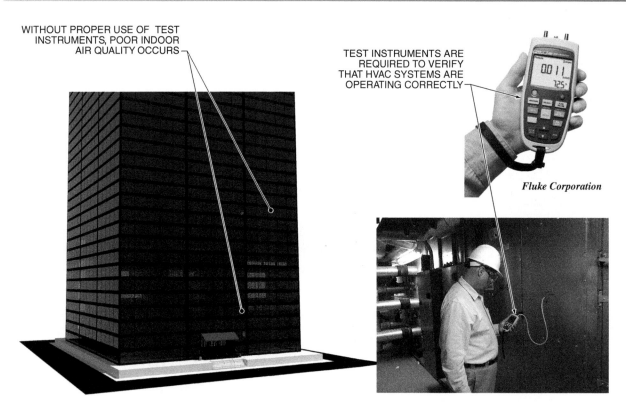

WITHOUT PROPER USE OF TEST INSTRUMENTS, POOR INDOOR AIR QUALITY OCCURS

TEST INSTRUMENTS ARE REQUIRED TO VERIFY THAT HVAC SYSTEMS ARE OPERATING CORRECTLY

Fluke Corporation

Figure 7-1. Proper verification of air-handling systems is necessary to prevent poor IAQ.

HVAC system balancing is the regulation of the flow of air in a system through the use of acceptable industry testing and balancing procedures. The designed flow is expressed in cubic feet per minute (cfm), air changes per hour (ACH), or other relevant measurements.

A certified HVAC mechanical contractor performs the initial testing, adjusting, and balancing of an HVAC system. After installation, ongoing TABs are performed by stationary engineers who have been trained in the proper procedures and techniques of using IAQ test instruments. Ongoing TABs can be useful for identifying and fixing problems in a facility as problems arise.

Air System Test Instruments

BLADE FAN

INLET FLOW

OUTLET FLOW

SPAR

BLADE

AXIAL-FLOW BLOWER

MOTOR

ROTATION

BLOWER BLADE

CENTRIFUGAL BLOWER

DISCHARGE OPENING

SQUIRREL CAGE

MOTOR

BELT

SHAFT

INLET VANES

PROPELLER FAN

MOTOR

PULLEYS

BELT

SHAFT

HUB

BLADES

ROTATION

PHOTO-CONTACT TACHOMETER

PHOTO TACHOMETER

445.3 RPM

MEMORY

CONTACT TACHOMETER

TEMPERATURE AND HUMIDITY METER

RELATIVE HUMIDITY

54.7 %RH

21.8 °C

AIR TEMPERATURE

ANEMOMETER

AIRFLOW RATE

24.6°

28.2

Figure 7-2. Test instruments can measure various aspects of air-handling systems including air temperature, humidity, and airflow rate.

General Airflow Testing, Adjusting, and Balancing Principles

Testing, adjusting, and balancing principles are used when testing air-handling systems in a building. TAB principles include using quality test instruments, taking measurements at the proper points in an HVAC system, converting measurements to different units, and documenting results. IAQ testing must be performed with properly calibrated high-quality test instruments. **See Figure 7-3.** Attempting to test a complex mechanical air-handling system with poor-quality test instruments yields poor and unreliable results. All test instruments must be calibrated and checked on a routine basis to ensure accuracy. **See Appendix.**

IAQ Test Instrument Certificates of Calibration

Fluke Corporation

Figure 7-3. Test instruments must be properly calibrated before performing tests.

The accuracy and range of an air-handling system test instrument must be verified (certified) by a testing laboratory traceable to the National Institute of Standards and Technology. Some air-balancing test instruments such as some manometers do not require calibration (certification).

Measurements must be taken at the proper points in an HVAC system for readings to have any value. Organizations such as the American Society of Heating, Refrigeration and Air-Conditioning Engineers (ASHRAE) and the National Environmental Balancing Bureau (NEBB) determine the proper IAQ measurement locations.

Measurement readings are often converted from one unit to another. **See Figure 7-4.** Velocity pressure is often converted to velocity (in feet per second) and velocity is often converted to flow (in cubic feet per minute). Unit conversions must be done accurately and properly. Some IAQ test instruments perform the conversions electronically while others require the stationary engineer to manually enter the conversions.

The results of all IAQ testing must be documented for future reference. To prevent problems with personnel not being able to find information, documentation must be stored in a secure, long-term, and accessible format. Computerized data loggers and readers can transfer data directly from test instruments into computer databases. The records may then be transferred from the computer into other long-term storage media.

Airflow Testing, Adjusting, and Balancing

While air-handling system testing, adjusting, and balancing is useful when evaluating indoor air quality issues, the most important of the three components is airflow balancing. The most common application of airflow balancing is keeping all building spaces at the same temperature. To understand testing, adjusting, and balancing of air-handling systems, the basic airflow measurements used for indoor air quality must be understood.

An air-balancing procedure begins with having architectural, mechanical, electrical, and HVAC control; original and as-built drawings; testing and balancing reports from the base building; and any system or equipment specifications ready for use. All of these documents must be organized and filed according to the established PM procedures of the company. Drawings and specifications are critical to the correct testing and balancing procedures being used for an application, and are required for reference during the process.

When testing begins, all base-building equipment information and original system test reports must be available for comparison purposes. Having the equipment information and test reports in-hand is the only way to determine if an air-handling system is correctly balanced.

It is important that the stationary engineer checks all equipment for proper operation. It is also important to check that all equipment shown in the base building or as-built drawings is actually in the system with the correct size motors, pulleys, belts, controls, dampers, and filters. The HVAC system must be complete and running to full design specifications before system performance measurements can be taken and recorded. The HVAC system design must be checked for compatibility with the current building configuration. **See Figure 7-5.** Checking for compatibility is especially important in older buildings, buildings that have been retrofitted, and buildings not being used as originally intended.

INDOOR AIR QUALITY UNIT CONVERSIONS		
MULTIPLY	**BY**	**TO OBTAIN**
Volume		
Cubic feet	1728	Cubic inches
Cubic feet	7.4805	Gallons
Pressure		
Inches of water	0.0361	Pounds per square inch
Inches of water	0.0735	Inches of mercury
Inches of mercury	0.4912	Pounds per square inch
Pounds per square inch	2.036	Inches of mercury
Pounds per square inch	2.307	Feet of water
Pounds per square inch	14.5	
Pounds per square inch	27.67	Inches of water
Pounds per square inch	0.06804	Atmospheres
kpa	100	
Miscellaneous		
Pounds	453.5924	Grams
Pounds	0.4536	Kilograms
HP (steam)	42,418	Btu
Therm	100,000	Btu
Btu	0.252	Calories
Cubic feet per minute	7.481	Gallons per minute

Atmosphere Equivalent		
29.921 in. Hg	=	1 Atmosphere
406.782 in. WC	=	1 Atmosphere
14.696 psi	=	1 Atmosphere

Figure 7-4. *Measurement readings often undergo unit conversions. Some instruments can perform conversions electronically.*

IAQ Fact

Sustaining acceptable indoor air quality involves more than maintaining the heating, ventilation, and air conditioning (HVAC) system for a building. A number of items interact with one another to affect IAQ, such as the building site, building occupants, area climate, contamination sources, HVAC equipment used, and the techniques used to construct the building.

Fluke Corporation

Measurements are taken between rooms to understand the airflow that occurs when doors are opened.

HVAC System Compatibility with Remodeled and Retrofitted Buildings

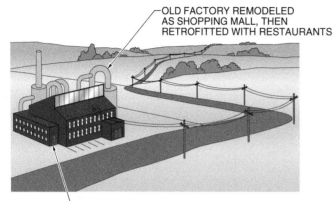

OLD FACTORY REMODELED AS SHOPPING MALL, THEN RETROFITTED WITH RESTAURANTS

HVAC SYSTEM MUST BE CHECKED FOR COMPATIBILITY WITH CURRENT BUILDING AND USAGE

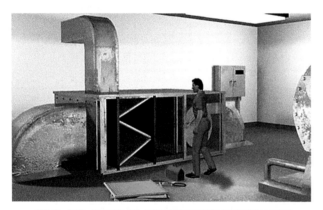

OLD AIR SYSTEM EQUIPMENT REWORKED

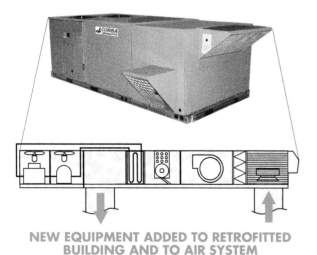

NEW EQUIPMENT ADDED TO RETROFITTED BUILDING AND TO AIR SYSTEM

Figure 7-5. When beginning the TAB process, the HVAC system design should be checked to ensure it is compatible with the building configuration.

Reviewing HVAC system prints and physically checking the system components provides an understanding of the air system. After confirming that an air-handling system has been installed to design specifications, a more detailed equipment check must be performed. **See Figure 7-6.** A detailed equipment check includes the following actions:

• Check and record air handler information for comparison to existing PM records including motor nameplate information, overloads, fan size and type, rotation, clearances, alignment, shaft and pulley diameters, bearing size and type, drives and belt tension, filters, and coils.

• Set automatic dampers to open all duct and outlet dampers 100%.

• Check the thermostat setting.

• Ensure that all windows and doors are closed.

• Record all fan amperage, voltage, and revolutions per minute (rpm) readings.

• Record pitot tube traverse for total fan airflow and compare the readings to the fan curve.

• Perform a comparative fan reading based on amperage readings.

• Record fan intake and discharge static pressures and temperatures.

• Record static pressure and temperature drops across system filters and coils.

• Check for any stratification of air in building spaces.

System diffusers (outlets) and branch ducts must always be balanced to design specifications. When necessary, the diffuser measurement and branch line static pressure should be rechecked on larger systems where a fan supplies air to a whole floor or several floors. The most suitable velometer must be used to record the airflow readings on the balancing report next to the design flow for each diffuser or register. The combined total CFM for all diffusers and registers must not vary more than 10% from the CFM supplied by the fan. After completing the testing, all holes are sealed and plugged in the ductwork and any insulation that was removed is placed back into position.

Detailed Equipment Check

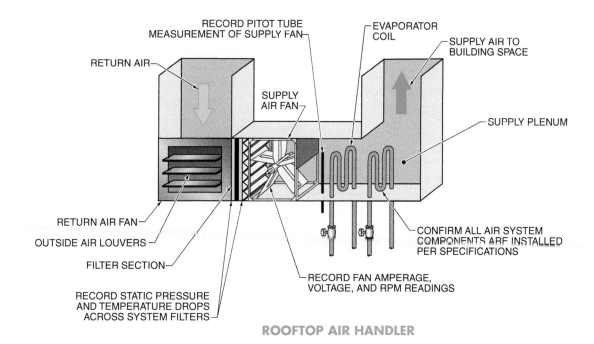

RECORD PITOT TUBE
MEASUREMENT OF SUPPLY FAN

EVAPORATOR
COIL

SUPPLY AIR TO
BUILDING SPACE

RETURN AIR

SUPPLY
AIR FAN

SUPPLY PLENUM

RETURN AIR FAN

OUTSIDE AIR LOUVERS

FILTER SECTION

CONFIRM ALL AIR SYSTEM
COMPONENTS ARE INSTALLED
PER SPECIFICATIONS

RECORD STATIC PRESSURE
AND TEMPERATURE DROPS
ACROSS SYSTEM FILTERS

RECORD FAN AMPERAGE,
VOLTAGE, AND RPM READINGS

ROOFTOP AIR HANDLER

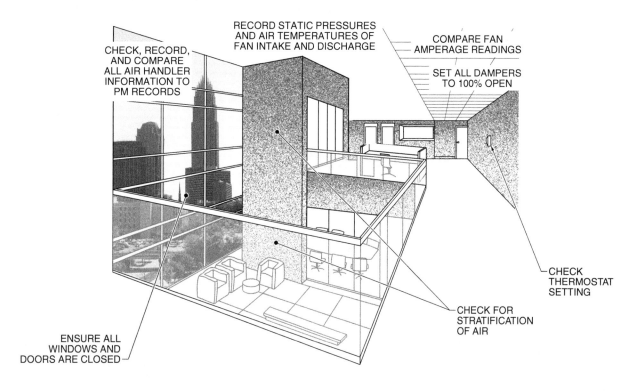

CHECK, RECORD,
AND COMPARE
ALL AIR HANDLER
INFORMATION TO
PM RECORDS

RECORD STATIC PRESSURES
AND AIR TEMPERATURES OF
FAN INTAKE AND DISCHARGE

COMPARE FAN
AMPERAGE READINGS

SET ALL DAMPERS
TO 100% OPEN

CHECK
THERMOSTAT
SETTING

CHECK FOR
STRATIFICATION
OF AIR

ENSURE ALL
WINDOWS AND
DOORS ARE CLOSED

Figure 7-6. A detailed equipment check is performed after confirming that all components of an air-handling system are installed to design specifications.

Duct Pressures

Duct pressures are measured to determine the airflow inside an air-handling system and are measured in inches of water column (in. WC). One inch of water column equals 0.03613 pounds per square inch (psi). This means that 27.7 inches of water equal 1 psi. Duct pressure is measured in static, velocity, and total pressure.

Static pressure is the pressure that acts against the sides of a duct and is the same throughout a cross section of the duct. **See Figure 7-7.** *Velocity pressure* is the pressure caused by airflow pushing against an object in the duct. Unlike static pressure, velocity pressure is not the same throughout a duct. *Total pressure* is static pressure and velocity pressure added together.

Airflow Calculations

Total pressure can be expressed in formula form as:

$$P_T = P_S + P_V$$
where
P_T = total pressure (in psi)
P_S = static pressure (in psi)
P_V = velocity pressure (in psi)

Total pressure and static pressure are usually measured with an insertion pitot tube. A *pitot tube* is a pressure-measuring instrument used to measure velocity pressures. A pitot tube is inserted into the airstream in the direction of flow and measures both static and total pressures.

When two pitot tubes are being used, one is the static pressure probe, and the second, which is upstream from the first, is the total pressure probe. The total pressure probe is bent or scooped to force moving air into the tube. The static tube is connected to the low pressure port of differential pressure transmitter, and the total tube is connected to the high pressure port of the transmitter. The velocity pressure can be found by subtracting static pressure from total pressure. Once the velocity pressure is found, the value can be converted into cubic feet per minute.

Once the average velocity pressure has been found, the velocity of the air can be determined by using the standard formula:

$$FPM = 4005 \times \sqrt{P_V}$$

where
FPM = velocity of the air (in fpm)
4005 = constant
P_V = average velocity pressure (in in. WC)

Duct Pressures

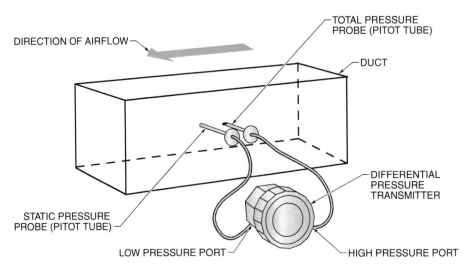

Figure 7-7. Total pressure is static pressure and velocity pressure added together.

For example, a duct yields an average velocity pressure of 0.25 in. WC. What is the air velocity in feet per minute? **See Figure 7-8.**

$$FPM = 4005 \times \sqrt{P_V}$$

$$FPM = 4005 \times \sqrt{0.25}$$

$$FPM = 4005 \times 0.5$$

$$FPM = \textbf{2002.5 fpm}$$

Knowing the air velocity allows the quantity of air to be found. If the duct area and air velocity are known, the quantity of air can be found by applying the following formula:

$$Q = A \times V$$

where

Q = quantity of airflow (in cfm)

A = area of the duct (in sq ft)

V = velocity of the air (in fpm)

For example, the insertion pitot tubes are in a rectangular duct with a size of 2′ × 3′. The velocity of the air was found to be 2002.5 fpm. What is the airflow in cubic feet per minute?

$$Q = A \times V$$

$$Q = (2 \times 3) \times 2002.5$$

$$Q = 6 \times 2002.5$$

$$Q = \textbf{12,015 cfm}$$

Fluke Corporation

Pitot tube test instruments are the most popular method used to measure pressures within ducts.

Airflow Velocity and Volume Calculations

DIRECTION OF AIRFLOW

FAN

DUCT (3′ × 3′)

TOTAL PRESSURE (442.72″ WC)

STATIC PRESSURE (442.47″ WC)

TOTAL PRESSURE	442.72″ WC
STATIC PRESSURE	− 442.47″ WC
VELOCITY PRESSURE	0.25″ WC

Figure 7-8. Once the average velocity pressure of a duct is known, the velocity of air can be calculated.

AIR-HANDLING SYSTEM DUCT MEASUREMENTS

There are various methods for determining the flow rate of air in ducts and grills or from diffusers and registers. Two methods of measuring airflow in ducts require direct air measurements using test instruments such as velometers or capture hoods. Other methods of determining airflow use estimated air measurements that involve measuring the temperature of the air.

Direct Air Measurements

Direct air measurements are measurements of air quantities with a test instrument for which no math or conversions must be performed. Many of the measurements and calculations require that the duct diameter be known. The *duct diameter* is the distance across the middle from side-to-side of either a rectangular or round duct. **See Figure 7-9.**

One measurement is not enough to get an accurate reading of pressure or velocity in a duct because pressure and velocity change throughout a duct. Many HVAC systems do not have correct spots to take measurements because of the duct's physical location, accessibility, obstructions in the duct, or length of the duct. Traversing is used to acquire measurements that can be used in calculations.

Traverse. A *traverse* is the process of taking multiple measurements across the area of a duct and averaging the values. When performing a duct traverse, it is important to ensure that turbulence in the airflow does not affect the measurements taken in the duct. Measurements are usually taken in a straight run of the ductwork that is away from openings or bends to reduce airflow turbulence. Measurements should be taken at least 7.5 duct diameters downstream or 3 duct diameters upstream from any turns or obstructions to airflow.

For the measurement of a rectangular duct, a minimum of 25 measurements is usually required. The number of test positions is carefully regulated and the exact number of test positions depends on the area (size) of the duct. **See Figure 7-10.** When a duct is less than 30″ wide, five data points must be taken along each side ($5 \times 5 = 25$ measurements). For duct sizes 30″ or 36″, six data points must be taken along the side. For ducts that are larger than 36″, seven data points must be taken along that side.

For the measurement of a round duct, the duct is divided into concentric circles, each containing an equal area. **See Figure 7-11.** An equal number of readings must be taken from each circular area to obtain the best average. When the duct has a diameter of 10″ or less, three concentric circles (4 to 6 measuring points per diameter), are used. For ducts with a diameter greater than 10″, four or five concentric circles (8 to 10 measuring points per diameter) are used. The preferred method is to drill three holes in the duct, at 60° angles from each other. Three traverses are taken across the duct, averaging the velocities obtained at each measuring point. An alternative method is to drill two holes 90° from each other.

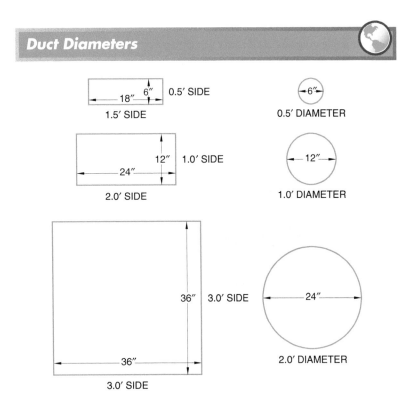

Duct Diameters

18″ 6″ 0.5′ SIDE
1.5′ SIDE

6″ 0.5′ DIAMETER

24″ 12″ 1.0′ SIDE
2.0′ SIDE

12″ 1.0′ DIAMETER

36″ 3.0′ SIDE
36″ 3.0′ SIDE

24″ 2.0′ DIAMETER

Figure 7-9. Duct diameter is the distance from side-to-side across the middle of a rectangular or round duct.

Traversing Rectangular Ducts

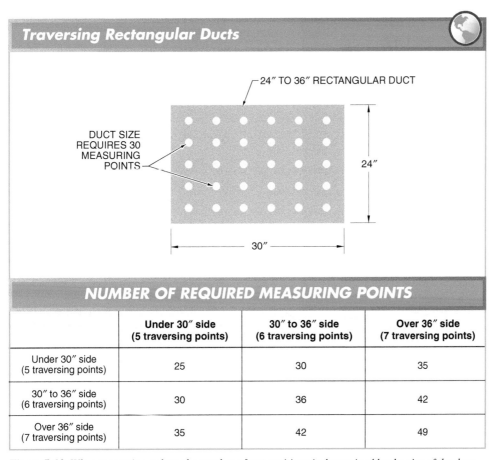

NUMBER OF REQUIRED MEASURING POINTS

	Under 30″ side (5 traversing points)	30″ to 36″ side (6 traversing points)	Over 36″ side (7 traversing points)
Under 30″ side (5 traversing points)	25	30	35
30″ to 36″ side (6 traversing points)	30	36	42
Over 36″ side (7 traversing points)	35	42	49

Figure 7-10. When traversing a duct, the number of test positions is determined by the size of the duct.

Traversing Round Ducts

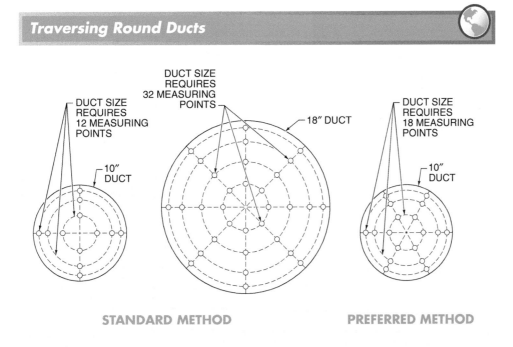

STANDARD METHOD   PREFERRED METHOD

Figure 7-11. Traversing a round duct requires dividing the duct into circles, each containing an equal area.

AIR-HANDLING SYSTEM TEST INSTRUMENTS

The *National Environmental Balancing Bureau (NEBB)* is an organization that sets procedural standards for testing, adjusting, and balancing environmental systems. NEBB has established criteria and frequency recommendations for the calibration of HVAC system test instruments.

Manometers

A *manometer* is an air-handling system test instrument that measures low air pressures, usually in inches of water column (in. WC). Manometers are often used with pitot tubes. When used with a pitot tube, manometers can measure air velocity pressure, which can be converted to velocity and then to airflow. Manometers can also be used in other applications, such as measuring the pressure differential between two rooms, the pressure drop across an air filter, or to determine the pressure an HVAC system is operating at. Two of the main types of manometers used are inclined-tube manometers and digital manometers.

Inclined-Tube Manometers. An *inclined-tube manometer* is a differential pressure-measuring test instrument used mainly in applications where a manometer is permanently attached. **See Figure 7-12.** A red or blue-colored oil mixture measures the difference between two pressures, which is indicated by markings on a clear tube. To acquire accurate readings, the tube must be properly mounted and leveled. An inclined-tube manometer must be zeroed and any bubbles removed from the liquid. When a manometer is moved the gauge must be reattached, leveled, and zeroed again. Inclined-tube manometers are difficult to use as portable instruments due to the leveling and zeroing requirements.

Inclined-Tube Manometer Procedures. Before measuring a pressure differential with an inclined-tube manometer, it must be checked to ensure that it is level and secure.

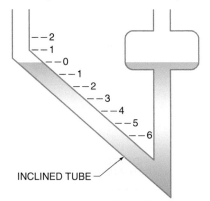

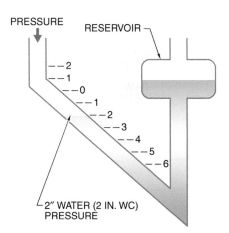

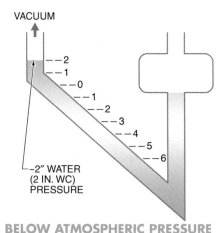

Figure 7-12. Inclined-tube manometers are mainly used in applications where the manometer is permanently attached.

See Figure 7-13. When a manometer is empty, the manometer must be filled with indicating fluid. The plugs for the high- and low-pressure sides are removed and the indicating fluid is poured into the high-pressure side of the tube. When all bubbles have dissipated, the zero-adjustment plunger is used to bring the fluid level to zero on the scale.

To measure a pressure differential, the high-pressure side fitting is connected to the environment to be measured, such as the inside of an HVAC duct. The low-pressure side is left open to the atmosphere. The inclined-tube manometer indicates a pressure measurement in inches of water column (in. WC).

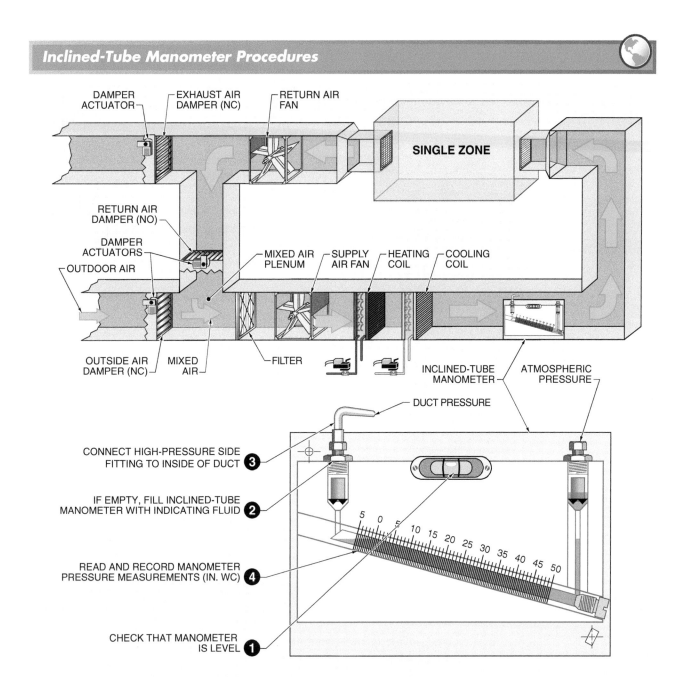

Figure 7-13. Inclined-tube manometer procedures for measuring the pressure differential in a duct include several steps.

Digital Manometers. A *digital manometer* is a pressure and differential pressure-measuring test instrument usually used in portable applications. Similar to the liquid inclined-tube manometer, digital manometers measure pressure differential and can use pitot tubes. Digital manometers make excellent portable pressure-checking test instruments because they do not require leveling. Depending on the model, some digital manometers may require zeroing. **See Figure 7-14.**

the pressure in a space with a digital manometer, tubing is connected between one of the meter inputs and the environment to be measured. The other input may be left open to atmospheric pressure. The meter then displays the pressure differential between the two inputs.

Like many other digital meters, a digital manometer may include a MIN/MAX function. This function records and displays the minimum and maximum measured values during the period that the MIN/MAX function was active.

Thermoanemometers

A *thermoanemometer* is a permanently installed air-handling system test instrument used to measure air velocity through a duct. Thermoanemometers are also known as hot-wire anemometers, as they use a principle of heat transfer to determine air velocity. Thermoanemometers can accurately measure a wide range of air velocities, from a minimum of 10 fpm to a maximum of 10,000 fpm.

The probe portion of a thermoanemometer is mounted within a duct. The external portion of the instrument passes a current through a very fine wire located in the probe. The probe is heated by the wire's resistance to current flow, but is cooled by the air passing over it. **See Figure 7-16.** The amount of airflow affects the amount of cooling of the probe. The temperature of the probe affects the current flow (since temperature affects the resistance of most metals), so this information can be used to determine the air velocity.

All thermoanemometers rely on the same principle of heat transfer and the resulting electrical relationships, but they measure these changes caused by different air velocities with slightly varying methods. Each method is based on the holding of one variable constant (voltage, current, or temperature) while measuring changes in the other variables. Therefore, thermoanemometers include constant-voltage, constant-current, and constant-temperature types.

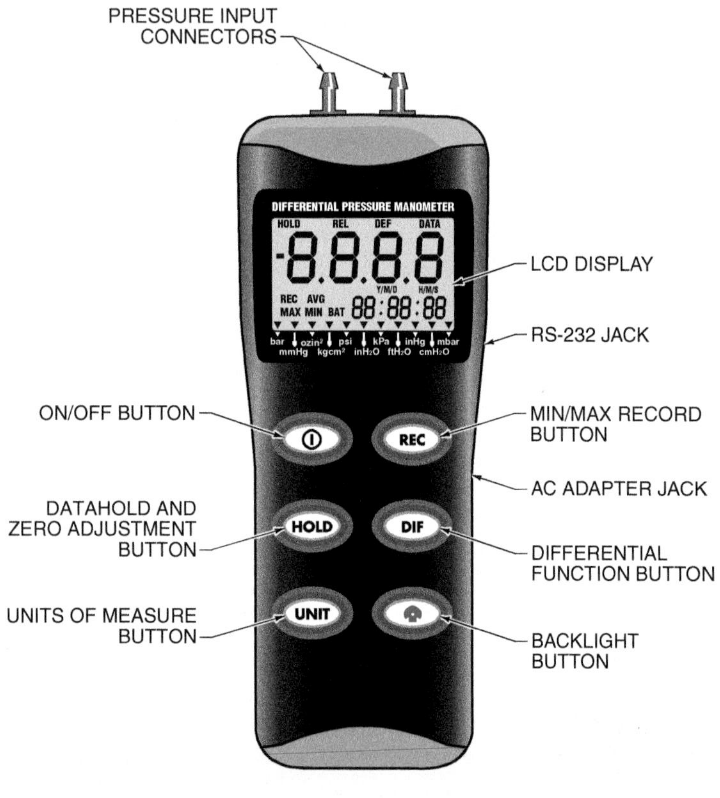

Figure 7-14. Digital manometers are portable test instruments used to measure pressure and pressure differential in air systems.

Digital Manometer Procedures. Digital manometers provide fast and continuous pressure readings. Digital manometers must first be zeroed, or calibrated to the ambient conditions. **See Figure 7-15.** To measure

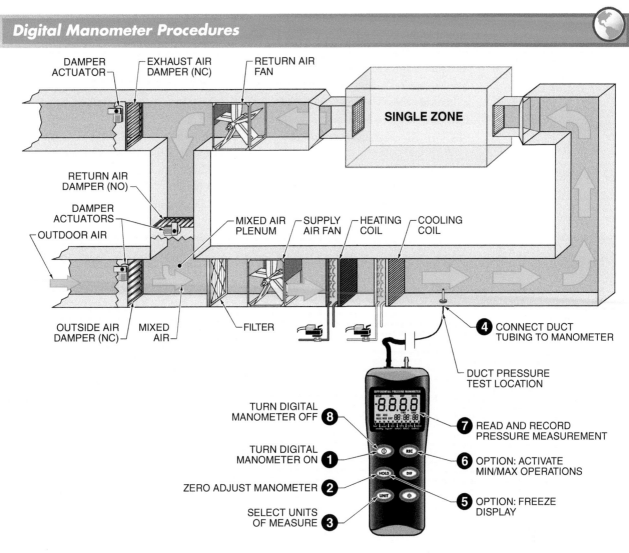

Digital Manometer Procedures

Figure 7-15. *Using a digital manometer to measure the pressure in a space requires several steps.*

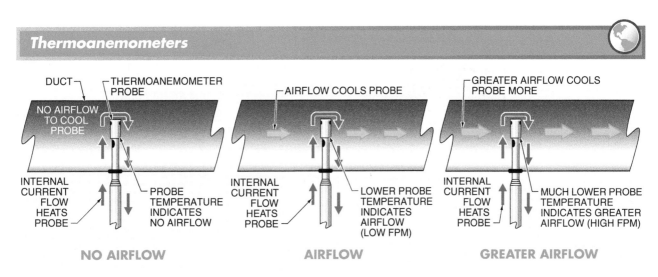

Thermoanemometers

Figure 7-16. *Thermoanemometers are permanently installed instruments that measure air velocity in a duct.*

Thermoanemometer Procedures. A thermoanemometer is composed of the probe, which is inserted into the environment to be measured, and a handheld electronic meter. **See Figure 7-17.** The two are connected by a flexible cable. The meter unit controls the measurement functions and calculates and displays the air velocity readings.

A special cover on the probe protects the thin wire when the probe is not used. It is also left in place when the meter is zeroed in order to calibrate the meter for a zero air velocity condition. The cover is then slid out of the way to expose the wire when ready to take measurements. It is important not to touch the wire while sliding the cover.

The probe is then inserted into the environment to be measured, such as through a hole in an HVAC duct. The probe must face a certain direction, as the orientation of the hot wire affects its cooling characteristics. A short time is required to allow the probe temperature to stabilize; then, air velocity measurements can be taken.

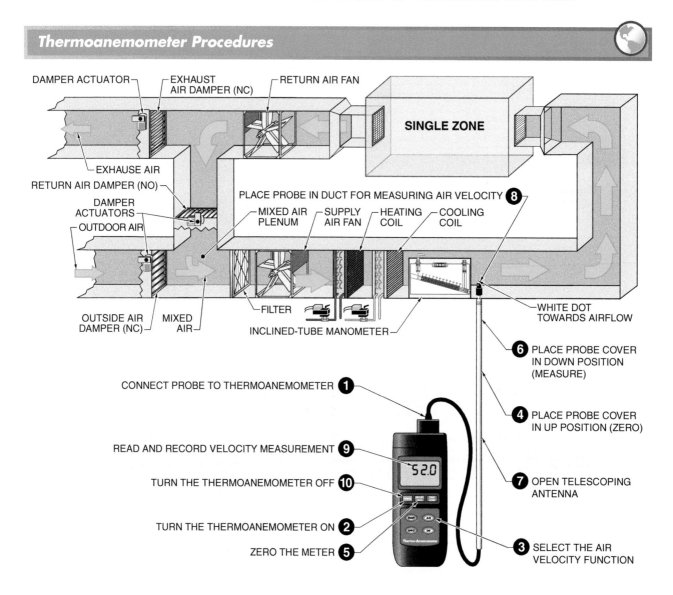

Figure 7-17. The proper procedure must be followed when using a thermoanemometer to measure air velocity.

Calculating Airflow Rate

To determine the square foot area of a duct opening, apply the following formula:

$$A = \frac{W \times H}{144}$$

where

A = area (in sq ft)
W = width (in in.)
H = height (in in.)
144 = constant

Then determine the airflow rate by applying the following formula:

$$Q = A \times V$$

where

Q = air volume (in cfm)
A = area (in sq ft)
V = air velocity (in fpm)

For example, a thermoanemometer in a 36″ × 24″ duct has an average velocity reading of 600 fpm. Find the airflow rate in cubic feet per minute. **See Figure 7-18.**

1. Find the duct area in sq ft.

$$A = \frac{W \times H}{144}$$

$$A = \frac{36 \times 24}{144}$$

$$A = \frac{864}{144}$$

$$A = \mathbf{6\ sq\,ft}$$

2. Find the airflow rate.

$$Q = A \times V$$
$$Q = 6 \times 600$$
$$Q = \mathbf{3600\ cfm}$$

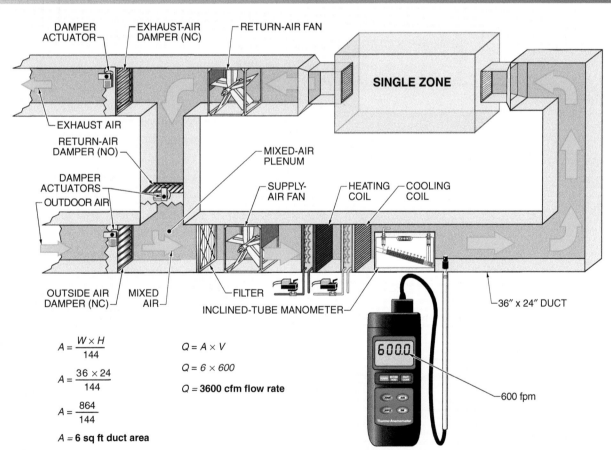

Calculating Airflow Rate

$$A = \frac{W \times H}{144}$$

$$A = \frac{36 \times 24}{144}$$

$$A = \frac{864}{144}$$

$A = \mathbf{6\ sq\ ft\ duct\ area}$

$$Q = A \times V$$

$$Q = 6 \times 600$$

$Q = \mathbf{3600\ cfm\ flow\ rate}$

Figure 7-18. To calculate the airflow rate of a duct, first find the duct area then solve for the airflow rate.

Mechanical-Vane Anemometers

Mechanical-vane anemometers have long been used in the refrigeration and air conditioning industry to measure airflow (in cfm). Mechanical-vane anemometers use a mechanical action that is pushed by the airflow. The two major types of mechanical-vane anemometers are rotating-vane and deflecting-vane anemometers.

Rotating-Vane Anemometers. A *rotating-vane anemometer* is a test instrument that measures airflow rate at a specific location and is similar to a small propeller or windmill in design. **See Figure 7-19.** Rotating-vane anemometers are commonly used by stationary engineers to measure airflow inside large air-handling units and on large banks of filters. Airflow hits the vanes, which causes them to turn. The rotation speed is proportional to the rate of airflow. Electronic devices in the unit measure the speed of the spinning vanes and convert this into an airflow rate.

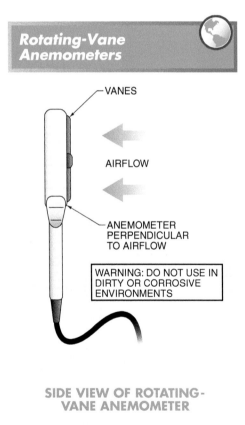

SIDE VIEW OF ROTATING-VANE ANEMOMETER

Figure 7-19. Rotating-vane anemometers are similar in design to a propeller or windmill.

Rotating-vane anemometers are available as a single-piece meter or as a two-part meter with a separate sensor and display unit that are connected by a cable. This allows the smaller sensor portion of the meter to be placed into tighter spaces, if necessary. Rotating-vane anemometer sensors are also available as accessories for digital multimeters. These accessories output small voltage signals that can be measured and displayed on the multimeter as values that are directly relatable to airflow units. Four-inch diameter vanes are the most common size for a rotating-vane anemometer, though other sizes are available.

To properly measure airflow rate, the anemometer must be oriented so that the vanes are perpendicular to the airflow. Other orientations result in inaccurately low airflow readings. Rotating-vane anemometers must be handled with care because they are easily damaged. In addition, they must not be used in dirty or corrosive environments.

Rotating-vane anemometers are available in both mechanical and digital versions. Digital versions have advanced features such as data storage, printing, and retrieval. Other features available with digital rotating-vane anemometers include:
- minimum, maximum, and averaging capabilities
- air temperature as well as airflow measurement
- direct conversions from air velocity and area to volume per minute

Rotating-Vane Anemometer Procedures. Rotating-vane anemometers are commonly used to measure airflow rate into a grill. **See Figure 7-20.** If a volume flow rate, such as cubic feet per minute (cfm) is required, this is selected from the meter's units function. Since volume flow rate is the product of air velocity and the area of the opening, the area of the grill or diffuser must be entered into the meter. The stationary engineer should accurately measure the size of the opening, calculate the area, and enter this number into the meter using its area function. This information is saved in the meter's memory.

The rotating-vane anemometer is then placed into the airstream, ensuring that it has the proper orientation in relation to the airflow. Airflow values may vary somewhat between the edges and the center of a grill or diffuser. Therefore, it is often best practice to use the meter's averaging mode as the anemometer sensor is traversed (moved) across the opening to obtain an average airflow reading for the grill or diffuser. This mode collects the measurements that the stationary engineer makes at several points and saves them in memory. When the last measurement is complete, the meter displays the average of the values.

Rotating-Vane Anemometer Procedures

7 PLACE ROTATING-VANE SENSOR INTO AIRSTREAM

ROOFTOP HEATING AND COOLING UNIT

ROTATING-VANE SENSOR IN POSITION TO TRAVERSE INTAKE

OUTSIDE-AIR INTAKE

REPEAT STEPS 8 THROUGH 10 **11**

AIRFLOW

1 CONNECT ANEMOMETER SENSOR CABLE TO METER

READ AND RECORD FLOW RATE MEASUREMENTS **12**

10 MEASUREMENT DISPLAYED

TURN ROTATING-VANE ANEMOMETER ON **2**

13 TURN ROTATING-VANE ANEMOMETER OFF

ENTER SQ FT AREA OF GRILL OR DIFFUSER **4**

3 SELECT DESIRED AIRFLOW UNITS

SAVE AREA ENTRY TO METER MEMORY **6**

5 USE THE UP, LEFT, AND RIGHT BUTTONS TO CHANGE AREA DIGITS

SELECT ANEMOMETER AVERAGING MODE TO TRAVERSE INTAKE **8**

9 TAKE MEASUREMENT AND PRESS AVG BUTTON

Figure 7-20. The airflow rate in a grill can be measured with a rotating-vane anemometer.

Deflecting-Vane Anemometers. A *deflecting-vane anemometer* is a test instrument that indicates the airflow rate with a pointer or arm that swings on a suspension. Deflecting-vane anemometers are also known as a swinging-vane anemometers. **See Figure 7-21.** Deflecting-vane anemometers are used in the field because of their portability, direct readings of air velocity, ease of use, and because they work without batteries or another power supply. In a deflecting-vane anemometer, there is little friction to disturb the accuracy. The flow through the test instrument is created by the velocity pressure at the point of measurement. Deflecting-vane anemometers are ideal for small terminal units with grills and diffusers that are difficult to reach.

Deflecting-Vane Anemometer Procedures. The primary advantage of a deflecting-vane anemometer is its simplicity. **See Figure 7-22.** It is portable, easy to use, and requires no batteries or other power supplies. This makes it an ideal field test instrument.

The deflecting-vane anemometer must first by zeroed by turning an adjustment screw until the pointer indicates zero, but then it is ready for taking measurements. The deflecting-vane anemometer is placed into the airstream to be measured that the flow rate is read directly from the indicator. If multiple measurements traversing a grill or diffuser are needed, the anemometer is moved as needed across the opening. The reading must be recorded by the stationary engineer, who must then calculate the average.

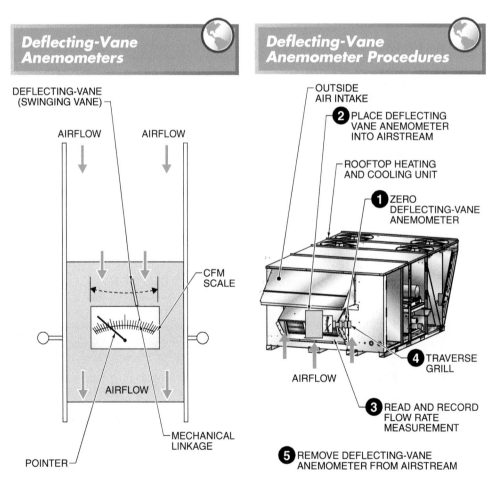

Figure 7-21. Deflecting-vane anemometers are used because of their portability, direct readings, and ease of use.

Figure 7-22. Deflecting-vane anemometers can be used to measure the airflow rate in a particular space.

Capture Hoods

In addition to ducts, the volume of air emerging from grills and diffusers is also an important measurement in ventilation and air conditioning systems. Traditionally, airflow readings were obtained by using a deflecting-vane anemometer with a diffuser probe. Then, a correction factor known as the K factor was used to properly calculate the airflow rate (in cubic feet per minute). Today, capture hoods are the test instrument of choice for diffuser airflow measurements. Capture hoods are also known as balometers.

Capture hoods are easy to use and provide quick direct readouts in cfm units. Capture hoods consist of a fabric skirt attached to a frame and a base unit with attached manifold. **See Figure 7-23.** The hood captures all air exiting from a diffuser or grill and causes the air to flow over a sampling manifold. The sampling manifold has evenly spaced inlets to average the flow rate. Various-sized fabric skirts and frames are available for different diffuser and grill configurations. Capture hoods are used on both supply and exhaust outlets. The two major categories of capture hoods available are analog and digital.

Capture Hoods

DIFFUSER (NOT SHOWN)

TUBULAR METAL FRAME (UNDER FABRIC SKIRT)

FABRIC SKIRT

BASE UNIT

MANIFOLD (NOT SHOWN)

TSI Incorporated

Figure 7-23. *Capture hoods may be used for airflow measurements from a diffuser or grill.*

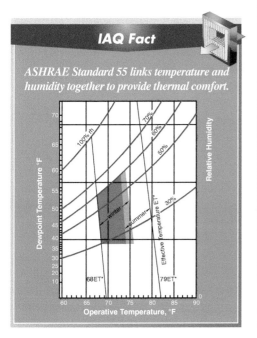

IAQ Fact

ASHRAE Standard 55 links temperature and humidity together to provide thermal comfort.

TSI Incorporated

Capture hoods are stored disassembled and must be carefully assembled. The calibration must also be checked before each use.

Analog Capture Hoods. An *analog capture hood* causes air to flow over a sensing manifold through a mechanical range selector (user adjustable) and finally through a deflecting-vane anemometer. **See Figure 7-24.** Analog capture hoods do not require any batteries or power supplies and can be used on many special types of diffuser configurations. Different hood sizes and configurations are available for low-flow and low-weight handheld applications so that the capture hood can be used in small, hard-to-reach applications.

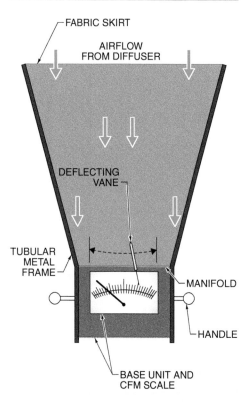

Figure 7-24. Analog capture hoods use a mechanical selector to force air over a sensing manifold.

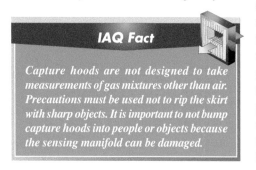

IAQ Fact

Capture hoods are not designed to take measurements of gas mixtures other than air. Precautions must be used not to rip the skirt with sharp objects. It is important to not bump capture hoods into people or objects because the sensing manifold can be damaged.

Analog Capture Hood Procedures. Analog capture hoods are very simple to operate. The procedure is nearly the same as for a deflection-vane anemometer, except that the analog capture hood has a large fabric skirt that is held snugly against the grill or diffuser while reading measurements. **See Figure 7-25.**

Before it is set in place, the analog capture hood is zeroed by turning an adjustment screw with a screwdriver until the pointer indicates zero. If the analog capture hood has multiple range settings, it is set for the highest possible reading first. If the resulting reading is low, the range selector is adjusted downward to the next lowest scale. This scale adjustment makes it easier to make a precise reading with the analog scale and pointer. The fabric hood is held over the grill or diffuser so that the opening is snug against the ceiling or wall and does not allow air to escape around the edge. All of the air must flow through the anemometer to produce accurate readings. The airflow rate measurement is read from the pointer's position on the analog scale.

Digital Capture Hoods. A *digital capture hood* is a capture hood with a thermoanemometer sensor to provide digital readings of airflow volume. **See Figure 7-26.** The ranging on digital capture hoods is automatic instead of manual. The scale automatically changes as the measurements change, which provides measurements with the highest possible precision. The measurements are shown on a liquid crystal display (LCD) or light-emitting diode (LED) display.

Digital capture hoods have many desirable features, including the fact that they can measure a wide range of airflow measurements, add and average airflow and temperature measurements, store readings to be printed, or upload readings to a computer. Some digital capture hood models allow reverse flow measurement by flipping a switch. Others allow the base plate electronics to be disconnected and removed from the base and used as a separate handheld unit with various probes.

Analog Capture Hood Procedures

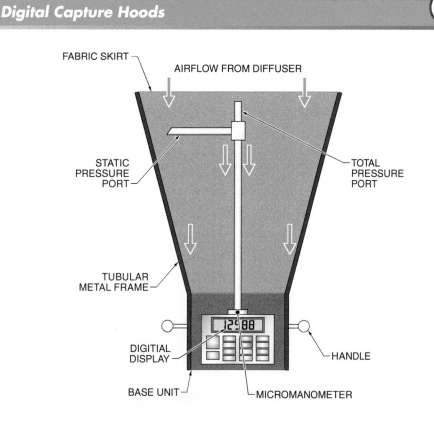

Figure 7-25. *Measuring the airflow rate from a diffuser with an analog capture hood requires several steps.*

Digital Capture Hoods

Figure 7-26. *Digital capture hoods have many of the same features as analog capture hoods.*

Some digital capture hoods can be used to calibrate other flow sensors by connecting the sensors to the electronics of the hood. The digital capture hood then writes the actual value to the sensor, adjusting it for any inaccuracies. Sensor adjustment is often used for variable-air-volume (VAV) box applications.

Digital Capture Hood Procedures. An airflow volume measurement with a digital capture hood is taken in a very similar way to measurements taken with the analog versions, except that the electronics must first be turned on and set to the correct operating mode. **See Figure 7-27.** Since it requires a power source, the stationary engineer must ensure that the unit has charged batteries or that it is plugged into the appropriate AC adapter.

When ready to measure airflow, the digital capture hood is held against the grill or diffuser and the airflow volume measurement is shown on the electronic display. The unit may include a feature to store one or more measurements in memory so that they can be recorded later on IAQ logs.

OUTDOOR AIR QUANTITIES

For IAQ purposes, the measurement of the outdoor air entering a building is the most important measurement that can be performed. There are four possible methods for determining the percentage of outdoor air being brought into a building. Two methods require direct air measurements using test instruments such as velometers and capture hoods. The other two methods of determining outdoor air percentage use estimated air measurements that involve measuring the temperature of the air or the amount of CO_2 in the air.

Direct Outdoor Air Measurements

Outdoor air quantities can be evaluated by directly measuring the airflow with test instruments that include the correct unit display. A common technique is to use a velometer, anemometer, or pitot tube to traverse both the supply and return ducts. Trained personnel must always take measurements in a straight run of a duct, away from duct openings or bends that induce airflow turbulence. A velometer provides direct cubic feet per minute readings. Anemometer and pitot tube readings must be converted into cubic feet per minute. **See Figure 7-28.** Using the cubic feet per minute readings, it is necessary to subtract the return air readings from the supply air readings to arrive at the total cubic feet per minute of outdoor air.

Digital Capture Hood Procedures

6 READ AND RECORD AIRFLOW MEASUREMENT

4 PLACE DIGITAL CAPTURE HOOD AGAINST DIFFUSER

9 TURN DIGITAL CAPTURE HOOD OFF

8 REPEAT STEPS 4 THROUGH 6 AS REQUIRED

AccuBalance

TSI

TURN DIGITAL CAPTURE HOOD ON **2**

5 TAKE AIRFLOW MEASUREMENT

1 INSTALL BATTERIES OR CONNECT AC ADAPTER

SAVE MEASUREMENT **7**

TSI Incorporated

3 SET HOOD TO MEASURE DIFFUSER AIRFLOW

Figure 7-27. Digital capture hoods can measure airflow rate to a space using a few procedural steps.

IAQ Fact

Climbing up ladders with capture hoods is dangerous. Two people must be used for capture hood work. One person scales the ladder, while the other person hands the capture hood to them.

Direct Measurements to Determine Outdoor Air CFM

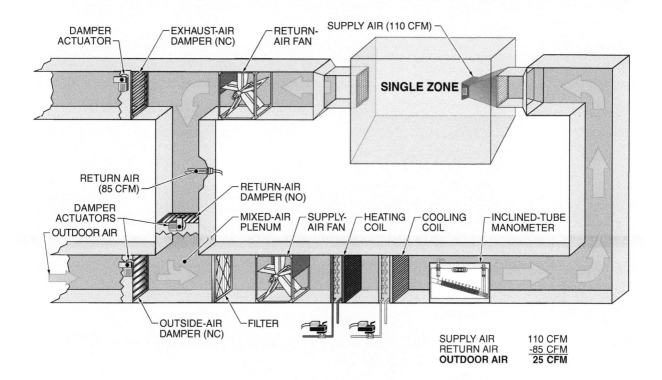

Figure 7-28. *A velometer provides direct cubic feet per minute readings, while anemometers and pitot tubes require a stationary engineer to convert readings into cubic feet per minute.*

Alternate Direct Measurements

An alternate technique depends on the accessibility of the outdoor air intake grill. When the outdoor air intake grill is accessible, one traverse of the grill can be performed to determine the total amount (in cubic feet per minute) of air being introduced into an HVAC system. **See Figure 7-29.**

Estimated Outdoor Air Measurements

The EPA recommends using either thermal balance (temperature) or CO_2 measurements to estimate outdoor air quantities. Temperature measurements must be taken over a long enough period of time to allow ample time for temperatures to even out in building spaces. However, temperature measurements should not take longer than

30 min, because temperatures can fluctuate excessively.

In the airflow formula, the difference of the mixed-air temperature subtracted from the return-air temperature $(70-50)$ is divided by the difference of the outdoor-air temperature subtracted from the return air temperature $(70-40)$. The resulting quotient (0.67) is multiplied by 100 to arrive at the percentage of out air (67%) being brought into the facility.

To ensure that the calculations to determine the percentage of outdoor air being brought into a facility is accurate, the following two conditions must be met:

• The outdoor air, return air, and mixed air streams must be accessible for measurement.

• There must be at least a several degree temperature difference between indoor air temperature and out air temperature.

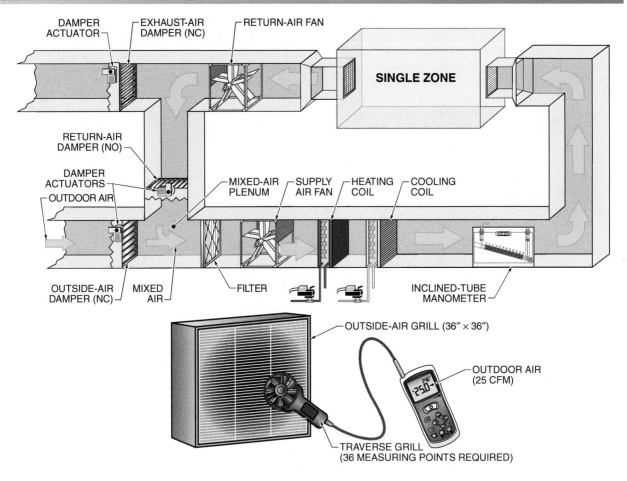

Figure 7-29. One technique for determining outdoor air quantities involves traversing the outside air duct.

Outdoor Air Percentage Calculations

The percentage of outdoor air being used in a system is found using the equation:

$$OA = \frac{T_R - T_M}{T_R - T_O} \times 100$$

where
OA = percentage of outdoor air
T_R = temperature of the return air (in °F)
T_M = temperature of the mixed air (in °F)
T_O = temperature of the outdoor air (in °F)
100 = constant

For example, when the mixed-air temperature of a facility is 50°F, the return air temperature is 70°F, and the outdoor air temperature is 40°F, what is the percentage of outdoor air being brought into the facility? **See Figure 7-30.**

$$OA = \frac{T_R - T_M}{T_R - T_O} \times 100$$

$$OA = \frac{70 - 50}{70 - 40} \times 100$$

$$OA = \frac{20}{30} \times 100$$

$$OA = 0.67 \times 100$$

$$OA = \mathbf{67\%}$$

Calculating Outdoor-Air Percentage Using Temperature

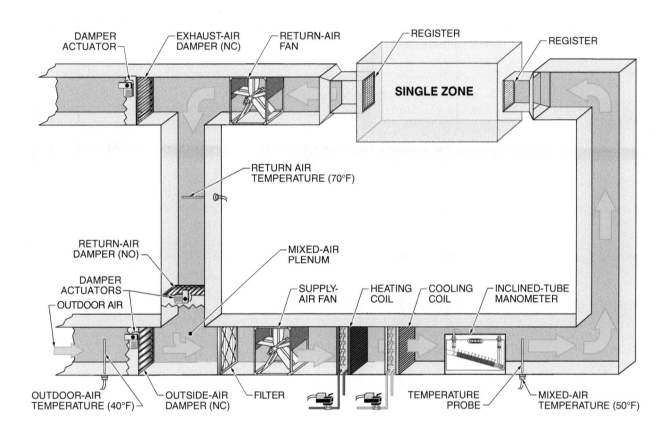

NOTE: DO NOT TAKE TEMPERATURE MEASUREMENTS USING MORE THAN 30 MIN IN TIME

$$QA = \frac{T_R - T_M}{T_R - T_O} \times 100 \qquad QA = \frac{20}{30} \times 100$$

$$QA = \frac{70 - 50}{70 - 40} \times 100 \qquad \begin{aligned} QA &= 0.67 \times 100 \\ \boldsymbol{QA} &= \textbf{67\% OF OUTDOOR AIR} \end{aligned}$$

TO ENSURE THE CALCULATIONS FOR DETERMINING THE PERCENTAGE OF OUTDOOR AIR BEING BROUGHT INTO A FACILITY IS ACCURATE:

• OUTDOOR AIR, RETURN AIR, AND MIXED AIRSTREAMS MUST BE ACCESSIBLE FOR MEASUREMENT
• THERE MUST BE AT LEAST A SEVERAL DEGREE TEMPERATURE DIFFERENCE BETWEEN INDOOR AIR TEMPERATURE AND THE OUTDOOR AIR TEMPERATURE

Figure 7-30. Calculating the outdoor air percentage of a facility can be done if the outdoor-air, mixed-air, and return-air temperatures are known.

Outdoor Air Estimate Using Carbon Dioxide Gas

A *tracer gas* is a gas that is identifiable in an airstream and can be measured with a degree of accuracy. The CO_2 level must be obtained for the return and supply air ducts and the outdoor air intake. To be accurate enough, the outdoor air concentration of CO_2 should be constant over the time required to perform the test measurements. Also, the measurements must be taken when the indoor CO_2 concentrations are likely to be highest and well above the outdoor concentrations such as late morning or late afternoon.

When calculating the percentage of outdoor air by hand the return, supply, and outdoor air measurements must be taken in the following sequence with no delay between each step:

1. return-air duct CO_2 measurement
2. supply-air duct CO_2 measurement
3. outside-air intake CO_2 measurement
4. supply-air duct CO_2 measurement second reading
5. return-air duct CO_2 measurement second reading

The two return-air duct concentrations are averaged together. The same is done with the two supply-air duct concentrations. Averaging two readings lessens the impact of fluctuations that can occur in these locations on the calculations.

The error rate of this calculation increases as the difference between CO_2 measurements decreases. Normal fluctuations of CO_2 and accuracy of measurement also affect the calculations. Some HVAC systems use CO_2 monitors that increase airflow or open outside-air dampers to increase outside airflow when CO_2 setpoint levels are exceeded.

Outdoor-Air Estimate Procedures. Some meters include a function that simplifies the estimation of outdoor air percentage by walking the user through the required sequence of measurements. **See Figure 7-31.** It saves each measurement in memory and calculates the estimate automatically when finished. Some meters have multiple sensing methods such as temperature and CO_2.

When an estimation method is chosen, such as using CO_2 levels, the meter prompts the stationary engineer to the location of the first measurements, the return-air airstream. The meter should be held with the sensor perpendicular to the airstream and the test is initiated. When this measurement is complete, the value is stored in memory and the meter prompts the user to the next testing location, the supply-air airstream. The same procedure is repeated for the supply-air airstream and the outdoor-air airstream, and then the return air and supply air are tested each a second time. When the final measurement is taken, the meter automatically calculates the estimate for the percentage of outdoor air being drawn into the building and displays the result.

The measured CO_2 levels should be recorded in the IAQ log. When finished, the testing procedure should be repeated to confirm the first results. If the second estimate is significantly different from the first estimate, the test should be repeated at least one more time or a different meter or testing method should be used.

Exhaust-and Return-Air Duct Measurements

In addition to measuring the volume of supply air to a space, it may also be desirable to measure the exhaust or return air in a duct or from a specific space. A capture hood can be connected to a return air grill to measure the airflow in reverse fashion (capture hood has reverse flow switch). **See Figure 7-32.** In a return duct, a manometer with pitot tube is used to perform a traverse. The reading then may be either read out directly or converted to cubic feet per minute, depending on the test instrument(s) being used.

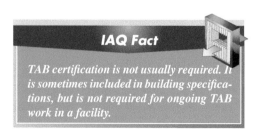

IAQ Fact

TAB certification is not usually required. It is sometimes included in building specifications, but is not required for ongoing TAB work in a facility.

Calculating Outdoor Air Percentage Using CO_2

PLACE THE CO_2 METER PERPENDICULAR TO THE RETURN-AIR AIRSTREAM ❸

❺ PLACE THE CO_2 METER PERPENDICULAR TO THE SUPPLY-AIR AIRSTREAM

DAMPER ACTUATOR

EXHAUST-AIR DAMPER (NC)

SINGLE ZONE

RETURN-AIR FAN

RETURN-AIR DAMPER (NO)

DAMPER ACTUATORS

OUTDOOR AIR

MIXED-AIR PLENUM

SUPPLY-AIR FAN

HEATING COIL

COOLING COIL

OUTSIDE-AIR DAMPER (NC)

MIXED AIR

FILTER

INCLINED-TUBE MANOMETER

❼ PLACE THE CO_2 METER PERPENDICULAR TO THE OUTDOOR-AIR AIRSTREAM

❹ MEASURE THE RETURN AIR CO_2 CONTENT

❽ STORE THE MIXED-AIR READING AND MEASURE THE OUTDOOR AIR CO_2 CONTENT

STORE THE RETURN-AIR READING AND MEASURE THE MIXED AIR CO_2 CONTENT ❻

❾ STORE THE OUTDOOR-AIR READING. METER CALCULATES PERCENTAGE OF OUTDOOR AIR

SET THE METER TO MEASURE PERCENTAGE OF OUTDOOR AIR ❷

❿ END THE PERCENTAGE OF OUTDOOR-AIR MODE

⓫ READ AND RECORD PERCENTAGE OF OUTDOOR-AIR CALCULATION

TURN THE CO_2 METER ON ❶

⓬ REPEAT STEPS 3 THROUGH 9

⓭ TURN THE CO_2 METER OFF

FLUKE *975 AIRMETER*

74.3 °F 25.7 %RH
0 CO   6.2 CO_2

Fluke Corporation

Figure 7-31. *The outdoor air percentage is often calculated using an air meter that measures carbon dioxide.*

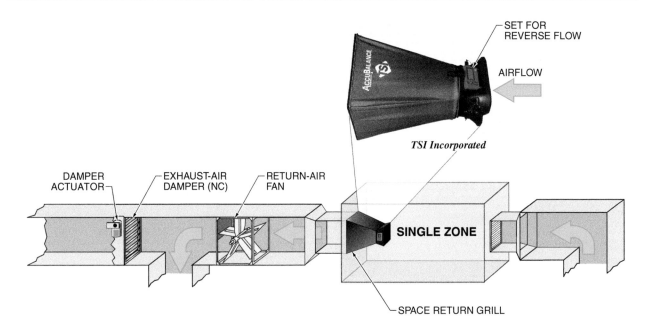

Figure 7-32. Connecting a capture hood (capture hood has reverse flow switch) to a return-air grill can be done to measure airflow in a reverse manner.

Air Volume

Air volume measurements are most commonly made to confirm that airflow out of each diffuser and register meets design specifications. To make this comparison, a schematic duct system layout or an outlet air balance report is needed from the most recent written test and balance report. **See Appendix.**

Air velocity meters such as flow capture hoods or anemometers are used to measure air volume. The equipment manufacturer's specifications should be consulted to ensure that the test equipment being used is appropriate for the measurements to be taken. All readings are recorded on an outlet air balance report.

Also required is the diffuser manufacturer's suggestions for the number of measurements required and the exact position for each measurement. All of the velocity readings are averaged, then multiplied by the K factor for the particular type of diffuser in use. The *K factor* is a number that represents the actual open area and the airflow characteristics of a specific diffuser. The result of the K factor equation is the cubic feet per minute flow out of a diffuser. The K factor is also available from diffuser manufacturers.

Ventilation Air Requirements

Governing agencies, such as ASHRAE, have published standards regarding the proper amount of ventilation air to be introduced into a building. The amount of ventilation air is measured in cubic feet per minute. The rate of air required is broken down by factors such as the size of the facility, functions being performed in the facility, and the number of people in the facility (density). The rate is then listed as cubic feet per minute per person or cubic feet per minute per square foot of the building floor space. **See Figure 7-33.** A generally accepted rule of thumb for the amount of ventilation air to bring into a facility is 20 cfm/person. ANSI/ASHRAE Standard 62.1-2007, *Ventilation for Acceptable Indoor Air Quality,* must be consulted for specific application information.

VENTILATION AIR REQUIREMENTS*				
	OUTDOOR AIR RATE†	AREA OUTDOOR AIR RATE‡	DEFAULT VALUES	
			OCCUPANT DENSITY§	COMBINED OUTDOOR AIR RATE†
Office Building				
Office space	5	0.06	5	17
Reception area	5	0.06	30	7
Telephone/data entry	5	0.06	60	6
Main entry lobbies	5	0.06	10	11
Miscellaneous Spaces				
Computer room (not printing)	5	0.06	4	20
Electrical equipment rooms	—	0.06	—	
Telephone closets	—	0.00	—	
Public Assembly Spaces				
Lobbies	5	0.06	150	5
Libraries	5	0.12	10	17
Retail				
Mall common areas	7.5	0.12	40	9

* Partial ASHRAE Table 6.1
† cfm/person
‡ cfm/sq ft
§ #/1000 sq ft

Sources: American Society of Heating, Refigerating, and Air-Conditioning (ASHRAE)

Figure 7-33. *ASHRAE publishes standards concerning ventilation air requirements.*

Zone Air Circulation

Another important consideration is zone air circulation. Air must circulate completely through occupied zones of a building, which flushes contaminants from the space. Contaminant flushing cannot happen if airflow becomes stratified or is blocked from reaching certain locations. **See Figure 7-34.**

Temperatures should be relatively even throughout a space. Wide variations in temperature within a space indicate poor air circulation. The vertical temperature difference between the floor and ceiling of an occupied zone should not exceed 5°F. ASHRAE sets maximum local air speeds at 50 fpm (0.25 meters per second [m/s]) in the summer and 30 fpm (0.15 m/s) in the winter. Higher air speeds can cause complaints from occupants about drafts.

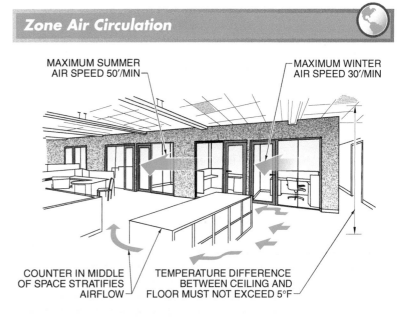

Zone Air Circulation

MAXIMUM SUMMER AIR SPEED 50'/MIN

MAXIMUM WINTER AIR SPEED 30'/MIN

COUNTER IN MIDDLE OF SPACE STRATIFIES AIRFLOW

TEMPERATURE DIFFERENCE BETWEEN CEILING AND FLOOR MUST NOT EXCEED 5°F

Figure 7-34. *Air should circulate completely through occupied zone of a building to flush out contaminants.*

Open Area Measurements

Large, open areas (vertical multistory spaces and large floor spaces) experience air movement in the form of drafts. Drafts lead to discomfort for the occupants of building spaces. In some industrial applications, such as labs, paint spray booths, and commercial applications such as multifloor office areas, drafts can interfere with airflow patterns that remove fumes and chemicals from the air. Drafts can be measured by placing digital anemometers at different points in the space. **See Figure 7-35.** Anemometers should be set to log the measurements over time, which can then be viewed and evaluated on a computer. The data can then be printed out for evaluation by staff.

Fan Testing, Adjusting, and Balancing

Fans in air-handling systems must be tested before being placed on permanent operation. When a fan is not operating properly, the malfunctioning fan affects the entire air-handling system. Fan operation is broken down into two major categories: electrical and motor drive operation. Electrical measurements are taken at the fan motor just as in the case of a pump motor. Both motor current and voltage for each leg are measured under proper operational conditions. **See Figure 7-36.** The current and voltage readings are then checked against motor nameplate and system characteristics data to ensure that the readings are correct for normal fan or operation.

Open Area Measurements

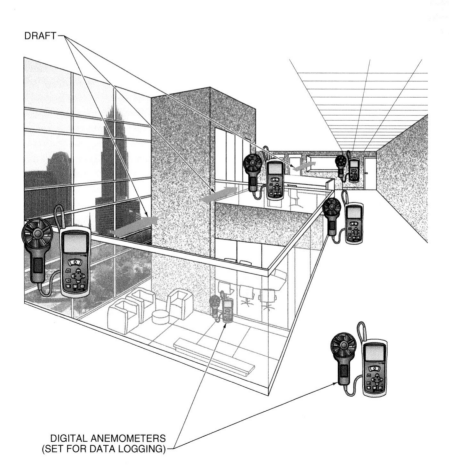

DRAFT

DIGITAL ANEMOMETERS
(SET FOR DATA LOGGING)

Figure 7-35. Open area measurements may be used for purposes such as measuring drafts.

Fan Voltage and Current Measurements

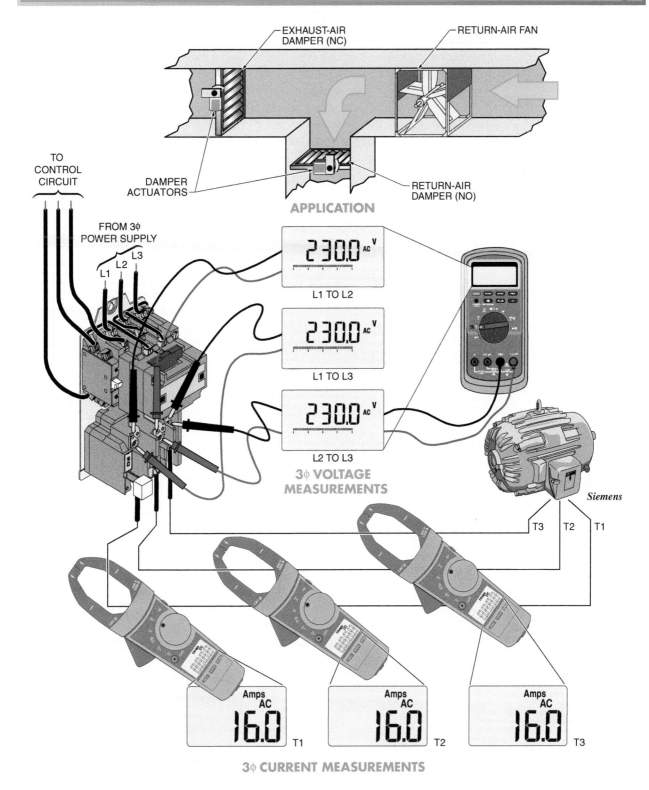

Figure 7-36. *Motor current and voltage measurements of a fan are necessary to ensure the fan is operating properly.*

Motor current is an indication of the amount of work being performed by a fan. A fan or blower that is performing less work draws less current. A fan may draw less current because the system pressure drop is less than the design pressure drop. A low-system pressure drop can be caused by improperly adjusting and balancing the system.

Mechanical readings are also taken in addition to the electrical measurements. One of the major differences between fan and pump systems is the use of belt drives on fan systems. The belts must be checked for proper tension and alignment prior to startup. Fan rotation (direction and rotations per minute), vibration and alignment readings, and inlet and discharge pressures must all be checked and verified. All of these readings are important when testing fan performance.

A fan is often tested at 100% of its rated capacity to ensure that enough airflow is available under maximum load. The capacity readings are checked against the fan curve to ensure that the fan operating point is correct. **See Figure 7-37.** The fan control sequences are usually tested as well.

When a variable-speed drive (VSD) is used, the drive operation and sequence are also verified. A fan variable-speed drive is adjusted as needed in order to meet proper operating parameters. Variable-speed drive adjustments are made in accordance with the manufacturer's procedures, using either the front panel keypad menu or a personal computer connected to the drive. The parameter that typically must be changed is the output frequency (motor revolutions per minute) of the drive required to maintain a specific static pressure in the discharge duct.

Adjusting Airflow Systems

When adjusting an airflow system, the fan is often the first device adjusted. The system fan must produce the correct volume of air and static pressures at the correct speed. The fan is adjusted until it reaches the desired place on its performance curve. Fans are adjusted to this operating point by changing the characteristics of the fan drive, which is either a mechanical or an variable frequency drive.

Mechanical Fan Drives. A mechanical fan drive consists of belts and pulleys. **See Figure 7-38.** The pulleys the belt rides on can have their diameter adjusted to change the rotational speed of the fan. The fan system belts also affect fan performance and must be checked for proper tension and alignment. Variable-speed belt drives are sometimes used to change fan speed. A different ratio of pulley diameters changes the rotational speed of the belt and output shaft of the variable-speed drive.

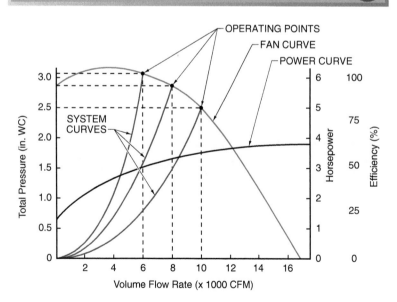

Fan Curve with Operating Points

Figure 7-37. Fan capacity readings are checked against a curve to ensure the fan operating point is correct.

IAQ Fact

For quality assurance, air-handling units must be designed, fabricated, and installed in compliance with NFPA Standard 90A, Standard for the Installation of Air Conditioning and Ventilating Systems. *Air-handling units must also be labeled by Underwriters Laboratories Inc. and factory tested in accordance with ARI 430,* Standard for Central-Station Air-Handling Unit. *Finally, air-handling units are labeled by the Air-Conditioning and Refrigeration Institute.*

Mechanical Fan Drives

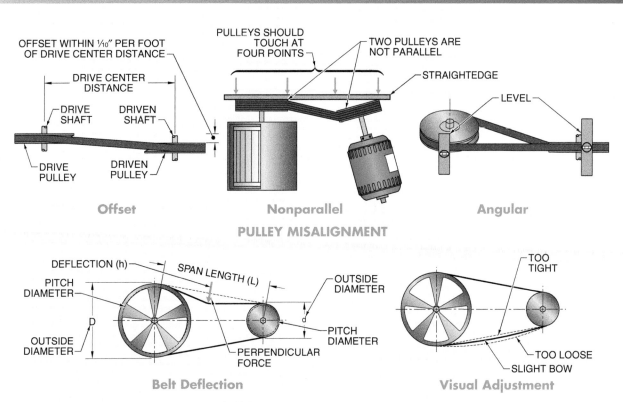

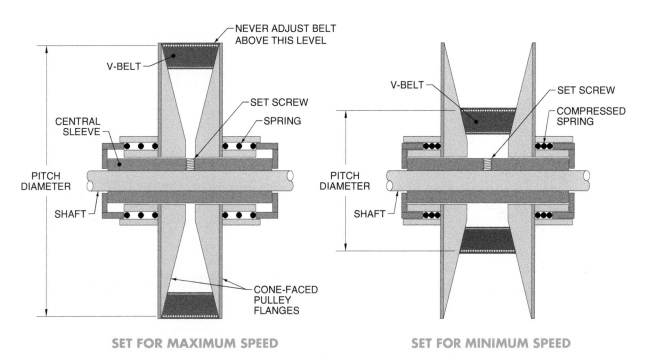

Figure 7-38. *A mechanical fan drive consists of belts and pulleys.*

Variable-Frequency Fan Drives. Fans that use modern variable-frequency drives offer far more flexibility and adjustments. **See Figure 7-39.** The variable-frequency drive built-in keypad can be used to change many of the motor operational characteristics such as the time required to ramp up during a soft start, maximum voltage and current, horsepower, and the speed of the motor.

System Component Testing and Adjusting

Variable-air-volume (VAV) systems often operate under a partial load due to the reduced need for cooling on a cool day. In this case, the stationary engineer is responsible for ensuring that the fan system produces the correct conditions at 100% load. A variable-air-volume system automatically changes fan speed (up or down) to the flow rate required.

After the system supply fan is adjusted, the return and exhaust fans must also be tested and adjusted. **See Figure 7-40.** An improperly functioning return or exhaust fan system causes as many or more problems as a poorly operating supply fan. After fan testing and adjusting are performed, the airflows in each supply and branch duct must be adjusted as well.

Adjustments are performed on the dampers located in the ductwork. The dampers are either opened to increase flow, or closed to reduce flow. The main duct dampers are usually adjusted first and then the dampers in the smaller branches of the variable-air-volume system are adjusted.

IAQ Fact

A chief engineer is the most important person a TAB technician can work with.

Electric Motor Drive Fan Drives

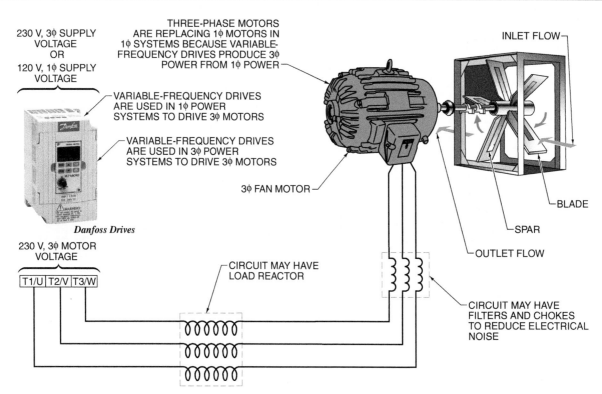

Figure 7-39. Fans that use electric motor drives offer more flexibility than mechanical fan drives.

Balancing Airflow Systems

The balancing of airflow systems is performed by adjusting the fan and/or dampers in the ducts until the proper airflow readings in that section of the building are reached. The mechanical contractor provides tables, known as schedules, which indicate the desired airflow for each duct section or room diffuser. The actual installation of an airflow system is often different than the design installation. The differences usually affect the duct schedules and the measured values. Any discrepancies must be corrected and recorded in the preventive maintenance and balancing records.

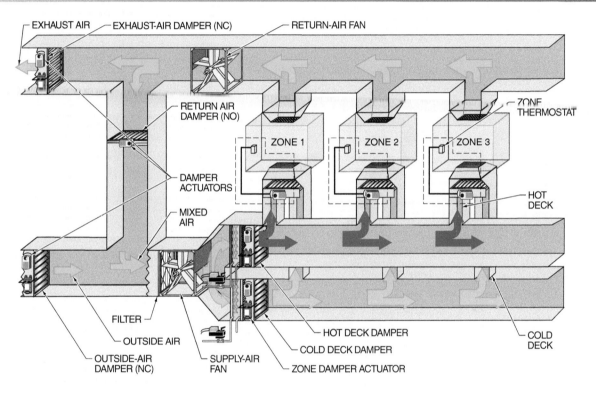

System Component Testing and Adjusting

Figure 7-40. After a system supply fan is adjusted, return and exhaust fans are tested as well.

Case Study – Using Various Test Instruments

Complaint History

A new store, which occupies the 1st, 2nd, and 3rd floors of a high-rise building, is having cooling problems in their 3rd floor display area on a hot summer day. The 3rd floor has two zones. One zone includes the store office areas, sales area, and display area where the complaint originated. Certified technicians from the National Environmental Balancing Bureau (NEBB) tested, adjusted, and balanced the system prior to the commissioning of the direct digital control (DDC) system. Building testing, adjusting, and balancing (TAB) technicians assisted in all aspects of preparing the system for commissioning.

Investigation

The room and diffuser temperatures in the zone are measured. The rooms have a thermocouple temperature of 72°F. Using an IR thermometer, technicians find that the zone diffusers have a temperature of 55°F (indicating that the system is calling for cooling) and that ceiling tiles located about 7′ from the diffusers have a temperature of 58°F. The temperature of desk surfaces in the zone that receives sunlight is 90°F.

Also, while checking the main fan, the technicians find that the motor and electric motor drive system show no signs of excessive temperatures. A laptop computer is connected to the system network; all zone temperatures are close to the setpoint, and the duct pressures are correct.

Using the IR thermometer again, the technicians check the temperatures of the system contactors. Pole 3 on compressor contactor #2 has a 30° increase in temperature compared to all other readings. When a voltage meter is used, L1 to L2, L1 to L3, and L2 to L3 all measure about 475 V. A clamp-on amperage meter indicates that L1, L2, and L3 all have the correct reading of 8.25 A.

Resolution

In the office area, the ceiling diffusers are exhibiting a prime example of surface effect. The air is clinging to the ceiling, reducing throw and drop velocity. The diffusers are adjusted to get the air away from the ceiling where the air can mix and throw to the outside walls.

After looking at pole 3 on compressor contactor #2 more closely, it is found that the insulation of the wire for the T3 pole has been captured under the lug, resulting in a poor connection. The wire is backed off, the compressor is placed back on-line, and the temperature measurements indicate normal operation.

IAQ Testing, Adjusting, & Balancing Definitions

- An *analog capture hood* causes air to flow over a sensing manifold through a mechanical range selector (user adjustable) and finally through a deflecting-vane anemometer.
- A *deflecting-vane anemometer* is a test instrument that indicates airflow rate with a pointer or arm that swings on a suspension.
- A *digital capture hood* is a capture hood that has the same basic components as an analog capture hood including a skirt, frame, and averaging grid but has a thermoanemometer sensor.
- A *digital manometer* is a pressure and differential pressure-measuring test instrument usually used in portable applications.
- *Direct air measurements* are measurements of air quantities with a test instrument for which no math or conversions must be performed.
- The *duct diameter* is the distance across the middle from side-to-side of either a rectangular or round duct.
- *HVAC system adjustment* is the process of changing the settings of operating devices to obtain the final setpoint of control or air-handling system balancing.
- *HVAC system balancing* is the regulation of the flow of air in a system through the use of acceptable industry procedures.
- *HVAC system testing* is the process of using specific IAQ test instruments to measure variables within an air-handling system to evaluate mechanical equipment and system performance.
- An *inclined-tube manometer* is a differential pressure-measuring test instrument used mainly in applications where a manometer is permanently attached.
- The *K factor* is a number that represents the actual open area and the airflow characteristics of a specific diffuser.
- A *manometer* is an air-handling system test instrument that measures low air pressures, usually in inches of water column (in. WC).
- The *National Environmental Balancing Bureau (NEBB)* is an organization that sets procedural standards for testing, adjusting, and balancing environmental systems.
- A *pitot tube* is a pressure-measuring instrument used to measure velocity pressures.
- A *rotating-vane anemometer* is a test instrument that measures airflow rate at a specific location and is similar to a small propeller or windmill in design.
- *Static pressure* is the pressure that acts against the sides of a duct and is the same throughout a cross section of the duct.
- A *thermoanemometer* is a permanently installed air-handling system test instrument used to measure air velocity through a duct.
- *Total pressure* is static pressure and velocity pressure added together.
- A *tracer gas* is a gas that is identifiable in an airstream and can be measured with a degree of accuracy.
- A *traverse* is the process of taking multiple measurements across the area of a duct and averaging the values.

**For more information please refer to ATPeResources.com
or http://www.epa.gov.**

1. What is the purpose of testing, adjusting, and balancing an HVAC system?

2. Explain the purpose of performing airflow balancing within a building.

3. Describe the process for performing a traverse of a round duct.

4. How do inclined-tube manometers and digital manometers differ?

5. Name the main set of standards that lists ventilation requirements for buildings.

HYDRONIC SYSTEM BALANCING AND BUILDING COMMISSIONING

8

chapter

Class Objective:
Define the benefits of testing and balancing and describe a general hydronic system balancing procedure.

On-the-Job Objective:
Justify testing and balancing procedures by their positive impact on IAQ at the facility.

Learning Activities:
1. List ways testing and balancing positively impact a hydronic system's IAQ.
2. Give an example of how a lack of system testing and balancing could lead to an IAQ problem.
3. Briefly describe the steps applied in a general hydronic system balancing procedure.
4. Describe the process of building commissioning.
5. **On-the-Job:** Determine if proper hydronic system testing and balancing procedures are being applied in the facility to improve IAQ.

Introduction

A hydronic system is a system that uses water as a heat-transfer medium in heating and cooling applications. Testing and balancing procedures are used to evaluate and monitor hydronic system performance and to provide the desired comfort level at a minimum cost in energy.

The mitigation of an IAQ problem often involves taking measurements to determine when a hydronic system is functioning improperly. To provide the proper water flow to the desired heating or cooling coil when there is a problem, it is important to understand how to quickly and effectively use the data collected from water flow test instruments.

TESTING, ADJUSTING, AND BALANCING OF CHILLED AND HOT WATER HYDRONIC SYSTEMS

To properly test, adjust, and balance a hydronic cooling and/or heating system, the circulation pump or pumps in the system must first be tested. A poorly operating circulation pump affects the operation of the entire hydronic system. Circulation pump testing is required when a new system is installed. Electrical measurements are taken off the motor for the pump. **See Figure 8-1.** Both the motor current and voltage for each leg of the three phases are measured under normal operational conditions. The readings are then checked to ensure that the readings match the motor nameplate and system characteristics.

Circulation Pump Electrical Testing

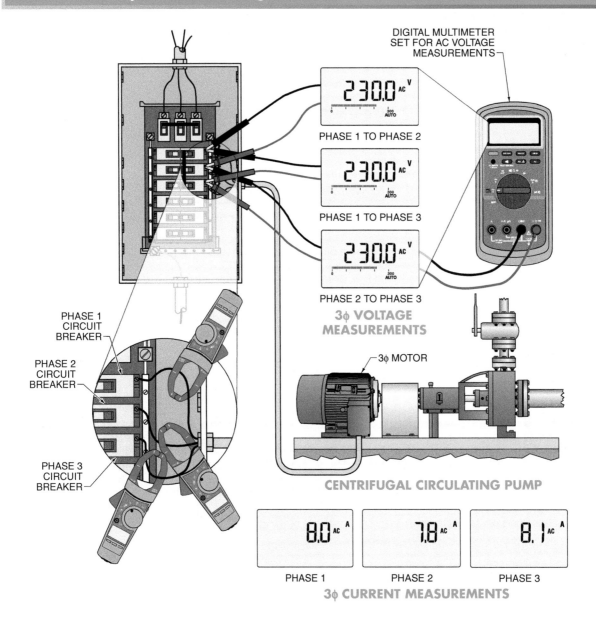

Figure 8-1. Circulation pump testing is required when a new system is installed.

Motor current is an indication of the amount of work being performed by a pump. A pump that is performing less work draws less current. This may indicate that the system pressure is less than the designed pressure. The wrong pressure may be caused by balancing valves that are not properly adjusted, or a failure in the automatic valves to properly open or close.

In addition to electrical measurements, mechanical readings are taken as well. Mechanical readings include pump rotation readings, vibration and alignment readings, pump suction and discharge pressures, and the measurement of water flow rates (in gallons per minute). All of these measurements are important when testing pump performance. All readings are checked against the pump curve to ensure that the pump operating point is correct.

Usually, a pump is tested at full capacity with the balancing valves wide open. It is normal for the water flow rate to be higher than desired when the balancing valves are 100% open. The balancing valves are progressively closed, increasing each balancing valve's resistance to the water flow and decreasing the flow rate. Then the pump control sequences are tested as well.

When a variable-frequency drive is being used, the drive operation and sequence must be verified. The variable-frequency drive for the pump may need to be adjusted to meet the proper operating parameters. Pump variable-frequency drive adjustments are done in accordance with the manufacturer's procedures, using either the front panel keypad menu or a computer connected to the drive. **See Figure 8-2.**

Circulation Pump Variable-Frequency Drive Testing

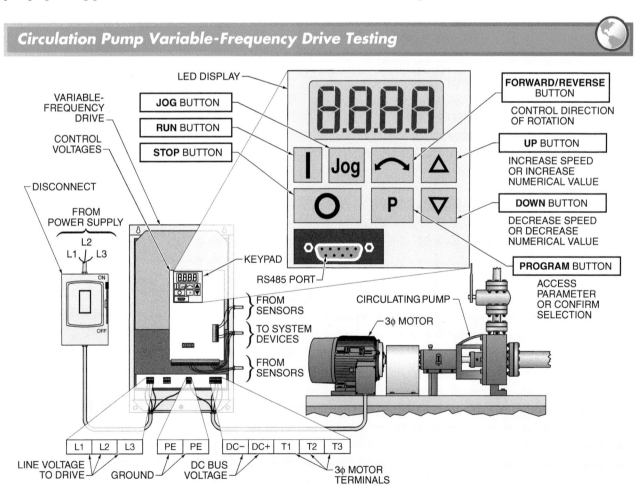

Figure 8-2. Variable-frequency drives need to be tested and adjusted to fit the correct operating requirements.

Hydronic System Test Instruments

Many IAQ test instruments provide data storage, output connections to a computer, a connection to a printer for printing paper copies, and an LCD or LED display. IAQ test instruments with digital readouts have become the most common type of instruments used today. Backlit models that provide illumination in poorly lit areas are available as well. Electronic test instruments require some type of power supply such as batteries or a connection to 120 V electricity from a wall socket. The most common test instruments used to measure hydronic system performance are electronic tachometers, digital multimeters (DMMs), clamp-on ammeters, hydronic manometers, and mechanical gauges.

Electronic Tachometers. An *electronic tachometer* is a test instrument that measures the rotational or linear speed of a moving object. Rotating speeds are displayed in revolutions per minute (rpm) and linear speeds are displayed in feet per minute (fpm). Electronic tachometers optically measure a mark made out of tape or chalk as the mark revolves or travels with the motor or pump shaft. **See Figure 8-3.** Some electronic tachometers also contain a contact tachometer on one end that can be applied directly to a shaft to measure RPM.

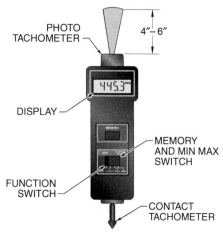

PHOTO-CONTACT TACHOMETER

LASER-CONTACT TACHOMETER

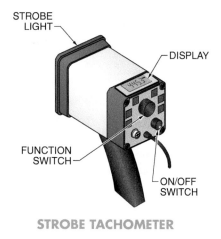

STROBE TACHOMETER

Fluke Corporation

Different test instruments are used to collect data from the various sections and equipment of a hydronic system.

Figure 8-3. *Electronic tachometers measure the speed of a moving object.*

Electronic Tachometer Procedures. To measure the rotational speed of a rotating shaft with an electronic laser tachometer, the laser tachometer should be within 6′ of the reflective tape. Some models allow as much as 15′ of distance. Most laser tachometers require a minimum distance of 2″ for accurate readings.

The reflective tape must first be attached to the shaft being measured before any readings are taken. Once the shaft of the piece of equipment is rotating, the electronic laser tachometer is pointed at the reflective tape to determine the speed (in revolutions per minute) of the shaft. **See Figure 8-4.**

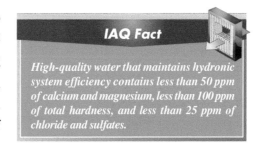

IAQ Fact

High-quality water that maintains hydronic system efficiency contains less than 50 ppm of calcium and magnesium, less than 100 ppm of total hardness, and less than 25 ppm of chloride and sulfates.

⚠ **WARNING**

Do not directly view or direct the laser of a laser tachometer at an eye. Low-power visible lasers do not normally present a hazard to people, but they may present a hazard if viewed directly for a period of time.

Electronic Tachometer Procedures

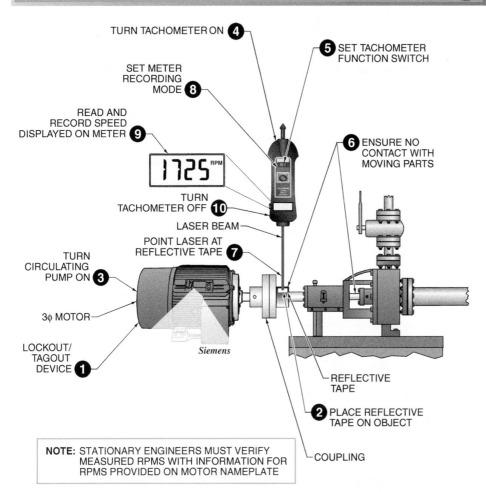

TURN TACHOMETER ON **4**

5 SET TACHOMETER FUNCTION SWITCH

SET METER RECORDING MODE **8**

READ AND RECORD SPEED DISPLAYED ON METER **9**

6 ENSURE NO CONTACT WITH MOVING PARTS

 1725 RPM

TURN TACHOMETER OFF **10**

LASER BEAM

POINT LASER AT REFLECTIVE TAPE **7**

TURN CIRCULATING PUMP ON **3**

3φ MOTOR

Siemens

LOCKOUT/ TAGOUT DEVICE **1**

REFLECTIVE TAPE

2 PLACE REFLECTIVE TAPE ON OBJECT

COUPLING

NOTE: STATIONARY ENGINEERS MUST VERIFY MEASURED RPMS WITH INFORMATION FOR RPMS PROVIDED ON MOTOR NAMEPLATE

Figure 8-4. The speed of a rotating shaft can be measured by an electronic tachometer.

Digital Multimeters. A *digital multimeter (DMM)* is a test instrument that can measure two or more electrical properties and displays the measured properties as numerical values. Digital multimeters can record measurements, and the digital display makes it easier to read displayed values. **See Figure 8-5.**

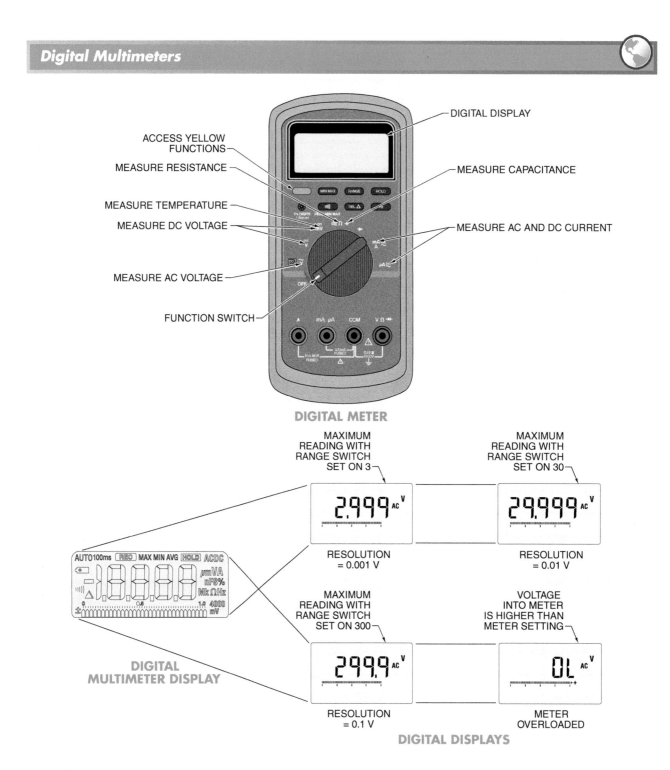

Figure 8-5. Digital multimeters record measurements of electrical properties.

Digital Multimeter Procedures. To measure the line voltage to an variable-frequency drive with a digital multimeter (DMM), the test leads must be in the proper jacks and the function switch must be set to AC voltage. **See Figure 8-6.**

A true-rms (root-mean-square) DMM is a DMM that can measure sinusoidal and nonsinusoidal waveforms. A nonsinusoidal waveform is a waveform that has a distorted appearance compared with a pure sine waveform.

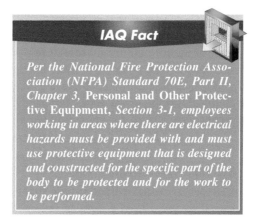

IAQ Fact

Per the National Fire Protection Association (NFPA) Standard 70E, Part II, Chapter 3, Personal and Other Protective Equipment, Section 3-1, employees working in areas where there are electrical hazards must be provided with and must use protective equipment that is designed and constructed for the specific part of the body to be protected and for the work to be performed.

Digital Multimeter Procedures – Voltage

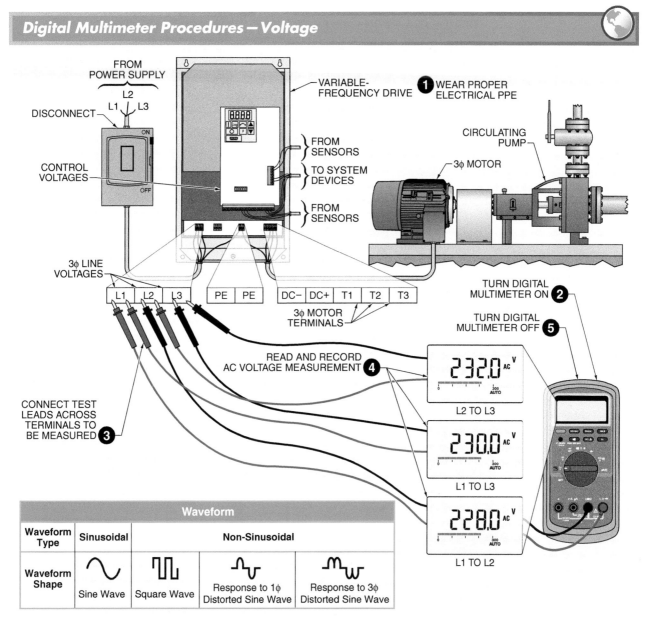

Waveform			
Waveform Type	Sinusoidal	Non-Sinusoidal	
Waveform Shape	Sine Wave	Square Wave	Response to 1φ Distorted Sine Wave · Response to 3φ Distorted Sine Wave

Figure 8-6. Digital multimeters are relatively easy to use to measure the line voltage of a variable-frequency drive.

Clamp-on Ammeters. A *clamp-on ammeter* is a test instrument that measures the current in a circuit by measuring the strength of the magnetic field around a single conductor. **See Figure 8-7.** A clamp-on ammeter is typically used to measure current in circuits with over 1 A of current and in applications where the jaws of the ammeter can be placed around one conductor.

Clamp-on Ammeter Procedures. To measure the current from a variable-frequency drive to a 3-phase motor with a clamp-on ammeter or a multimeter with a clamp-on attachment, it is important to ensure that only one conductor at a time is in the center of the jaws and aligned with the alignment marks. **See Figure 8-8.**

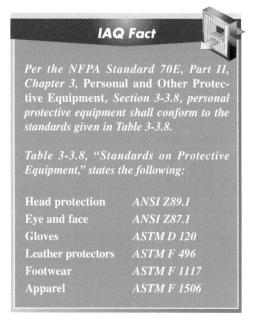

IAQ Fact

Per the NFPA Standard 70E, Part II, Chapter 3, Personal and Other Protective Equipment, Section 3-3.8, personal protective equipment shall conform to the standards given in Table 3-3.8.

Table 3-3.8, "Standards on Protective Equipment," states the following:

Head protection	*ANSI Z89.1*
Eye and face	*ANSI Z87.1*
Gloves	*ASTM D 120*
Leather protectors	*ASTM F 496*
Footwear	*ASTM F 1117*
Apparel	*ASTM F 1506*

Clamp-on Ammeters

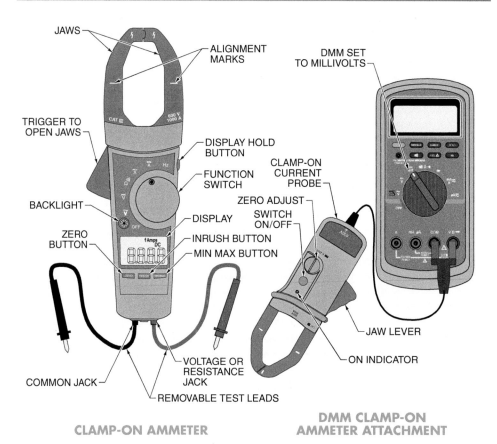

CLAMP-ON AMMETER

DMM CLAMP-ON AMMETER ATTACHMENT

Figure 8-7. Clamp-on ammeters measure current in a circuit.

Clamp-on Ammeter Procedures

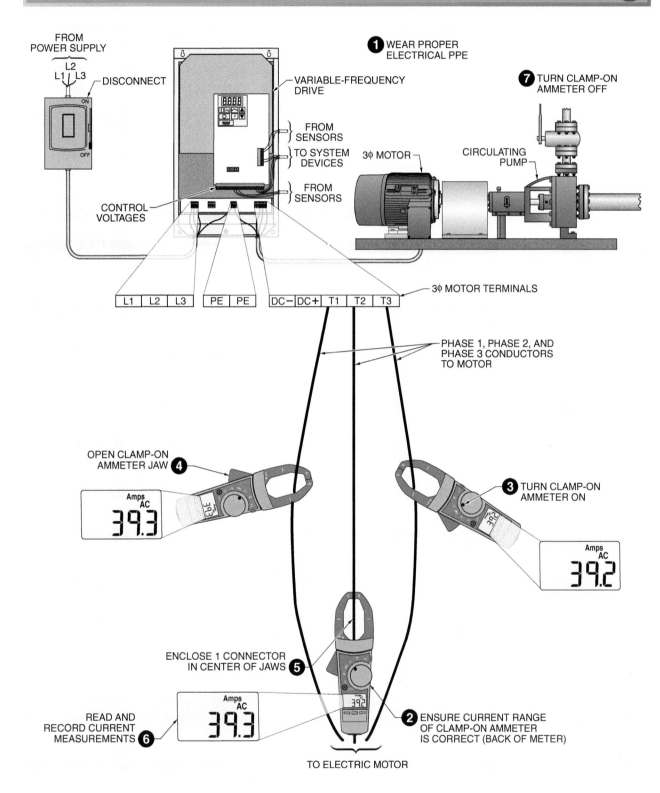

Figure 8-8. *The procedure for using clamp-on ammeters to measure current involves several steps.*

Hydronic Manometers. A *hydronic manometer* is a test instrument that measures water pressures (differential and gauge) to determine the flow rate and to balance a hydronic system. **See Figure 8-9.** Facilities that use hydronic systems for heating and/or cooling are usually balanced for terminal flow using differential pressures. Hydronic manometers are used to check pump performance and verify terminal flow and the flow through balancing valves, venturis, and orifice plates.

Hydronic Manometers

HIGH-PRESSURE PORT

LOW-PRESSURE PORT

VALVE HANDLE

LCD DISPLAY

KEYPAD

ON/OFF BUTTON

SHUTOFF VALVES

LOW-PRESSURE HOSE (BLUE)

HIGH-PRESSURE HOSE (RED)

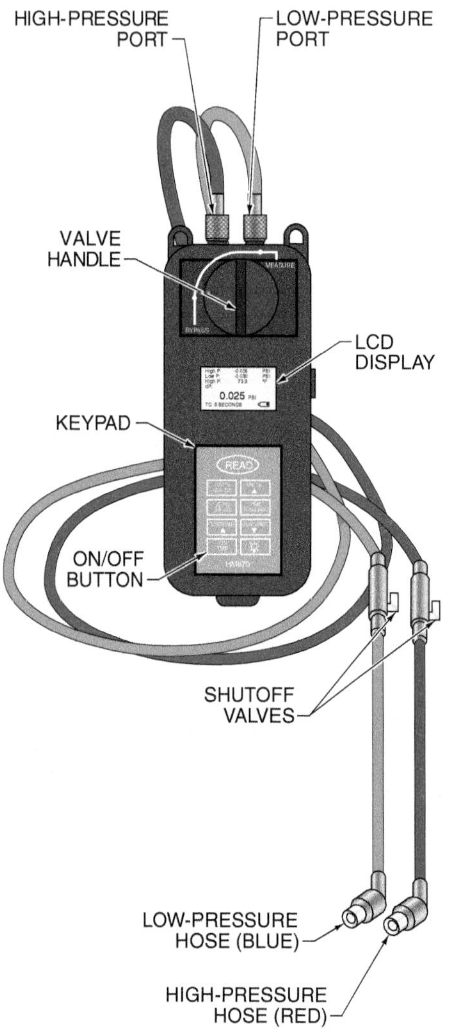

Figure 8-9. Hydronic manometers measure water pressure to help balance a hydronic system.

⚠ WARNING

- *Observe all safety precautions and wear appropriate personal protective equipment, including gloves and eyewear, when working with systems that are pressurized or have high temperatures.*

- *Use caution when disconnecting a hydronic manometer from a pressurized system. Air or water discharged under pressure poses a potential risk of serious injury.*

- *Use caution when using hydronic manometers near electrical equipment. Water spray associated with purging, connecting, or disconnecting hoses poses a potential risk of injury and/or damage to equipment.*

- *Never connect a hydronic manometer or accessories to systems that exceed the maximum pressure specifications of the manometer.*

- *Never use a hydronic manometer or accessories on potable water systems or other systems that are used to convey fluids for human or animal consumption.*

- *Never use a hydronic manometer to measure the pressure of volatile, flammable, or otherwise hazardous fluids or gases.*

- *When using hydronic manometers, verify all hose connections are secure prior to taking any pressure measurements. Loose connections will result in the discharge or pressurized air or water, which poses a potential risk for serious injury.*

- *Always drain and dry the hoses and internal piping of a hydronic manometer after each use. This limits the possible growth of hazardous microorganisms in the instrument.*

IAQ Fact

Most balancing valves have a stop on their stem (mechanical memory), which permits reopening a balancing valve to its correct position after servicing or troubleshooting a system.

Hydronic Manometer Procedures. To measure pressures in a hydronic system with a hydronic manometer, the two hoses of the manometer are connected to the low- and high-pressure test ports. **See Figure 8-10.**

Hydronic Manometer Procedures

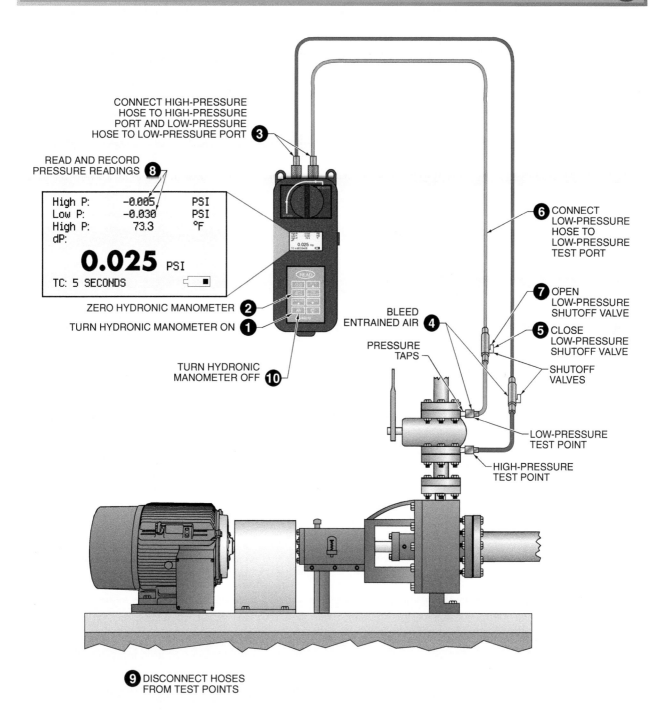

CONNECT HIGH-PRESSURE HOSE TO HIGH-PRESSURE PORT AND LOW-PRESSURE HOSE TO LOW-PRESSURE PORT ❸

READ AND RECORD PRESSURE READINGS ❽

High P: −0.005 PSI
Low P: −0.030 PSI
High P: 73.3 °F
dP:
0.025 PSI
TC: 5 SECONDS

ZERO HYDRONIC MANOMETER ❷
TURN HYDRONIC MANOMETER ON ❶

TURN HYDRONIC MANOMETER OFF ❿

❻ CONNECT LOW-PRESSURE HOSE TO LOW-PRESSURE TEST PORT

❼ OPEN LOW-PRESSURE SHUTOFF VALVE

❺ CLOSE LOW-PRESSURE SHUTOFF VALVE

SHUTOFF VALVES

BLEED ENTRAINED AIR ❹

PRESSURE TAPS

LOW-PRESSURE TEST POINT

HIGH-PRESSURE TEST POINT

❾ DISCONNECT HOSES FROM TEST POINTS

Figure 8-10. Hydronic manometer procedures are used to check pump performance and verify terminal flow.

Mechanical Gauges. Mechanical gauges (Bourdon tubes and Schrader-types) are available in various numerical pressure ranges such as 0 psi to 10 psi, 0 psi to 60 psi, or 100 psi to 600 psi. **See Figure 8-11.** Bourdon tube gauges are the gauge design with the highest accuracy, but Bourdon tube gauges are susceptible to pressure surge damage. Schrader gauges are not as accurate as Bourdon tube gauges, but Schrader gauges are not damaged by pressure surges. Pressure gauges are also available in various units of measurement such as pounds per square inch (psi), absolute pressure (psia), inches of water (in. WC), or inches of mercury (in. HG).

Flow-Balancing Valves

A *flow-balancing valve* is a special valve that is usually attached to the output side of individual heating and/or cooling coils (terminal units) in the main piping runs in order to control flow. Flow-balancing valves usually consist of a manual valve with an indicator that shows the percentage that the valve is open. **See Figure 8-12.** Flow-balancing valves are also known as balancing cocks and have a Schrader-type pressure tap on both sides of the valve. Hydronic manometers or mechanical pressure gauges are attached to the pressure taps on both sides of flow-balancing valves and provide either an electronic or analog readout of the pressure.

Mechanical Gauges

POUNDS PER SQUARE INCH (PSI) GAUGES

ABSOLUTE PRESSURE GAUGE

WATER COLUMN GAUGE

VACUUM GAUGE

Figure 8-11. Mechanical gauges measure pressure and are available in a variety of ranges.

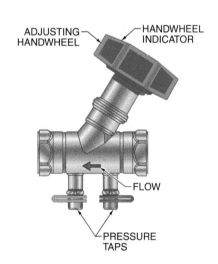

Manual Flow-Balancing Valve

ADJUSTING HANDWHEEL

HANDWHEEL INDICATOR

FLOW

PRESSURE TAPS

Figure 8-12. Flow-balancing valves often have a manual valve with an indicator to show how much the valve is open.

The readout in differential pressure is compared to a chart, usually provided by the valve manufacturer, indicating the pressure measurement versus the percentage that the valve is open. This chart provides readouts in gallons per minute (gpm) at a specific pressure difference and at the percentage that the valve is open. **See Figure 8-13.**

Determining Balancing Valve Flow with Chart Procedures

Flow coefficient values (Cv) at various handwheel settings

Hand-wheel Setting	½"* DN† 15	¾" DN 20	1" DN 25	1¼" DN 32	1½" DN 40	2" DN 50	2½" DN 65	3" DN 80	4" DN 100	5" DN 125	6" DN 150
1	0.21	0.39	0.56	0.92	1.39	2.32	3.2	6.4	9.3	11.6	20.9
1.5	0.29	0.56	0.75	1.28	1.97	3.25	4.6	8.7	12.8	19.7	29
2	0.37	0.7	0.89	1.53	2.38	4.18	5.9	11	15.7	25.5	38.3
6	1.36	2.44	4.18	6.84	10.3	19.4	59.2	56.8	115	174	285
6.2	1.44	2.6	4.47	7.25	11	20.4	62.6	61.5	122	183	298
6.4	1.52	2.76	4.76	7.66	11.8	21.5	66.1	66.1	129	194	311
6.6	1.62	2.96	5.1	8.12	12.5	22.5	69.6	70.8	135	204	322
6.8	1.74	3.16	5.45	8.58	13.2	23.5	73.1	75.4	140	215	332

* North American nominal size in in.
† ISO nominal size in mm

1 DETERMINE BALANCING VALVE PRESSURE DROP IN FEET OF WATER
(6.37 PSI PRESSURE DROP = 14.7 FT. WC)

2 IDENTIFY VALVE SIZE (½")

3 IDENTIFY PERCENT VALVE OPEN (HANDWHEEL SETTING-OPEN 6.8 TURNS)

4 DRAW HORIZONTAL LINE ON CHART FROM VALVE SIZE; OPEN SETTING
TO Cv SCALE (1.7)

5 DRAW A LINE CONNECTING THE PRESSURE DROP OF VALVE THROUGH
THE Cv SCALE AND EXTEND TO GPM SCALE (4.5)

6 READ AND RECORD GPM SCALE VALUE

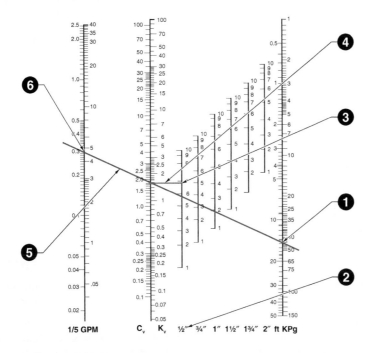

Figure 8-13. *The flow rate through balancing valves can be determined using charts and graphs.*

During the initial facility balancing, the stationary engineer confirms proper flow amounts for each device based on design drawings provided by the mechanical engineering firm. Design drawings contain tables, also known as schedules, that provide the design flow for hydronic systems. A stationary engineer may also be asked to put check marks or initials on the tables as each device is tested.

Adjustments are performed by rotating a handwheel or by using a wrench to change the percentage that a valve is open, which affects the pressure differential across the valve. When the measured pressure difference and the percentage a valve is open are combined, the flow in gallons per minute is determined. Hydronic system balancing is accomplished by matching the measured flow against that provided by the valve schedule.

Hydronic System Flow Calculations

When reviewing a flow schedule, a chilled water coil requires a specific amount of flow of water (in gallons per minute) for proper operation. The following example demonstrates the procedure to test the balancing valve for proper flow:

1. Attach a digital or mechanical manometer across the balancing valve.

2. Obtain a chart from the manufacturer that indicates the pressure measurement versus the percentage that the valve is open.

3. By referring to the chart provided, determine what the pressure drop should read on the gauge when the valve is 20% open.

Once the proper balancing valve flow is reached, the total British thermal units delivered by the water flow can be computed by using the following formula:

$$Q = 500 \times GPM \times \Delta T$$
where
Q = heat flow (in Btu/hr)
500 = constant
GPM = gallons per minute
ΔT = temperature differential (in °F)

For example, a coil is balanced to obtain 50 gpm at full flow. The temperature of the water entering is 42°F and the temperature of the water leaving is 52°F. **See Figure 8-14.**

How many British thermal units are removed by the chilled water flow?

$$Q = 500 \times GPM \times \Delta T$$
$$Q = 500 \times 50 \times (52 - 42)$$
$$Q = 500 \times 50 \times 10$$
$$Q = \textbf{250,000 Btu/hr}$$

For the conversion, 12,000 Btus are equal to 1 ton. Apply the following formula to convert the British thermal units per hour to tons of cooling:

$$Q = \frac{250,000}{12,000}$$
$$Q = \textbf{20.83 tons of cooling}$$

Adjusting Water Flow Rate. Automatic valves and balancing valves are sized by pressure drop. The pressure drop is represented by the flow coefficient (C_v) of the valve. The flow coefficient is the amount of water flow through a valve that causes a pressure drop of 1 psi. To find the water flow rate (Q) value, apply the following formula:

$$Q = C_v \times \sqrt{PD}$$
where
Q = water flow rate (in gpm)
C_v = flow coefficient
PD = pressure drop across the valve (in psi)

For example, an automatic control valve has a flow coefficient of 50. The pressure drop across the valve is 5.25 psi. Find the water flow rate through the valve. **See Figure 8-15.**

$$Q = C_v \times \sqrt{PD}$$
$$Q = 50 \times \sqrt{5.25}$$
$$Q = 50 \times 2.29$$
$$Q = \textbf{114.5 gpm}$$

IAQ Fact

Organizations that have standards for hydronic system balancing include the following: the American Society of Heating, Refrigerating, and Air-Conditioning Engineers (ASHRAE), the American Society of Mechanical Engineers International (ASME), the National Environmental Balancing Bureau (NEBB), and Sheet Metal and Air Conditioning Contractors' National Association (SMACNA).

APPLICATION: Determining British Thermal Units From Chilled Water

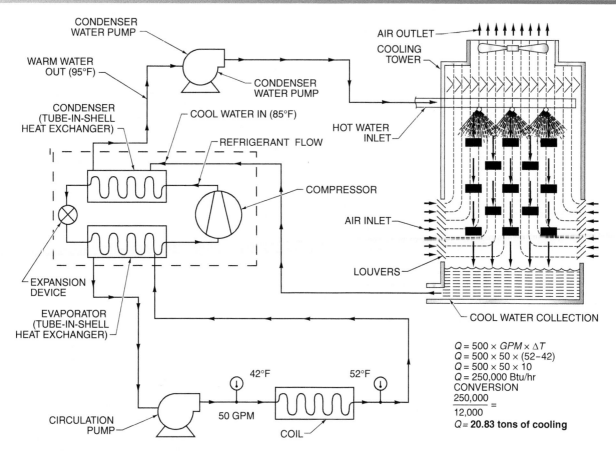

$Q = 500 \times GPM \times \Delta T$
$Q = 500 \times 50 \times (52-42)$
$Q = 500 \times 50 \times 10$
$Q = 250{,}000$ Btu/hr
CONVERSION
$$\frac{250{,}000}{12{,}000} =$$
$Q =$ **20.83 tons of cooling**

Figure 8-14. *If the gallons per minute of flow and the temperature differential of the water in a coil are known, the amount of British thermal units removed can be determined.*

APPLICATION: Determining Balancing Valve Flow Rate

$Q = C_v \times \sqrt{PD}$

$Q = 50 \times \sqrt{5.25}$
$Q = 50 \times 2.29$
$Q =$ **114.5 gpm**

Figure 8-15. *The water flow rate through a valve can be determined if the flow coefficient and pressure drop are known.*

COMMISSIONING OF COMMERCIAL BUILDINGS' HVAC SYSTEMS

At the conclusion of a construction project, the mechanical contractor turns over the operational control of a building to the owner and staff. The turnover process may be hampered by problems, such as documentation being lost or stored in the wrong format. In other cases, the training that the contractor is required to provide to building staff has not been performed properly. Another problem that may be encountered occurs when alterations to the mechanical system were made without proper documentation or without informing building staff. The fact that HVAC systems are installed at the lowest possible cost and that today's mechanical and control systems are more complex than ever also contribute to the problem. Ultimately, installing mechanical systems at the lowest possible cost often leads to many problems, including the following:

- building occupation delays due to schedule overruns
- litigation costs
- high vacancy levels due to uncomfortable building conditions
- equipment not operating properly long after installation
- costly corrections
- inability to perform adequate building operation and maintenance
- high warranty claims and associated costs
- possible indoor air quality issues

Building system commissioning is the systematic process of ensuring and documenting that all building systems perform together properly according to the original design intent and the owner's operational needs. Commissioning is necessary but is not automatically included as part of the typical HVAC design and installation process. Commissioning is critical for ensuring that the HVAC design created is actually installed and performs properly to ensure the comfort of building occupants.

Building commissioning is performed by a commissioning agent. A *commissioning agent (CA)* is a mechanical contractor who acts as a third-party representative to ensure that the mechanical systems of a building are performing properly. In general, the CA represents the owner's best interests.

Principles of Commissioning

Building commissioning is an extensive process. The process begins before a building is constructed and continues until after the building has been on-line for at least a year. In some cases, the timeline may be longer. The general phases of the commissioning process are as follows:

1. predesign
2. design
3. construction
4. turnover/acceptance
5. post-turnover/acceptance

Predesign. During the predesign phase, the CA meets with the owner on an ongoing basis to determine what the owner wants and expects from the commissioning process. Depending on the job, the CA spends time gathering data for future use but does not have a highly active role.

Design. The major role of the CA during the design phase is to determine the scope of the project and the systems to be installed, and to document the process, which includes the documentation required for and from each contractor. The overall commissioning plan starts in this phase.

Construction. The commissioning plan usually is completed during the construction phase. All of the relevant data such as drawings, submittals, operation and maintenance manuals, and project scheduling data have been obtained and compiled. The CA creates detailed functional performance test plans for each installed system and each piece of equipment involved in the commissioning process. At this point, the CA is on site, observes all construction activities, and records any details that would lead to equipment or system performance problems. The CA works with the various on-site contractors to ensure that all start-up tests are performed, all checklists are completed, and any problems are corrected.

Turnover/Acceptance. Using the functional performance test plans created in the construction phase, the CA observes and verifies the proper operation of equipment, systems, and controls per contract documents during the turnover/acceptance phase. When necessary, the CA also verifies that corrections are made and ensures the presence of several sets of complete operation and maintenance manuals. Actual performance testing is usually carried out by a number of different contractors. The CA can intervene at any time in the process to ensure compliance with project documentation. At the time of turnover, all documentation is wrapped up and a final commissioning report, supported by relevant job documents, is submitted to the owner.

Post-Turnover/Acceptance. In the post-turnover/acceptance phase, building operations and/or maintenance staff ensure that building HVAC systems are functioning properly. Building staff are also responsible for handling ongoing adjustments to the HVAC system, changing occupancy and use patterns, and maintaining and documenting history and any changes to the facility or HVAC system. The CA can be involved in establishing the documentation requirements for the post-turnover phase, reviewing performance, and recommending improvements.

Building Commissioning

Building commissioning is usually considered a process that applies to new building construction but this is not always the case. Types of building commissioning include ongoing commissioning and retro-commissioning. *Ongoing commissioning* is building commissioning that is centered on analyzing and optimizing heating, ventilation, and air conditioning (HVAC) system operation and control for existing building conditions. *Retro-commissioning* is building commissioning that focuses on analysis of old systems, old buildings, and systems that have been upgraded with new technology. Ongoing commissioning

is a process designed to resolve operating problem areas, improve comfort, optimize energy use, and identify possibilities to upgrade older building systems used in commercial, industrial, and central plant facilities. **See Appendix** and **CD-ROM.**

Ongoing commissioning may provide energy savings of up to 20% on average, with a short payback period of three years or less. The following strategies contribute to long-term commissioning savings:
• ongoing monitoring of energy savings and strategies used
• improving the system reliability and building comfort by optimizing system operation and control schemes based on actual occupancy and other actual usage patterns in the building
• reducing turnover and improving the performance of the operating staff's skills by encouraging them to participate in the commissioning process

Building Commissioning Costs

Building owners often object that the process of building commissioning costs too much. However, final costs vary depending on the project size and the number and complexity of the systems involved. Building systems to be commissioned may include HVAC/mechanical, lighting, life safety, electrical, and plumbing systems. A general industry rule of thumb is that equipment commissioning costs 2% to 5% of the equipment cost. Commissioning an entire building is often less than 1% of the total costs involved in the construction of the building. **See Figure 8-16.** The added cost may seem to be a burden, but there are direct and indirect costs resulting from the decision to not commission these systems. These costs include:
• expensive change orders to modify the system after installation
• legal costs
• uncomfortable and complaining tenants
• unacceptably high vacancy rates
• high operational energy costs

BUILDING COMMISSIONING COSTS	
Commissioning Scope	**Commissioning Cost**
Entire building (HVAC, controls, electrical, and mechanical)	0.5% to 1.5% of total construction cost
HVAC and automated control system	1.5% to 2.5% of mechanical system cost
Electrical systems	1.0% to 1.5% of electrical system cost
Energy efficiency measures	$0.23 to $0.28 per square foot

Figure 8-16. While building commissioning can be expensive, often the cost is less than 1% of the total construction costs.

Case Study – Replace Chiller Compressor or Parts of a Building

Event

Some HVAC jobs can be quite complicated, such as the case of a 20-story office building that contained some HVAC-related problems. The 750-ton centrifugal water chiller operated with R-11 refrigerant. Because of refrigerant phase-out legislation, the owners of the office building wanted to replace the chiller with one that used a more environmentally safe refrigerant. However, after reviewing the work that would be involved to remove the old chiller, the owners decided that the cost of a lifting crane and the cost of cutting away and repairing part of the building were too great.

Investigation

After further discussion, a conversion retrofit of the old chiller made more sense. By replacing the old compressor, building freight elevators could be used, which would eliminate the need for a lifting crane. It was decided to replace the old compressor with a new high-efficiency centrifugal compressor and motor. The new compressor type would be chosen to efficiently operate with the existing evaporator and condenser heat exchanger. The new compressor is also environmentally friendly because it uses R-123 refrigerant.

Resolution

The retrofit was easier to accomplish and also cost less than a new chiller. At the time of the compressor retrofit, a new control system was installed on the chiller and system, which optimized cooling tower and ventilation fan operations.

Hydronic System Balancing and Building Commissioning Definitions

- *Building system commissioning* is the systematic process of ensuring and documenting that all building systems perform together properly according to the original design intent and the owner's operational needs.
- A *clamp-on ammeter* is a test instrument that measures the current in a circuit by measuring the strength of the magnetic field around a single conductor.
- A *commissioning agent (CA)* is a mechanical contractor who acts as a third-party representative to ensure that the mechanical systems of a building are performing properly.
- A *digital multimeter* is a test instrument that can measure two or more electrical properties and displays the measured properties as numerical values.
- An *electronic tachometer* is a test instrument that measures the speed of a moving object.
- A *flow-balancing valve* is a special valve that is usually attached to the output side of individual heating and/or cooling coils (terminal units) in the main piping runs in order to control flow.
- A *hydronic manometer* is a test instrument that measures water pressures (differential and gauge) to determine the flow rate and to balance a hydronic system.
- *Ongoing commissioning* is building commissioning that is centered on analyzing and optimizing heating, ventilation, and air conditioning (HVAC) system operation and control for existing building conditions.
- *Retro-commissioning* is building commissioning that focuses on analysis of old systems, old buildings, and systems that have been upgraded with new technology.

**For more information please refer to ATPeResources.com
or http://www.epa.gov.**

Review Questions

1. What is an electronic tachometer used for?

2. When are clamp-on ammeters typically used?

3. Describe the purpose of a hydronic manometer.

4. What is another name for a flow-balancing valve?

5. List the general phases of the commissioning process.

Fluke Corporation

Class Objective
Define common IAQ prevention strategies and how they are used.

On-the-Job Objective
Put together a detailed list of prevention activities currently used in the facility.

Learning Activities:
1. List four safe practices related to chemical use.
2. Identify the components of and uses for a material safety data sheet (MSDS).
3. Explain how green products are used in construction activities.
4. **On-the-Job:** Put together a plan to begin implementing necessary prevention strategies that are not currently in place.

Introduction

One of the ways to minimize or eliminate indoor air quality issues is to engage in practices that will prevent them as much as possible. The overall amount of effort and cost required to prevent many common IAQ issues is low, so it makes sense to implement preventive activities as much as possible. Preventive activities help avoid the high costs of correcting IAQ problems or the legal costs of an IAQ-related lawsuit. Some prevention methods may be common knowledge; however, others require a more formal approach. Typical prevention areas include chemical safety and storage, accident prevention and response, use of green products, random air samplings, and visual inspections.

CHEMICALS

Many different types of chemicals are present in commercial buildings. Chemicals are used for many normal day-to-day functions that take place in the building. **See Figure 9-1.** The type and amount of these chemicals will depend upon the function and activities that take place in the building.

When not stored or used properly, chemicals can be a major source of indoor air problems. Chemicals are used inside buildings by individual tenants for producing products and services, as well as for building operations, cleaning, and maintenance. Special care must be taken in the application of all pesticides, paints, chemicals, and cleaning materials used inside a building.

COMMON CHEMICALS AND CHEMICAL SOURCES

CHEMICALS
Adhesives, air fresheners, cleansers, disinfectants, drain cleaners, lubricants, paints and coatings, pesticides, solvents, waxes and polishes

CHEMICAL SOURCES
Cafeterias, medical offices, science laboratories, tobacco products, dry cleaning, vacuuming, wet mops

Figure 9-1. A wide array of chemicals may be found in commercial buildings and other facilities. Chemicals are used for many daily activities.

Pesticides

Pesticides are used to eliminate or control pest infestation. Pesticides are also used as plant regulators, defoliants, and drying agents. Biocide substances are used to control microorganisms in cooling towers, fountains, and swimming pools.

Pesticides must always be used according to the manufacturer's instructions. Many are toxic substances and must be handled accordingly. Different guidelines may apply for chemicals depending on whether they are being used, transported, or stored. The Federal Insecticide, Fungicide, and Rodenticide Act (FIFRA) is a law set up to regulate pesticides in order to protect consumers and the environment. The FIFRA requires certification for applicators of restricted-use pesticides. Some states also require that individuals who apply general pesticides in commercial facilities be trained and certified. It is important to know a state's requirements and which pesticides are labeled for restricted use.

As a general guideline, pesticides are applied when a building or space is unoccupied. The space must then be flushed with large amounts of outdoor air before the space can be reoccupied. Occupants must be given advance notification of pesticide application, including the date of application and the type of product to be used. In particular, individuals who will be present during pesticide use must be notified in advance, while individuals with compromised immune systems or allergic reactions should be encouraged to leave the area.

Manufacturer instructions for pesticide application must be followed. Only the amount required can be applied and only in the necessary locations. Pesticide application must be avoided around or on any item or surface that individuals contact through ingestion, inhalation, or skin contact. This includes food preparation surfaces such as countertops and sinks.

It is advisable to select the least toxic pesticides available by reading the labels and comparing the potency of various ingredients and the toxicity warnings. Products are labeled for toxicity from most toxic to least toxic. By comparing the material safety data sheets (MSDSs) of similar products, the user can identify the safest and least toxic ingredients. Collection and storage procedures of all pesticides can be found on the MSDSs and from the manufacturers' instruction labels.

Practicing integrated pest management (IPM) is an alternative to using pesticides. *Integrated pest management* is a type of pest management where an environmentally sensitive and common-sense approach to pest management is used. IPM seeks the maximum use of natural and nonchemical means of pest control.

For commercial facilities, IPM uses a combination of biological, physical, habitat, and chemical control techniques. In IPM, prevention is always the first step and chemical control is used as a last resort. However, when IPM proves infeasible and chemical control is indeed needed, safe practices must be used. Alternatives to chemical control include the following:

- building monitoring
- identifying the pests to control
- determining the level of infestation
- controlling pests' means of entry and access to food sources
- understanding pest habitats
- making use of natural predators
- ensuring good sanitation practices

Performing a chemical inventory can help to identify chemicals that must be disposed of due to their toxicity level.

Painting

Paints and painting products are other contamination sources of indoor air quality within a facility. A protocol for painting within a building must be established and enforced, and followed by both in-house and contractor personnel. One suggested protocol includes using low-VOC emission and fast-drying paints whenever possible. Other safe practices are to paint during unoccupied hours and to keep the lids on paint containers that are not in use.

An additional preventive measure to follow when painting includes ventilating the building with significant quantities of outdoor air during and after painting. Also, it is necessary to ensure that several complete air exchanges occur in the painted space prior to occupancy. Using exhaust ventilation for storage spaces, while eliminating return air from storage spaces, reduces exposure to toxic fumes. Similarly, another safety guideline includes using higher than normal outdoor air ventilation for a period of time after re-occupancy.

A guideline for diluting and mixing paints or stains is to avoid spraying paint when possible. When finished with paints, personnel must store them in closed, airtight containers inside metal cabinets that are kept in a separate location from air distribution sources.

Chemical Inventory

Properly managing chemicals within a facility can prevent IAQ problems and reduce or eliminate unpleasant and irritating odors. Managing building chemicals also lowers the risk of other hazards, such as fire or costly damage from spills or accidents. Tenants may store chemicals for various purposes and processes. Frequently, the tenant's lease addresses the topic of how tenants must store and handle chemicals. However, oversight and enforcement should be part of the chemical inventory and storage program for the facility. The first step in any good chemical management plan is to take a physical inventory of all chemicals in a building or facility. **See Figure 9-2.**

A *chemical inventory form* is a form used to identify all chemicals found in a building and indicates how the chemicals are used. When taking a chemical inventory, personnel must correctly dispose of damaged, partially used, unlabeled, and old containers. A chemical inventory must note where pesticides are used, even when pest control is contracted to an outside company. Once a chemical inventory is complete, the list must be kept in a safe and accessible place, and updates are to be conducted annually.

Chemical Inventory

Chemical Inventory:_____ File Number:_____ Date:_____

Address:_____

Completed by:_____ Phone: _____

The inventory should include chemicals stored or used in the building for cleaning, maintenance, operations, and pest control. If you have an MSDS for the chemical, put a check mark in the right-hand column. If not, ask the chemical supplier to provide the MSDS, if one is available.

If the MSDS is not on file, enter the manufacturer's name and phone number in the third column (*) and order the MSDS.

Storage/Location: _____
(Use one form for each storage location)

Date	Chemical/Brand Name	Mfr/Phone*	Use	Quantity	MSDS on file?

Figure 9-2. Keeping a chemical inventory can assist the facility maintenance staff in quickly identifying detected odors and can help prevent IAQ problems.

Storage and Handling

Maintaining a current chemical inventory can assist facility personnel in chemical product management. By continuously updating the chemical inventory, facility managers can take note of redundant chemical storage and quickly identify chemicals that are no longer used within the facility. It is imperative to always store, handle, and use chemical products according to the manufacturer's recommendations. Most manufacturers provide written directions on the safe use of chemicals. Many manufacturers also provide hotlines that answer questions on chemical safety.

When purchasing chemical products, it is important to select a product with the least amount of chemical toxicity. By requesting the MSDS for products prior to their purchase, the stationary engineer can make an informed decision on what product is capable of performing the task with the least amount of risk from accidental chemical exposure. When chemicals are being used, increasing outside airflow into a building and/or isolating the space affected can prevent the spread of toxic fumes.

When storing chemicals, the minimum amount of chemicals should be kept in stock for immediate or near-term use, and they should be stored in a central area under negative pressure with the exhaust vented outdoors. Chemicals must not be stored near HVAC or electrical equipment, or near appliances that can heat up. **See Figure 9-3.** All chemical storage containers must be labeled, sealed, and maintained in good condition and in accordance with federal, state, local, and municipal regulations. Mechanical and electrical rooms should never be used for storage.

In order to safely handle and store the various chemicals in any facility, the stationary engineer should designate storage areas, list the potential uses of chemicals, and organize an inventory with an MSDS for each. While designating a chemical storage facility, the stationary engineer should determine what ventilation requirements are needed, if any. The stationary engineer should also evaluate the original purpose of the room being used to store chemicals and find out if there is adequate ventilation for the room. Whether the exhaust for the room truly exhausts air outdoors or returns it to the air-handler system is another concern.

Storing Chemicals

Figure 9-3. Chemicals should not be stored directly next to return-air ducts, HVAC equipment, or electrical systems.

Training personnel for the proper handling of chemicals is an important component in maintaining a safe indoor environment. Aerosols and chemical products that have a powerful smell should not be used in occupied spaces. Knowing the correct application for using each chemical, how to correctly store it, and what to do in case it is spilled or used incorrectly makes a big difference in the indoor environment.

When chemicals are transferred to storage, spills do occur. Chemical spills must be immediately cleaned up according to the chemical's MSDS recommendations. In addition, accumulated dirt, stacks of papers, boxes, and rags in storage areas are fire hazards and provide ideal breeding grounds for insects and molds. Therefore, such hazards must be removed and kept away from air handlers.

Accidents

Effective housekeeping and preventive maintenance programs do reduce the number and severity of accidents that impact IAQ. Whenever an accident occurs within a facility, facility operators must immediately investigate the potential impact on IAQ. During an investigation, the immediate threat to IAQ should be considered as well as any impact that may occur as a result of the cleanup or repairs that are initiated by the accident.

Housekeeping Programs

Comprehensive housekeeping programs are essential for establishing and maintaining good indoor air quality in buildings. A *housekeeping program* is a set of procedures that includes preventing excess dirt from entering an environment, using cleaning products that introduce the least amount of contaminants into the air, and implementing maintenance procedures with trained personnel. A housekeeping program involves a survey of the building, determining if any areas require special attention, and a number of other tasks. Avoiding the unnecessary release of any chemicals into a building's air supply minimizes the potential for contamination. Housekeeping programs usually require that a cleaning survey be performed.

IAQ Fact

Housekeeping programs extend the life of furnishings and prevent dirt accumulation that affects health, thus reducing overall costs.

A *cleaning survey* is a method by which sources of dirt and debris are identified. A comprehensive cleaning survey can assist in identifying ways to prevent infiltration. The section on the cleaning survey devoted to identifying sources and accumulation of dirt and debris may include some or all of the following:

- dirt potential of pedestrian traffic, loading docks, receiving areas, and garages
- surrounding vegetation, pollen, and proper drainage around the building
- interior sources such as printing and copy rooms, kitchens, eating areas, smoking areas, trash areas, storage areas, carpets and furnishings, and ceiling tiles
- dirt accumulation areas such as horizontal surfaces, corners and edges, furniture, high ledges, and windows with blinds

Moving trash items outdoors helps to control pests and minimizes odors.

A cleaning survey makes it easier to identify ways of preventing dirt and pollution from entering a building space and to identify the cleaning needs of each part of the building. Various parts of a building require different types of cleaning. Areas that may require different cleaning methods include entryways, lobbies, bathrooms, hallways, corridors, kitchens, cafeterias, and offices.

Special attention must be paid to objects that collect or produce dust or fibers, such as carpeting, draperies, upholstered furnishings, open shelving, old or deteriorated furnishings, bookshelves, and stacks of paper. It is important to keep these items clean and minimize dust. The use of deodorizers, fragrances, and air fresheners should be limited, as they add more chemicals into the air. Instead of using air freshener chemicals, the sources of odors should be controlled.

The creation of airborne dust and dirt can be limited by choosing the right equipment and operating it correctly. A feather duster is an example of a cleaning item that should not be used in a commercial environment because it spreads dust, creating an IAQ problem. The proper use of high-quality vacuum bags and dust catchers helps to control airborne dust and dirt. For example, HEPA vacuums are highly effective in removing small particles such as fungi, bacteria, molds, animal dander, dust particles, and pollen from the airstream flowing through the vacuum. In addition to vacuuming carpeted floors, all fabric surfaces should be vacuumed.

Other effective housekeeping practices include using trash basket liners and changing the liners frequently, sanitizing trash baskets, sealing and moving all trash outdoors, and properly storing trash in outdoor containers as quickly as possible. Finally, frequent trash collection minimizes odors caused by microbial growth while also aiding in pest control.

As a regular practice, all custodians and maintenance workers should be looking for the following sources of IAQ problems:

- unsanitary conditions
- water damage
- microbial growth (mildew smell)
- standing water
- unusual occupant activities
- changes in space use

Personnel should report any problem conditions to management and correct them before IAQ problems result.

A comprehensive housekeeping program can be developed based on the needs identified in the cleaning survey. Elements of a comprehensive housekeeping program, as recommended by the EPA and other governing agencies, include the following:

- using maximum extraction methods
- using low-polluting equipment and methods
- choosing low-polluting products
- properly mixing and storing housekeeping products in a ventilated room or closet
- training housekeeping personnel
- comparing contractors
- monitoring results

Low-Polluting, High-Extraction Cleaning Methods. When cleaning, it is important to use equipment and methods chosen for maximum extraction and minimum polluting capabilities. Deep cleaning carpets at regular intervals and avoiding carpet treatments with sticky residues are common methods for ensuring quality indoor air. When vacuuming, high-efficiency filtration bags are best. Other cleaning guidelines include using only floor machines with vacuum capability, using lint-free dusting cloths, avoiding dusters that do not capture dust (feather dusters), covering the top of dust mops with dust cloths to avoid passing dust over the mop, or spraying a product directly onto a cloth. Also, per the EPA, toggle-top chemical dispensers must be used.

Cleaning products should be chosen by referencing MSDS sheets to find environmentally preferable products with minimal amounts of volatile organic compounds (VOCs). Products with a moderate pH (between five and nine) are ideal. Another important practice is to minimize the use of ammonia, chlorine, volatile acids, and other products that are corrosive or that react with other cleaning products. Many aerosols and particle cleaners have particles that become airborne, so they should only be used on a limited basis because of health concerns.

Ventilation, Mixing, and Storage. Adequate ventilation protects building workers from hazardous conditions when mixing and applying chemicals, as does the use of personal protective equipment. Other important safety precautions include the proper labeling of chemicals, the clear designation of mixing areas, and the proper implementation of mixing procedures.

Training and Monitoring. Participation and communication between cleaning personnel, building occupants, and contractors help to ensure that safety procedures are being followed. Everyone should participate in the development, implementation, and refinement of the program. Building occupants must understand how their actions, such as leaving food debris in a work area, impact the indoor environment and the cleaning process.

Employers are required by OSHA standards to provide specific information and training to employees on hazardous chemicals found in their work area at the time of their initial assignment and whenever a new hazard is introduced into their work area. Proper training of housekeeping personnel before they participate in a cleaning operation contributes significantly to the correct implementation of cleaning procedures. One improper application of a chemical product or failure to use a dust-free wipe can contaminate an indoor space, creating IAQ problems.

In addition to training personnel on equipment, materials, and methods, training must also include ways to create and maintain a healthy environment. Housekeeping personnel should be capable of performing tasks in a manner that protects the indoor environment. Monitoring the results of housekeeping tasks ensures that a building is being kept clean as required. Similarly, it is important to keep good records of what is cleaned, how it is cleaned, and when it is cleaned.

Cleaning waste often has a detrimental impact on the outdoor environment. Therefore, the handling and disposing of waste according to applicable codes and regulations minimizes the risks to the indoor and outdoor environments. Regulations governing the proper disposal of nonmedical hazardous materials, such as chemical cleaners, must be followed.

MATERIAL SAFETY DATA SHEETS

A *material safety data sheet (MSDS)* is a form containing data regarding the properties of a particular substance. Material safety data sheets are used for cataloging information on chemicals, chemical compounds, and chemical mixtures. MSDSs are an important component of product stewardship and workplace safety. MSDSs are intended to provide workers and emergency personnel with procedures for handling or working with a substance in a safe manner. MSDSs include important information, such as physical data (i.e., melting point, boiling point, flash point, etc.), toxicity, health effects, first aid, reactivity, storage, disposal, protective equipment, and spill-handling procedures.

An MSDS for a substance is not primarily intended for use by the general consumer. MSDSs focus more on the hazards of working with the material in an occupational setting. Chemical cleaners intended for household consumers are usually low-strength formulas and generally do not pose the same level of danger that commercial products may.

The Occupational Safety and Health Administration's (OSHA) *Hazard Communication* regulation requires that the MSDS for potentially harmful substances handled in the workplace be on-site and available to employees. Under Section 311 of the Emergency Planning and Community Right-to-Know Act, it is required that the MSDS be made available to anyone in the event of an emergency. The information and format used in preparing MSDS data can vary from one manufacturer to another or one distributor to another.

MSDS Format. MSDSs contain a variety of information to alert users to the possible dangers of using a certain product. **See Figure 9-4.** Information listed on an MSDS can include the following:

- MSDS creation/revision date
- manufacturer's name, complete address, and an emergency telephone number
- Section 1—Product identification
- Section 2—Composition
- Section 3—Hazards identification
- Section 4—First aid measures

Material Safety Data Sheets (MSDSs)

MidSun Group, Inc.

Figure 9-4. Material safety data sheets provide an extensive listing of information on a product including potential hazards and permissible exposure limits.

- Section 5—Fire-fighting measures
- Section 6—Accidental release measures
- Section 7—Handling and storage
- Section 8—Exposure controls/personal protection
- Section 9—Physical and chemical properties
- Section 10—Stability and reactivity
- Section 11—Toxicological information
- Section 12—Ecological information
- Section 13—Disposal considerations
- Section 14—Transport information
- Section 15—Regulatory information
- Section 16—Other information

The creation date is important, as some chemical compounds change with age. Manufacturer contact information is necessary in the event of a big spill or accident. Chemical common names are listed under the product identification section for individuals handling the chemical who may not have an extensive background in chemistry.

Every chemical has a permissible exposure limit (PEL) established by OSHA. The PEL is a time based maximum allowable exposure limit. This information is important when chemicals are exposed to temperature and pressure variations. Firefighting personnel must have access to the MSDS for every chemical in a facility.

Housekeeping and maintenance personnel must be provided with hazard information, carcinogen levels, and the chemical points of entry in order to effectively clean spills and avoid problems. This information, along with exposure control measures, is also required by emergency responders to determine what personal protective equipment must be worn.

Additionally, MSDS information can be used to prevent IAQ problems, such as fumes caused by the reaction between chemicals and/or the physical injury caused by the toxicity of a chemical. An MSDS can also help in the process of solving problems when they occur by properly identifying spilled chemicals, instructing how to safely clean up a particular spill, and how to protect building occupants from a spill or chemical reaction. When encountering a product that does not have an MSDS, the manufacturer can be contacted to request one. **See Figure 9-5.** MSDS documentation can be managed manually with a paper-based system or from a computer database.

Sample MSDS Request Letter

Date

> Each hazardous chemical must have an MSDS. This form may be used to request an MSDS from the distributor or the manufacturer; telephone them for a complete address, if necessary.
>
> **Important:** Copies of each letter should be kept in the IAQ records.

Manufacturer/Distributor
Address
City, State Zip Code

Subject: Material Data Sheet

Please send us a material safety data sheet (MSDS) for the products listed below:

1.
2.
3.

The MSDS is for our hazard communication program required by the Hazard Communication Standard. Please make sure each MSDS meets the requirements of 29 CFR 1910.1200, OSHA Hazard Communication Standard.

Thank you for your assistance.

Sincerely,

Figure 9-5. For chemicals that do not have a material safety data sheet, one may be requested from the chemical manufacturer or downloaded from the manufacturer's website.

Paper-Based MSDS System. A *paper-based MSDS system* is a chemical tracking method that keeps paper MSDS files in three-ring binders at an accessible location. Files must contain an inventory of all chemicals, followed by the MSDS for each chemical in the chemical inventory. Using a three-ring binder allows the easy addition or removal of pages to reflect a changing building inventory. Many facilities issue numbered MSDS labels that are applied to all chemical containers for easy container-to-MSDS identification.

Computer-Based MSDS System. A *computer-based MSDS system* is a chemical tracking method that uses software to eliminate the difficulty of handling paperwork for MSDS storage, manual updates, and other MSDS maintenance issues. Some features available with MSDS computer software include:

- scanning or importing MSDSs into one central database, reducing search time
- access to updated MSDSs, which helps to be compliant with OSHA 29 Code of Federal Regulations (CFR) 1910.1200—*Hazard Communication* and EPA regulations
- documents accessible from any work location
- documents searchable by any combination of criteria, including material name, synonym, components, and vendor
- secure archive of MSDSs, containing the history and status of each MSDS and the reasons for making changes
- generation of request letters to manufacturers and suppliers for missing or incomplete MSDSs
- interface with computerized maintenance management software (CMMS) for sharing information between MSDS and preventive maintenance software programs

LEADERSHIP IN ENERGY AND ENVIRONMENTAL DESIGN

The Leadership in Energy and Environmental Design (LEED®) is a Green Building Rating System™ developed by the U.S. Green Building Council (USGBC). LEED provides a set of standards for environmentally sustainable construction. Since its inception in 1998, LEED has been used throughout the United States and in over 30 countries. The technical criteria proposed by LEED committees are publicly reviewed for approval by the membership organizations that constitute the USGBC.

LEED certification is available for existing buildings, renovation, and new construction. There are four levels of LEED certification that are determined by a point-based system. **See Appendix.** Points are assigned in one of six categories. The categories are:

- sustainable sites
- water efficiency
- energy and atmosphere
- materials and resources
- indoor environmental quality
- innovation in design

While building and/or renovation costs associated with LEED certification can be higher than normal construction, these costs may be offset in the long run. Buildings that are LEED certified may also qualify for state or local government incentives.

Existing Buildings

LEED for new buildings focuses on the construction of a building. LEED for existing buildings certifies building operations and maintenance and creates an agenda for maintaining high performance. LEED for existing buildings encourages continual inspection, reporting, and review of building practices and operations over the lifetime of the building.

LEED for existing buildings rewards good practices and provides an outline to reduce the use of energy, water, and other natural resources. A LEED outline may point out ways to improve indoor air and building inefficiencies. Establishing policies to meet LEED standards can also help create a more productive environment for employees, reduce environmental damage, and reduce costs associated with maintaining building operations such as HVAC.

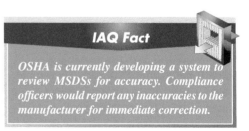

IAQ Fact

OSHA is currently developing a system to review MSDSs for accuracy. Compliance officers would report any inaccuracies to the manufacturer for immediate correction.

LEED® and IAQ

The USGBC is also working to create a design guide with tips on IAQ. The guide will illustrate a process to help building management and personnel to improve IAQ throughout an existing building, or create high-performance designs for a new building. The guide is being created with the help of several organizations. A few of the contributing organizations include the Environmental Protection Agency (EPA), the American Institute of Architects (AIA), and the Building Owners and Managers Association (BOMA).

Green Products

In the United States, the building environment accounts for approximately one-third of all energy, water, and material consumed and generates a similar proportion of pollution. Green products are products that have been selected for their environmentally friendly qualities. Green products are environmentally friendly by reducing pollution either indoors or outdoors. Examples of product qualities that are environmentally friendly include the following:

* reduced toxicity
* reusability
* energy efficiency
* eco-responsible packaging
* recycled content
* manufactured with minimal environmental impact
* minimal or no artificial materials

Many products are now available with little or no emissions or odors. VOC emissions are a primary IAQ problem. Green products with low emissions and odors are produced without known harmful chemicals and the manufacturing process does not cause pollution. The best green products are those that can be manufactured and disposed of with a little or no impact on the environment and are recyclable.

One example of a green product is Eco Spec® interior latex semi-gloss enamel paint. As opposed to conventional latex paint, Eco Spec does not contain VOCs or solvents, and is a low odor product. Eco Spec is GREENGUARD® Indoor Air Quality Certified.

The GREENGUARD Environmental Institute is an organization that produces the green certification for certain products. GREENGUARD compares standards for emissions set forth by agencies, such as the EPA, and chooses the lowest permissible level as its standard. All products must test below these levels in order to be certified by GREENGUARD. There are many GREEN-GUARD certified products available on the market including adhesives, lubricants, cleaners, furniture, carpet, paper products, degreasers, and more.

In general, green products are the least toxic construction products available for a job. Building materials that are considered to be green include plant materials like bamboo and straw, lumber from forests certified to be sustainably managed, stone, and recycled metal. Other materials that are considered to be green are usually non-toxic, reusable, renewable, and/or recyclable.

Every day, more businesses are using green products and construction practices. Governmental regulations, environmental design concerns, and skyrocketing energy costs are just some of the factors that have contributed to this wave of change. To increase environmental friendliness, building materials should be extracted and manufactured locally (near the building site) in order to minimize the energy required for material transportation.

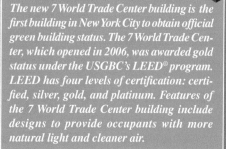

IAQ Fact

The new 7 World Trade Center building is the first building in New York City to obtain official green building status. The 7 World Trade Center, which opened in 2006, was awarded gold status under the USGBC's LEED® program. LEED has four levels of certification: certified, silver, gold, and platinum. Features of the 7 World Trade Center building include designs to provide occupants with more natural light and cleaner air.

Emission Levels

Low-impact, low-emission building materials are used wherever feasible. For example, insulation can be made from low VOC-emitting materials such as recycled denim, rather than insulation materials that may contain toxic materials, such as formaldehyde. To discourage insect damage, alternate insulation materials are treated with boric acid. Organic or milk-based paints can be used on green construction materials.

Architectural salvage and reclaimed materials are used when appropriate. Often, when older buildings are demolished, any good wood is reclaimed, possibly renewed, and then sold. Various parts of a building can be reused, such as doors, windows, mantels, and hardware, which reduces the consumption of new goods.

When possible, building materials can be gathered from the construction site itself. For example, when a new structure is being constructed in a wooded area, wood from the trees that were cut to make room for the building can be used as part of the building. However, when new materials are used, designers look for materials that are rapidly replenished.

Rapidly replenished materials are materials that are readily available for purchase and whose extraction from the environment has a minimal impact on the environment.

Fluke Corporation

Visual inspections help to prevent IAQ problems. During an inspection, a stationary engineer or technician may look for specific items such as extreme temperatures and ventilation problems.

Bamboo, a rapidly replenished material, can be harvested for commercial use after only six years of growth. Cork oak is another rapidly replenished material in which only the outer bark is removed for use, so the tree is preserved.

INSPECTIONS

Inspections and recordkeeping are important preventive IAQ activities. Regularly scheduled inspections help to identify possible IAQ issues before they become widespread and/or cause health issues for building occupants.

In addition to a formally scheduled PM program, every stationary engineer should employ two other pollution prevention strategies. The pollution prevention strategies include walk-through visual inspections and regularly scheduled random air-samplings.

Structural Problems

During an inspection, it is important to look for any structural problems. Water and air leakage are two main causes of IAQ problems. Repair work orders must be written immediately to fix all water leaks and to completely clean all signs of mold growth. Moisture and dirt cause mold growth, which can lead to serious IAQ complaints. Air infiltration into a building can permit excessive carbon monoxide or other gases to enter and accumulate in occupied spaces. Equally important is to check for vapor barriers and seals between parking garages, in mechanical rooms, in laundries, and in the rest of the building or facility.

Visual Inspections

Walk-through visual inspections are a regular part of a stationary engineer's job. However, visual inspections usually concentrate on equipment and facility operation. Nevertheless, visual inspections provide a prime opportunity to prevent IAQ problems. When walking through a building or facility, potential IAQ problems to look for include the following:

- HVAC equipment or ductwork problems
- lack of ventilation
- new equipment being installed
- odors
- leaks (air and water)
- extreme temperatures
- unusual conditions

To prevent small problems from becoming major IAQ problems, minor repairs and adjustments to the HVAC equipment and the facility must be made. HVAC checklists and Pollutant, Source, and Pathway surveys can help to identify areas that require inspection.

IAQ Recordkeeping

IAQ records usually consist of an IAQ log and visual inspection checklists. A *visual inspection checklist* is a checklist that is used to record any repairs needed or possible problems that are identified during a visual inspection. The HVAC checklist and the Pollutant, Source, and Pathway survey forms are examples of visual inspection checklists. The forms are customized as needed for the specific applications in the building. Because of the complexity of IAQ work, good recordkeeping is essential.

Records can show a distinguishable pattern over time for the diagnosis of an IAQ problem or help a stationary engineer work out a hypothesis on IAQ issues. Records also help outside consultants when they are needed. When an IAQ complaint leads to legal proceedings, the records show the efforts made to solve the problem.

IAQ Log. An IAQ log complements various types of information related to HVAC maintenance and operation. An IAQ log is an ongoing record of all air-sampling readings, with a brief comment section for any system adjustments that are made. **See Figure 9-6.** Advanced test instruments can electronically store test results, which later can be downloaded to a computer.

It is essential to follow established protocols for each type of air sampling, especially when more than one person is taking measurements. Measurements taken under varying methods and conditions can easily lead to false conclusions or interpretations, so it is important to be consistent, follow protocols, and make sure that all test instruments are properly calibrated.

> **IAQ Fact**
>
> *The most important parts to IAQ recordkeeping is that the recordkeeping must be easy-to-use and kept up-to-date.*

IAQ LOG

Store your measurement protocol and instrument documentation with this log.

Measurement performed: _____ Baseline readings: _____

Instrument used: _____ OA temperature: _____ Location: _____

Date performed: _____ OA relative humidity: _____ Location: _____

Performed by: _____

Time	Location	Space RH	Space Temperature	CO	CO₂	Measurement	Systems Adjustments or Comments

Figure 9-6. An IAQ log contains a record of air-sampling readings and system adjustments for easy review.

HVAC Checklist

An HVAC checklist provides a convenient way to identify and list items needing unscheduled repair during a visual inspection. An *HVAC checklist* is a document listing any HVAC equipment and maintenance procedures that might have a negative impact on a building's indoor air quality. **See Appendix.** For instance, some of the items on the HVAC checklist include humidifiers, outdoor air inlets, screens and filters, and duct cleaning.

Humidifiers must be inspected periodically to ensure they are functioning and draining properly. When the drain of a humidifier is not operating correctly, the inside of the ductwork will get wet, becoming a breeding ground for mold. Outdoor air inlets are inspected to verify that they are clear of debris and that no outdoor air pollutants are being brought into the building. Screens and filters are inspected to ensure that they are removing dirt, debris, and solid contaminants. Ducts are examined for the presence of excessive dirt, which would indicate that the ducts require cleaning. This inspection should be conducted at least annually and any time there are changes in tenants or external environmental conditions.

A policy and procedure should be developed for handling problems that stem from air and particles such as truck exhaust, generator exhaust, outdoor tobacco smoke, landscaping pesticides, and grass cutting entering the facility. All facility loading docks should be considered a pathway for contaminants to enter. Pollutants can enter the building from the loading dock area directly into the facility or enter the building through the outdoor air intake.

Parking garages are another potential source of pollutants for the facility. In addition to vehicle exhaust, a parking garage may also have storm sewer sump pumps, sewer ejector pumps, and grease traps. These items must be maintained through a regular preventive maintenance program. However, it is a good IAQ practice to perform an inspection of these areas frequently.

Pollutant, Source, and Pathway Surveys

An HVAC checklist lists the components of an HVAC system to be inspected. The Pollutant, Source, and Pathway surveys list the sources of IAQ contaminants that must be inspected. This shows how random air sampling is a great preventive strategy in identifying pollutant sources before they cause a problem. **See Appendix.** Contaminants can be in the HVAC system and throughout building spaces.

Common Airflow Pathways. Contaminants travel with the airflow along pathways, sometimes over great distances. Pathways may lead from an indoor source to an indoor location or from an outdoor source to an indoor location. The location experiencing a contaminant problem may be close by, in the same area, or in an adjacent area as a contaminant source. The location may also be a great distance from and/or on a different floor from the contaminant source. Some of the tools that can aid in identifying pathways are thermal imagers, inert smoke, and airflow test instruments.

Knowledge of common building airflow pathways helps to track down the source and to prevent contaminants from reaching building occupants. Common airflow pathways include:
- stairwells
- elevator shafts
- vertical electrical or plumbing shafts
- receptacle outlets and building openings
- ducts and plenums
- flue or exhaust leakage into room spaces
- outside air intakes
- windows, cracks, and crevices
- substructures or slab penetrations

Common Pathway Effects. Pathways such as stairs, elevators, and vertical shafts allow air and contaminants to move within a room or through doorways and corridors to adjoining spaces. **See Figure 9-7.** The stack effect of certain vertical shafts encourages airflow by drawing air toward these shafts on the lower floors and away from the shafts on the higher floors, affecting the flow of contaminants. A *stack effect* is the flow of air that results from warm air rising, creating a positive pressure area at the top of a building and a negative pressure area at the bottom of a building.

Pathways also allow contaminants to easily enter and exit building cavities. Contaminants are commonly carried by the HVAC system throughout the occupied spaces of a building. Cooling tower exhaust should be examined to ensure that exhaust does not get drawn into outdoor air intakes. The discharge on the air side of the cooling tower contains chemical vapors and odors that could cause serious problems if introduced into the indoor environment. Cooling towers are also a potential source of contaminated water, including Legionella, so proper maintenance, water treatment, and visual inspections are critical.

Sewer gas leaking into an occupied space through floor drains, such as restroom and mechanical room floor drains, is another common problem for facilities. Leaks from sanitary exhausts or combustion flues can cause health problems. To prevent sewer gas from leaking, floor drain traps must be primed (contain water) to form a hydraulic seal in the plumbing trap. This prevents sewer gas from backing up through the plumbing into the facility.

Some contaminants should be addressed before they enter the air pathway. Plants and flowers can produce allergens and should be recognized as potential problems in the indoor environment. Plants should be chosen based on potential emissions. Some species are more likely to affect indoor air quality than others.

A plan should be established for selecting plants and flowers that produce the least amount of pollen, insects, and odors. Pesticides used to control insects for plants and flowers should also be chosen based on their ability to control pests effectively without creating another indoor air problem.

Used together, HVAC checklists and Pollutant, Source, and Pathway surveys provide a snapshot of the building during any given inspection. These forms help identify potential problems and aid in the investigation of IAQ complaints. Whenever a problem is spotted that requires attention, stationary engineers must record the problem on one of the inspection forms and issue a standard repair work order. Once a repair work order has been issued, the information from the work order must be marked on the checklists. The completed checklists are then stored as back-up documentation.

COMMON PATHWAYS AND THEIR EFFECTS	
COMMON PATHWAYS	**PATHWAY IAQ EFFECTS**
Stairwells and elevator shafts	Air and contaminants move within a room or through doors and corridors to adjoining spaces
Vertical electrical or plumbing shafts	A stack effect draws air toward chases on the lower floors and away from chases on the higher floors
Receptacles, outlets, and building openings	A negatively pressurized building draws air and outside pollutants into the building through any available opening
Ducts and plenums	Duct leakage generates unplanned airflow and energy loss
Flue or exhaust leakage into room spaces	Leaks from sanitary exhausts or combustion flues can cause serious health problems
Outside air intakes	Polluted air can enter the building through the air intake
Windows, cracks, and crevices	Contaminants easily enter building cavities and move from space to space
Substructures or slab penetrations	Radon, soil gases, and microbial contaminated air travel through substructures in the building

Figure 9-7. Contaminants can follow many different pathways. Each contaminant pathway can contribute to poor IAQ.

Case Study

Case Study – Fiberglass Exposure Court Case

Basis

In the late 1990s, a woman sued her company for exposure to fiberglass particles that she believed affected her health. The claimant had worked since 1990 as a secretary within a public services building for seven years with no health issues. In 1997, she changed positions and her workstation was moved to a different floor within the same building.

In June, the claimant developed a rash on her face and arms. The claimant was seen by a doctor and given medicine for the condition. When the claimant went on vacation, her symptoms cleared up. Upon returning to work, the rash returned as well. The claimant began to suspect that her skin condition was the result of contaminants in the workplace based on a memo sent to all employees in February 1998.

Further Evidence

The memo stated that the heating, ventilation, and air conditioning (HVAC) system had been contaminated during building renovation, allowing contaminants to enter the system. The HVAC system was cleaned, and a high amount of fiberglass particles were released into the air in the process. In March 1998, the facilities manager issued another memo stating that a new source of fiberglass particles had been found. The source was fiberglass blankets that had been placed above a conference room for noise suppression.

A few months later, the claimant's doctor received data saying that the woman tested negative for the presence of surface fiberglass. The claimant then returned to her regular workstation. The rash returned again after the claimant had been in the building for three hours.

Two doctors continued to examine the claimant's condition. The doctors could not conclude positively that the condition was caused by fiberglass exposure, but they did concede that the condition was probably related to fiberglass exposure. The claimant had no history of any skin condition prior to these episodes.

Between 1996 and 1998, tests done by the employer showed high concentration levels of fiberglass particles and other contaminants in the workplace. Finally, a biopsy was performed on the claimant. Doctors determined that, while there was no definite evidence of fiberglass particles, the changes in her skin could have been caused by a material such as fiberglass. They also noted that fiberglass particles do not penetrate the skin deeply and would not be present after a few weeks.

Ruling

After examining all of the evidence, including the doctors' statements, the court ruled in favor of the claimant. The court determined that the claimant had established that her skin condition was in large part due to her exposure to fiberglass particles.

Definitions

Prevention Definitions

- A *chemical inventory form* is a form used to identify all chemicals found in a building and indicates how the chemicals are used.
- A *cleaning survey* is a method by which sources of dirt and debris are identified.
- A *computer-based MSDS system* is a chemical tracking method that uses software to eliminate the difficulty of handling paperwork for MSDS storage, manual updates, and other MSDS maintenance issues.
- *Green products* are products that have been selected for their environmentally friendly qualities.
- A *housekeeping program* is a set of procedures that includes preventing excess dirt from entering an environment, using cleaning products that introduce the least amount of contaminants into the air, and implementing maintenance procedures by trained and monitored personnel.
- An *HVAC checklist* is a document listing any HVAC equipment and maintenance procedures that might have a negative impact on a building's indoor air quality.
- *Integrated pest management* is a type of pest management where an environmentally sensitive and common-sense approach to pest management is used.
- A *material safety data sheet (MSDS)* is a form containing data regarding the properties of a particular substance.
- A *paper-based MSDS system* is a chemical tracking method that keeps paper MSDS files in three-ring binders at an accessible location.
- *Rapidly replenished materials* are materials that are readily available for purchase and whose extraction from the environment has a minimal impact on the environment.
- A *stack effect* is the flow of air that results from warm air rising, creating a positive pressure area at the top of a building and a negative pressure area at the bottom of a building.
- A *visual inspection checklist* is a checklist that is used to record any repairs needed or possible problems that are identified during a visual inspection.

**For more information please refer to ATPeResources.com
or http://www.epa.gov.**

1. Name two ways that pesticides are used.

2. Explain how comprehensive housekeeping programs can minimize IAQ issues.

3. List six pieces of information that are on an MSDS.

4. What are green products and why are they used?

5. Identify three possible IAQ pathways and how they may contribute to poor IAQ.

PREVENTIVE MAINTENANCE

10
chapter

Fluke Corporation

Class Objective
Identify the reasons for and benefits of operating a preventive maintenance program.

On-the-Job Objective
Classify and organize work within a preventive maintenance program.

Learning Activities:
1. Define preventive maintenance.
2. List the economic benefits of a preventive maintenance program.
3. Give an example of how the lack of preventive maintenance could lead to an IAQ problem.
4. Define unscheduled maintenance and give several examples of equipment requiring this type of maintenance.
5. Identify one purpose of a repair work order.
6. **On-the-Job:** Describe examples of scheduled and unscheduled work in the facility.

Introduction
All building mechanical system components require some type of maintenance. *Maintenance* is the repair of any sort of mechanical or electrical device that is broken, as well as the implementation of routine actions that keep the device in working order or prevent trouble from arising. There are several types of maintenance approaches used in buildings.

OVERVIEW OF MAINTENANCE TYPES

An effective preventive maintenance program is one of the most important tools in preventing IAQ problems. Facilities with effective preventive maintenance programs generally have fewer IAQ problems. IAQ problems cannot be prevented or solved quickly without an effective preventive maintenance program and the records it generates.

Maintenance activities vary depending on the type of maintenance program in place. Maintenance may be performed before any breakdowns occur or only after a breakdown occurs. There are different types of maintenance methods utilized within industry. The various methods are referred to as predictive, proactive, reliability-centered, enterprise asset management, run-to-fail, and preventive maintenance.

Predictive Maintenance

Predictive maintenance (PdM) is maintenance that involves evaluating the condition of equipment by performing periodic or continuous equipment monitoring. PdM is also known as condition-based maintenance. Data collected through equipment monitoring is analyzed to check if values are within acceptable tolerances. **See Figure 10-1.**

The goal of PdM is to predict the future and to perform maintenance at a scheduled time when the maintenance activity is most cost effective and before the equipment fails. PdM is in contrast to maintenance-based time and/or operation count, where a piece of equipment gets maintained whether it needs it or not. It is most commonly used on expensive or critical equipment.

HVAC Equipment Analysis Using Predictive Maintenance

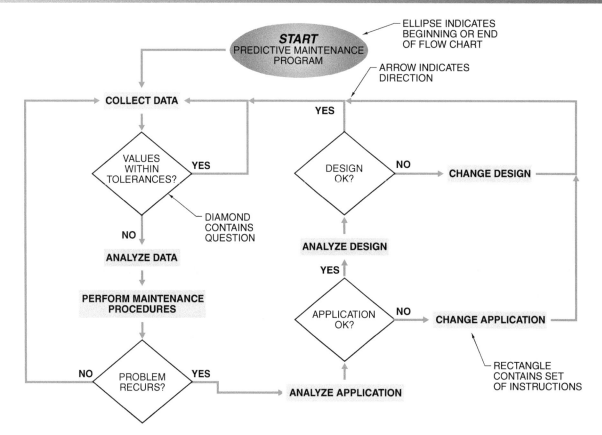

Figure 10-1. Predictive maintenance is most commonly used on expensive or critical equipment.

The predictive component of predictive maintenance is designed to predict the future condition of equipment. This approach uses principles of statistical process control to determine when in the future maintenance activities are appropriate. Most PdM inspections are performed while equipment is in service, thereby minimizing disruption of normal system operations.

Adopting PdM for the maintenance of equipment usually results in higher system reliability. PdM technologies used to evaluate equipment condition include infrared, vibration analysis, sound level measurements, oil analysis, refrigerant analysis, electrical current and voltage analysis, and other specific on-line tests. **See Figure 10-2.** In terms of resources, PdM requires a substantial investment in training and equipment. PdM may be labor intensive, ineffective in identifying problems that develop between scheduled inspections, and is not always cost effective.

Proactive Maintenance

Proactive maintenance is a maintenance strategy that stabilizes the reliability of equipment using specialized maintenance services. The central theme of proactive maintenance involves directing corrective actions aimed at the root causes of failure, not active failure symptoms, faults, or machine-wear conditions. A typical proactive maintenance regiment involves the following steps:

1. Set a quantifiable target or standard related to a root cause of concern (e.g., keeping a proper lubricant level for a fan).

2. Implement a maintenance program to control the root-cause property within the target level (e.g., routine monitoring of lubricant level).

3. Perform routine monitoring of the root-cause property using a measurement technique (e.g., oil level sensor) to verify the current level is within the target.

In other words, the goal of proactive maintenance is to find the cause of equipment failures. Maintenance procedures may call for a redesign or modification of equipment operation or components. Proactive maintenance extends the life of HVAC equipment and reduces equipment repair needs.

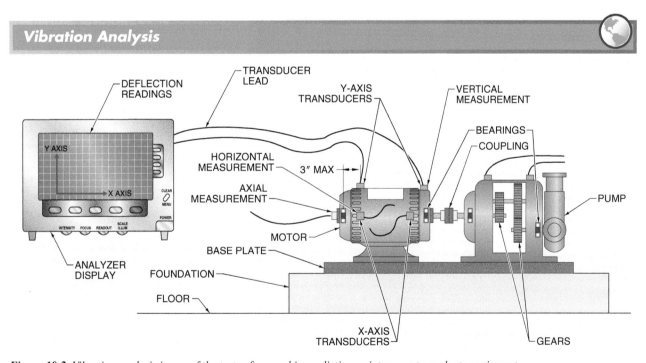

Figure 10-2. Vibration analysis is one of the tests often used in predictive maintenance to evaluate equipment.

Reliability-Centered Maintenance

Reliability-centered maintenance (RCM) is a maintenance plan used to create a cost-effective maintenance strategy to address the dominant causes of equipment failure. RCM is often used by the military and large maintenance organizations such as petrochemical, manufacturing plants, and other mission-critical facilities. RCM is a systematic approach to defining a routine maintenance program composed of cost-effective tasks that preserve important functions. The important functions of a piece of equipment that need to be preserved with routine maintenance are identified. The dominant failure modes and causes are then determined and the consequences of failure ascertained. Critical levels are then assigned to the consequences of failure.

Maintenance tasks are selected that address the dominant causes of failure. This process directly addresses maintenance-preventable failures. The result is a maintenance program that concentrates scarce economic resources on those items that would cause the most disruption if they were to fail.

TSI Incorporated

In reliability-centered maintenance, a stationary engineer identifies the important functions of a piece of equipment and then defines a maintenance program to preserve them.

IAQ Fact

Reliability-centered maintenance (RCM) is often chosen because it can help a business avoid added costs. For instance, RCM can help to reduce equipment failures and their associated costs. RCM can also help to reduce time-based maintenance work hours. RCM tends to vary in operation methods based on the type of facility and its age.

Enterprise Asset Management

Enterprise asset management (EAM) is the whole-life, optimal management of the physical assets of an organization to maximize value. EAM covers things such as the design, construction, commissioning, operations, maintenance, and decommissioning/replacement of a plant, equipment, and mission critical facilities. Enterprise refers to the management of the assets across departments, locations, facilities, and, in some cases, business units. By managing assets across the facility, organizations can improve utilization and performance, reduce capital costs, reduce asset-related operating costs, extend asset life, and subsequently improve the return on investments.

Run-To-Fail Maintenance

Run-to-fail maintenance is corrective, reactive, and unscheduled maintenance. In run-to-fail maintenance, no maintenance is performed until the equipment has failed and no longer functioning. It is the simplest maintenance strategy available. However, because it is the harshest on equipment, it is the least desirable strategy.

Run-to-fail maintenance, also known as repair, is used to get equipment working again. This maintenance approach is often much more expensive than any other approach. Indirect costs, such as safety problems and the unavailability of critical building systems, contribute to the high cost of run-to-fail maintenance. Run-to-fail maintenance may be more likely to cause IAQ problems the most because any interference with ventilation and/or cooling may allow the buildup of IAQ pollutants.

Preventive Maintenance

Preventive maintenance (PM) is the periodic inspection and servicing of a building and equipment for the purpose of preventing failures. The time period involved in performing PM may be based on a calendar (i.e., monthly or quarterly), or based on the amount of run time on the equipment (i.e., every 500 hours). PM uncovers conditions leading to equipment and production breakdowns or harmful depreciation. PM is performed to avoid harmful conditions or to address any issues while they are still minor. PM includes tests, measurements, adjustments, and parts replacement, performed specifically to prevent faults from occurring.

FUNDAMENTALS OF PREVENTIVE MAINTENANCE PROGRAMS

A preventive maintenance (PM) program is designed to provide the proper care to all building mechanical systems and components to keep them operating at peak performance according to manufacturers' specifications. In a PM program, all facility equipment is inventoried, surveyed, and the care that each piece of equipment requires is listed. PM work orders are then issued and the appropriate work completed.

Within a PM system, trained personnel monitor the results of maintenance work and adjust the type and amount of work to control PM costs. Files for the collection and coordination of IAQ records are also established and maintained. PM records enable people to track IAQ problems and prevent them altogether or solve them quickly. If IAQ problems result in legal action, PM records may provide evidence of the company's efforts to maintain good IAQ.

Preventive maintenance has many advantages, including the fact that if IAQ issues are documented and/or prevented, large-scale breakdowns may be avoided and fewer repairs may be needed. Preventive maintenance also provides better scheduling and workload distribution for equipment than other kinds of maintenance. Preventive maintenance provides greater safety for workers, more efficient equipment operation, saves energy, and prolongs the life of equipment. Eliminating many reactive maintenance costs may save up to 10 times the cost of the PM.

Economic Benefits

Preventive maintenance that includes infrared scans can lower energy costs. Clean heating and cooling coils transfer airflow more efficiently than dirty coils. Clean filters and ductwork lower the workload of fan motors, which lowers electricity consumption.

A well-maintained facility has fewer hot, cold, and airflow complaints, which translates to lower maintenance costs per square foot of occupied space. Lower costs mean a higher profit margin for building owners and management firms. The fewer the number of HVAC calls that require a response, the more effective the stationary engineers can be at providing other quality services for building occupants.

Well-maintained facilities can bring in higher lease rates. Some facilities have used their HVAC systems to attract tenants by highlighting in their marketing information the system's energy efficiency and ability to ensure tenant comfort. In some cases, prospective tenants are beginning to send HVAC experts to inspect the HVAC system before agreeing to the terms of the lease.

Extended equipment life and fewer unscheduled repairs are additional economic benefits of PM work. Extending equipment life by using infrared scans can also help maintain the value of expensive capital investments and offset labor costs. PM records help identify high-maintenance equipment, provide data for new equipment purchases, ensure better inventory control, and help to control the maintenance budget more effectively. By preventing IAQ problems, building and/or companies may avoid medical, production, and legal expenses associated with poor IAQ.

Resolving Problems

As equipment wears, it operates less efficiently. This creates problems. The specific problems that require proper preventive maintenance vary with each facility and include problems within mechanical and electrical systems.

Mechanical. A PM program can help identify and resolve mechanical problems. For example, motor failure can be avoided by replacing worn bearings and stretched belts. Tightening loose fasteners stops vibration and prevents fatigue failure of mounts and other structural components. Checking condensate drains prevents clogs and standing water in which bacteria might form.

Electrical. Preventive maintenance can help identify and resolve electrical problems. For example, resistance buildup causes a voltage drop and a resulting amperage increase. Tightening loose electrical connections can prevent resistance buildup. An increase in current would cause the motor to operate at a higher temperature, tripping a motor overload and/or burning out the motor. Locating a broken wire prevents the failure of an overloaded electric heat strip element. Removing a buildup of carbon prevents short circuits that would disable a system.

Air-Handling Units. When an air-handling unit (AHU) in an HVAC system is new, the system should perform as designed. If the equipment operates and receives no PM, it wears until it no longer performs within the manufacturer's original specifications.

For example, an AHU consists of fans, coils (heating/cooling), filters, and outdoor-air (OA) and return-air (RA) dampers. **See Figure 10-3.** During the cooling mode, the unit is designed to deliver 7000 cubic feet per minute (cfm) of supply air at 55°F, in order to cool the space to 70°F.

If the filters are dirty, the fan may only deliver 6800 cfm at 55°F, which causes the space temperature to be higher than 70°F. If the cooling coil is dirty, airflow is restricted and a poor heat transfer effect results. This causes the supply air temperature to rise above 55°F and the space temperature also rises above 70°F.

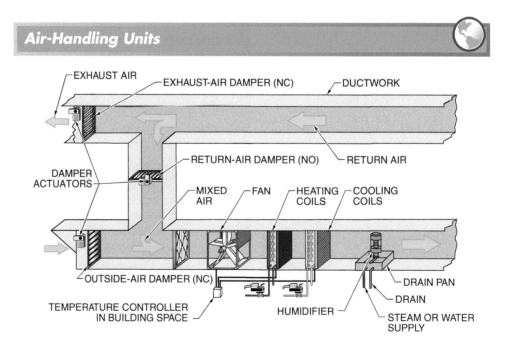

Air-Handling Units

Figure 10-3. *Air-handling units (AHUs) consist of fan(s), coils, dampers, ductwork, humidifiers, filters, and controls to condition and distribute air throughout a building.*

In the case of both dirty filters or coils, the fan motor requires more energy (amps and horsepower) to operate, and this could potentially damage the motor. Because of the lower airflow, lower amounts of dehumidification occur. IAQ contaminant levels may then rise. The air temperature becomes more difficult to control and this may cause IAQ complaints.

The wearing down of equipment is natural and unavoidable. PM work slows and helps correct the wearing process. The goal of PM is to keep HVAC equipment and systems working at the original specifications of the manufacturer.

HVAC systems must be installed, inspected, and balanced according to manufacturer's specifications. **See Figure 10-4.** An effective

PM program can help ensure that this important task is properly completed. A program for balancing the entire system and periodic rechecking of the system performance can eliminate future IAQ problems.

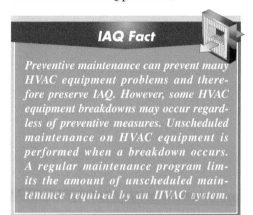

IAQ Fact

Preventive maintenance can prevent many HVAC equipment problems and therefore preserve IAQ. However, some HVAC equipment breakdowns may occur regardless of preventive measures. Unscheduled maintenance on HVAC equipment is performed when a breakdown occurs. A regular maintenance program limits the amount of unscheduled maintenance required by an HVAC system.

Air-Handling Unit Print

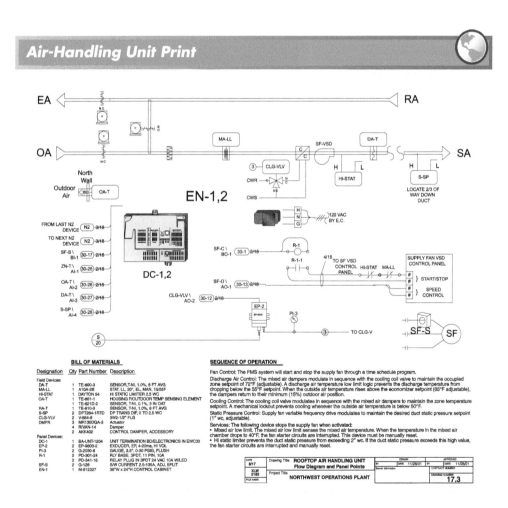

Figure 10-4. *Documentation for an air-handling unit may include a unit print, parts list, and sequence of operation.*

Fluke Corporation

Today, temperature measurements of HVAC electrical and mechanical equipment are included in all building preventive maintenance programs.

PREVENTIVE MAINTENANCE TYPES

Preventive maintenance is designed to prevent failures and maintain the safe and efficient operation of the equipment. The two types of preventive maintenance are scheduled work and unscheduled work. Both types of PM are aimed at keeping HVAC equipment in peak operating condition.

Scheduled Maintenance

Scheduled maintenance work is planned and scheduled through the PM system to prevent most emergency repairs and operating problems. Scheduled maintenance is controlled through the issuing of PM work orders.

Unscheduled Maintenance

A machine receives unscheduled maintenance when it is determined that work is required outside of normal PM or the choice is made to work on a machine only after it breaks down. Electronic equipment that is inexpensive and not critical can be run on unscheduled maintenance because of its reliability. In IAQ

work, most equipment should not be run on unscheduled maintenance. Unscheduled maintenance is controlled through the issuing of repair work orders.

Example of a PM Schedule

An inexpensive fan located in a paint locker is designed to exhaust 100% of the air to keep the locker under negative pressure so no fumes can escape. The fan fails and a chemical smell enters a nearby work area. Occupants complain, and two of them go home with headaches.

Replacing the fan takes less than half an hour and is relatively inexpensive, but the damage is already done. The occupants suspect that the air is bad, making it much harder to prove that an effective IAQ program is in place. As a result of this incident, the decision is made to inspect these fans quarterly.

ESTABLISHING A NEW PM SYSTEM

The work needed to organize and implement a new PM program must be documented. Documentation can be used to justify the use of new programs or equipment to support it. The future operation of a facility can be affected by presenting a well-documented case to management, using the facts to justify the additional work hours and money needed to perform PM tasks. Justification through documentation is one sure way that management can assess and properly allocate funding for these projects. Presentation of the facts and personal credibility may be an important factor in a management decision to proceed with a PM program.

PM Organizational Tools

The first step in beginning a PM program is choosing an electronic or a manual organizing tool. The organizing tool must help record, store, and retrieve the information that PM activities create. **See Figure 10-5.** Ideally, an organizing tool enables personnel to keep track of the cost of PM operations, the time spent on PM, and the relative dependability of the equipment. This information can be used to improve facility operation.

Manual vs Electronic Tools

PAPER FORM

COMPUTERIZED MAINTENANCE MANAGEMENT SYSTEM (CMMS)

Figure 10-5. *Items such as work orders can be produced manually or electronically using a computerized maintenance management system.*

Manual systems range from expensive, commercially produced organizers to simple three-ring binders and file folders. The type and sophistication of the organizing tool depends on the operating budget, the size of the facility, and the size of the maintenance crew. A small facility could be run using three-ring binders and file folders. One person can organize a manual or small computer system themselves by using and modifying PM forms. **See Appendix.**

Manual systems are generally cheaper than computer systems but offer fewer features. Large facilities often require a computerized system. These are known in the industry as computerized maintenance management systems (CMMSs).

Computerized Maintenance Management Systems. Computerized maintenance management systems use software programs that can be purchased ready-made, customized, or created by the user. Using a computer makes it easy to locate information quickly. Also, some facilities use handheld personal digital assistants (PDAs) to retrieve records and administer preventive maintenance tasks.

Do You Need a CMMS?

Follow a step-by-step process to determine if a computerized maintenance management system (CMMS) fits your operation.

Form a Team. A CMMS team should include a plant engineer, a maintenance manager, maintenance personnel, and representatives from departments that impact CMMS. One of these persons should be the project leader responsible for implementing the CMMS.

Identify Problems with Existing System. Create a list of problems you hope a CMMS can solve. The list should include categories such as labor productivity, equipment availability, inventory control, product quality, environmental controls, management support, and maintenance information.

Define CMMS Objectives and Benefits. Determine the objectives needed to solve the identified problems. Create a list of CMMS features that will solve the identified problems.

Conduct a Financial Analysis. Create a list of benefits and calculate the dollar savings generated from each benefit. Total software costs include system analysis, program installation, data entry, software acquisition, training, operations, and software maintenance. Hardware costs include PCs, a network, and possibly handheld devices.

Compute the Return on Investment. A return on investment (ROI) calculation justifies the use of a CMMS $(ROI = \dfrac{Savings - Costs}{Costs} \times 100)$.

Application

Computerized PM systems can track almost any type of information necessary. CMMS software costs can range from several hundred to thousands of dollars, plus annual maintenance fees. A basic computer system can be created with a store-bought database, spreadsheet, and wordprocessing software package. **See Figure 10-6.**

A CMMS software package maintains a computer database of information about an organization's maintenance operations. This information is intended to help maintenance workers do their jobs more effectively. For example, information gathered from a CMMS program can help determine which storerooms contain the spare parts needed for a certain task.

CMMS information also helps management make informed decisions. An example of this is calculating the cost of maintenance for each piece of equipment used by the organization, possibly leading to a better allocation of resources. The information may also be useful when dealing with third parties. For an organization involved in an IAQ liability case, the data in a CMMS database can serve as evidence that proper maintenance has been performed.

CMMS packages may be used by any organization that must perform maintenance on equipment and property. Some CMMS products have a general focus, while some focus on particular industry sectors, such as schools or health care facilities. Different CMMS packages offer a wide range of capabilities and cover a correspondingly wide range of prices. A typical package addresses some of the following topics:

- work orders
- scheduling jobs
- assigning personnel
- reserving materials
- recording costs
- tracking relevant information

CMMS programs include functions to keep track of PM inspections and jobs, including step-by-step instructions or checklists, lists of materials required, and other pertinent details. Usually, the CMMS software automatically schedules PM jobs based on schedules and/or meter readings. Different software packages use different techniques for reporting when a job should be performed.

CMMS asset management includes functions to record data about equipment and property including specifications, warranty information, service contracts, spare parts, purchase date, and expected lifetime. **See Figure 10-7.** CMMS inventory control includes functions to manage spare parts, tools, and other materials. The management of tools includes the reservation of materials for particular jobs, a record of material storage location, determination of when more materials should be purchased, tracking shipment receipts, and taking inventory.

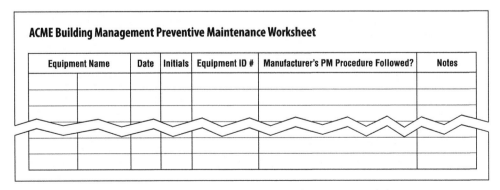

ACME Building Management Preventive Maintenance Worksheet

Equipment Name	Date	Initials	Equipment ID #	Manufacturer's PM Procedure Followed?	Notes

Figure 10-6. *Computer spreadsheets may be used to create simple PM paperwork for a custom program.*

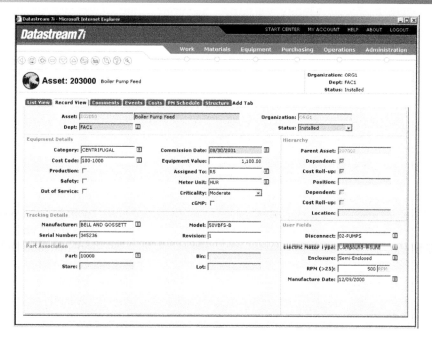

Figure 10-7. *Asset management features of a CMMS program keep track of a wide variety of equipment information.*

CMMS packages can produce status reports and documents giving details or summaries of maintenance activities. The more sophisticated the package, the more analysis functions are available. Many CMMS packages can be either web-based or LAN-based. Web-based packages are hosted by the company selling the product on an outside server. LAN-based packages imply that the company buying the software hosts the product on their own server.

CMMS packages are closely related to facility management system packages, also known as facility management software (FMS). For the purposes of many organizations, the two progams can be saved on the same computer and may also swap data between one another. Some advanced CMMS programs allow mechanical equipment to be barcoded automatically to indicate the proper PM information. After selecting a manual or CMMS organizing tool, a survey of the facility should be done to create the basic system files and master equipment records.

Preventive Maintenance Surveys

A facility PM survey consists of a complete inventory of all equipment in the facility and an evaluation of the equipment's condition. The stationary engineer should complete the facility PM survey during equipment installation, but this can also be done any time thereafter to help improve the overall building IAQ.

The goal of the inventory is to create a record for each piece of equipment listing all the information critical to maintaining that equipment. This information includes serial numbers, spare parts required, parts suppliers, and all equipment specifications. All essential parts can be ordered by using these forms. These records become the basis of an equipment history record file.

After the inventory is complete, equipment should be evaluated. The goal of evaluating equipment condition is to eventually restore the equipment to the manufacturer's specifications. After evaluating the equipment, repair work orders are scheduled to restore and/or maintain the equipment at peak operating condition.

Survey Process. The facility PM survey can be completed by entering information onto the preventive maintenance forms by hand or into the CMMS. **See Figure 10-8.** If a computerized system is used, the information can also be entered into computer terminals that are linked to a main computer server.

The acronyms from mechanical prints should be used to distinguish equipment properly on the survey forms. For example, an air supply fan might be called HV2. The acronym HV stands for heating and ventilating fan, and the number two is the unit number. This name should also be marked or painted on the unit after the equipment survey and equipment history records are completed and all of the equipment has been identified.

A survey should begin with the air-handling units supplying one section of the facility. Air-handling units are especially important to IAQ issues. Equipment can be surveyed while maintenance or repair is done. Surveying during maintenance prevents using excessive overtime hours.

During the survey process, the HVAC system should be inspected to determine if it was ever balanced. Reviewing the original balance reports and/or performing a composite evaluation of the airflow by obtaining total CFM output from each fan system can help determine this.

If no balance reports exist, the reports cannot be found, or the airflow is incorrect, then the entire system must be balanced according to the design specifications and blueprints. If it appears that the system was balanced, then the balancing reports must be used in conjunction with the as-built blueprints to ensure that the system is still arranged properly and uniformly, producing the same amount of air.

PM Survey Form

Form Number: _____ Date of Survey: _____ Name: _____

Date of Equipment Purchased: _____ Purchase Price: _____

Equipment Name: _____ System: _____

Location: _____ Floor: _____ Zone: _____

Installing Contractor's Information: _____

Address: _____

Phone: _____ Fax: _____

Manufacturer Name and Address: _____

_____

Phone: _____ Fax: _____

Model Number: _____ Serial Number: _____

Frame: _____ Type: _____ Duty Rating: _____ HP: _____

Full Load Amps: _____ Run Amps: _____ Voltage: _____ Phase: _____

Hertz (Cycles): _____ RPM _____ Max Ambient Temp.: _____ Service Factor: _____

Location of Disconnect: _____ Panel Fed From and Location: _____

Type of Fuses: _____ Number of Fuses: _____ Overload Protection: _____

Belt Size: _____ Number of Belts: _____ Belt Tension: _____

Sheave Type and Size: _____ ID: _____ OD: _____ Shaft Size: _____

Bearing Size: _____ Bearing Type: _____ Number of Bearings: _____

Greasable Bearing: _____ Nongreasable Bearing: _____ Condition: _____

Other Lubrication: _____

Filter Type(s): _____

Filter Size: _____ Number of Filter: _____ Filter Reading: _____

Figure 10-8. *A PM survey lists information about a piece of equipment and its current condition.*

Equipment History Records

Once the survey is completed, a system of files and records should be created that are accessible and provide the historical and reference information for all facility equipment. The three components to this system of records include an equipment history file, a master equipment list, and the operating manuals and manufacturers' data.

Equipment History Record File. From the PM survey, there should be a form for each piece of equipment in the facility. These forms are the start of a historical record, known as the equipment history record file, for each piece of equipment. **See Figure 10-9.**

Forms can be organized alphanumerically by the acronyms on the PM survey forms, which are based on the mechanical prints. For larger facilities, another option is to break the equipment file down further by floor/zone and then alphanumerically.

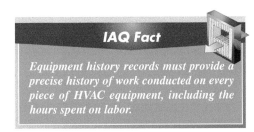

IAQ Fact

Equipment history records must provide a precise history of work conducted on every piece of HVAC equipment, including the hours spent on labor.

Equipment History Record File

Description:	Furnace #1	Voltage:	440	Warranty ID:	IT6P882
Asset ID:	FUR0001.00	Amperage:	600	Warranty Date:	02/17
Asset Type:	Furnace Systems	Wattage:	264000		
Parent ID:		Phase:	3		
Priority:	8 Active:☒	Elec Line:	10		
Manufacturer:	Brown Boveri, Inc.	Air Area:	COMP 6	YTD Labor Hr:	45.00
Model:	IT6P			YTD Downtime:	22.00
Serial Number:	OP2810C2B				
Vendor:	Lewis Systems, Inc.				
Vendor Address:	1862 Erie St.			TD Labor Hr:	740.00
	Cleveland, OH 55117	Counter UOM:		TD Downtime:	493.00
Vendor Phone:	216-555-1340	Current Counter:	1		
Asset Tag:	00509	Counter Rollover:	0		
Location:	4247 Piedmont Building 2	Meter UOM:			
	Floor-1 Room 23 CL15F	Current Meter:	1234550	YTD Labor Cost:	724.00
Department ID:				YTD Misc Cost:	1231.22
Cost Center:	Fixed Asset Repair			YTD Part Cost:	7701.19
Supervisor:	Jones, Fred			Total:	9656.41
		Meter Rollover:	0		
		Purchase Date:	02/17		
		Install Date:	08/05	TD Labor Cost:	9620.60
		Retire Date:		TD Misc Cost:	5631.80
		Install Cost:	97000	TD Part Cost:	14,942.00
		Replacement Cost:	97000	Total:	30,194.40

Comment: Manufact. warranty extremely strict. Document all hours worked and parts used.

- -

Report Totals:	YTD Labor Hr: 45.00	YTD Labor Cost: 724.00	TD Labor Cost: 9620.60
Assets: 1	YTD Downtime: 22.00	YTD Misc Cost: 1231.22	TD Misc Cost: 5631.80
	TD Labor Hr: 740.00	YTD Part Cost: 7701.19	TD Part Cost: 14,942.00
	TD Downtime: 493.00	Total: 9656.41	Total: 30,194.40

Figure 10-9. *An equipment history record file is a detailed list of all pertinent information about equipment including the manufacturer, serial number, and downtime information.*

Master Equipment List. As a quick reference to the equipment history record file, a master equipment list is created. The master equipment list includes the name, location, and a brief description of each piece of equipment. The list should be kept at the front of the equipment history record file to serve as a reference and as a table of contents to the file. A CMMS program can also be used to electronically duplicate this entire file for easier storage and referencing.

Operating Manuals and System Prints

Modern HVAC systems cannot be maintained and repaired without using operating manuals and system prints. A duplicate set of the operating manuals, manufacturers' data, and system prints should be made to serve as a working copy. The duplicate set of files should include the following:

- complete, updated facility design blueprints
- complete, chronological set of as-built blueprints of mechanical, electrical, and plumbing systems
- manufacturer's specifications for all system controls, including catalogs
- schematic duct layout from last testing and balancing work showing ductwork, fans, outlets, etc. **See Figure 10-10.**
- complete air balance reports and airflow specifications
- equipment operating manuals
- prints for each floor that details all occupant activities and room layout
- building commissioning reports for all HVAC equipment (if applicable)

The control blueprints should include a written description of how the system operates. The building commissioning report may not be available for the facility because these reports are relatively new to the building industry. Building commissioning is a procedure for documenting and verifying the performance of HVAC systems so that systems operate in conformity with the design intent. The commissioning applies to all phases of a project, from conception through occupancy. Commissioning also applies when changes in occupancy and space use or the normal deterioration of a building indicates a need for recommissioning.

This information should be stored in an appropriate filing system near the equipment history record file. The original set of documents should be kept as a permanent, clean copy that serves as a backup if the working copy is lost or damaged. The original copy should never be taken out of the file except for updates, as changes occur with the system, or for duplicating a new working copy. Both copies should be updated when equipment changes are made.

Review System Prints. After working copies of documents have been made, the system prints should be reviewed. When checking the system prints, it can be determined whether the HVAC system has been changed from its intended design. The HVAC system should be traced from the controlled devices (heat strips, valves, or dampers) for each loop, working backwards following the ductwork. When a difference between the real-world system and the prints exists, the changes must eventually be noted and dated on the print. Each revision must accurately reflect the real-world layout. Depending on the maintenance department, the procedure used to revise a print varies.

A review of system prints assumes that the system has been installed correctly. If the equipment was improperly installed according to the prints, refer to the facility PM survey. The survey should help to determine whether the installation or the prints are correct.

IAQ Fact

Electronic documents of large, complicated drawings are easier to use when compared with traditional paper documents. Specific parts can be enlarged on screen to display details. In addition, information on setup, operating, troubleshooting, and preventive maintenance procedures may be more easily included in the electronic document files.

Schematic Duct Layout

LOUVER   OA DAMPER   FILTER   HEAT COIL   COOLING COIL

9,800 CFM

48x24

RA DAMPER

36x30

2   1

(2) 48x24 REG
3000 CFM EA.

FD

500 CFM
6 FT LINEAR   158 CFM TROFFER   400 CFM 12x12 REG   SD

13   14   15   16

700 CFM EA
15x15 NECKS   SD

9   10   11   12

800 CFM EA
14″ Ø   SD

5   6   7   8

FD = FIRE DAMPER
SD = SMOKE DAMPER
VD = VOLUME DAMPER

600 CFM EA
12″ Ø   VD

1   2   3   4

Figure 10-10. *Schematic duct layouts are needed to perform maintenance or repair on HVAC equipment.*

Equipment that does not match the as-built drawings raises some serious questions. If the system has been installed contrary to the original design, a determination must be made whether the system can properly perform its intended function. HVAC systems go through many upgrades and retrofits through their years of service. System evaluation can become complex, depending on the circumstances, and may require consultation with the installation contractor. The system may have to undergo further modification depending on the results of the evaluation.

IAQ Fact

Many chillers that are 20 years old or older have been upgraded with new compressors and controls, even though they resemble the original design.

PM Charts

Once the prints and system are in agreement, the operating manuals and manufacturer's specifications help determine what maintenance each piece of equipment requires and the frequency needed. This information should be recorded on a PM chart for each piece of equipment. **See Figure 10-11.**

When completing the PM charts, certain pieces of equipment are more important to facility operation than others. Vital pieces of equipment require more frequent care in general. Equipment may also require more work during certain times of the year. For example, the cooling system requires special care just before the cooling season starts.

PM Work Orders

A *PM work order* is a form or ticket that indicates specific preventive maintenance (PM) tasks to perform at regularly scheduled intervals. **See Figure 10-12.** One PM work order for each interval of service is created by using the information in the PM charts for each piece of equipment. For instance, a piece of equipment that requires 20 tasks, spaced out quarterly, has four PM work orders. Each work order lists the specific tasks to be performed in that quarter. The notes on the PM chart regarding work to be completed in a specific month, such as boiler maintenance in July or August, help to space out the workload.

PM Chart
(PM Tasks to be Performed)
Annual: _____
_____
_____
_____

Biannual: _____
_____
_____
_____

Bimonthly: _____
_____
_____
_____

Monthly: _____
_____
_____
_____

Biweekly: _____
_____
_____
_____

Weekly: _____
_____
_____
_____

Figure 10-11. *A PM chart lists the maintenance needed and frequency of maintenance required for a piece of equipment.*

PM Work Order

☐ Weekly ☐ Biweekly ☐ Monthly ☐ Bimonthly

☐ Quarterly ☐ Biannual ☐ Annual

Date: _____ Name: _____

Equipment Name: _____

☐ Voltage: _____ A Phase _____ B Phase _____ C Phase _____

☐ Amperage: _____ A Phase _____ B Phase _____ C Phase _____

☐ Check and tighten all electrical connections: _____

☐ Check all fuses and overload protection: _____

☐ Inspect and lubricate bearings: _____

☐ Check belt condition and tension: _____

☐ Supply air temp: _____ Return temp: _____ Supply pressure: _____ Return pressure: _____

☐ Supply water temp: _____ Return temp: _____ Supply pressure: _____ Return pressure: _____

☐ Check and replace filters: _____ Filter reading before: _____ Filter reading after: _____

☐ Parts used: _____

☐ Comments: _____

_____

_____

_____

_____

_____

_____

Date completed: _____ Total hours: _____ Signature: _____

Figure 10-12. *Sample PM work orders include data such as frequency, date, and PM tasks appropriate for each piece of equipment.*

If it was determined as part of the base PM program start-up procedure that a balance of the entire HVAC system was necessary, it should be scheduled into the work orders at this point. If a complete system balance is necessary to establish a baseline from which to work, this should be a one-time procedure, barring any major renovations.

PM work orders are issued for work that is scheduled daily, weekly, monthly, quarterly, biannually, or annually. PM work order activities include:
- inspections for unusual conditions like excessive noise and heat
- inspections for leaks, rust, dirt, and mechanical problems
- regular and careful lubrication
- mechanical and electrical adjustments

- electrical and mechanical inspections and tests
- HVAC testing and balancing
- operational checks
- parts replacement
- coil cleaning and filter replacement

Some CMMS systems may also schedule a specific number of PM hours to individual personnel. This means that a stationary engineer may periodically work from a work order and check the general operation of all HVAC fans. The stationary engineer inspects belt tension, damper operation, filter cleanliness, and all mechanical supports such as motor mounts. When needed, stationary engineers also tighten the belts and change the filters during inspections.

Repair Work Orders. In addition to scheduled PM, unscheduled maintenance may also occur as a result of a PM inspection, which identifies repairs that cannot be completed during the inspection. In this case, a repair work order is established. **See Figure 10-13.** When repairs are completed, information about the repair is entered into the PM information system. This information should include what type of repair was done, what supplies were used, what caused the problem, and how long the job took.

IAQ Fact

The Environmental Protection Agency's document IAQ Maintenance and Housekeeping Programs *contains detailed information on the maintenance and housekeeping of HVAC systems.*

Master Schedule

Once PM work orders are complete, they should be sorted according to the time intervals that work is performed. For example, all annual, monthly, and weekly work orders that should be completed in January are grouped together. **See Figure 10-14.** When scheduling the next PM procedure, it is important to keep in mind the dates of last service, or of installation in a new facility.

Priority equipment or equipment that requires PM in a specific month should be scheduled first. The PMs ought to be spaced out so that excessive work is not created for any one month. This may mean that PM is performed on certain equipment sooner than required, but this usually does not occur after the start-up phase. CMMS systems can create these schedules automatically after a few decisions are made by the stationary engineer.

PM Repair Work Order

Department: _____   Work order number: _____

Date call received: _____   Time called: _____   Time arrived: _____

Requester: _____   Room number/location: _____

Equipment name: _____   Location: _____

Problem: _____
_____
_____

Cause and solution: _____
_____
_____
_____

Parts used: _____
_____
_____
_____

Employee name: _____   Time worked: _____

Employee name: _____   Time worked: _____

Employee name: _____   Time worked: _____

Employee name: _____   Time worked: _____

Employee name: _____   Time worked: _____

Signature: _____

Figure 10-13. *Repair work orders are issued for equipment that is not working properly.*

Work Orders

WORK ORDER PROJECTION – TASK BY WEEK

RF Industries

Week 2

Date	Task No.	WO No.	Equipment No.	Cost Center	Expense Class	Hours
7/12	BEARING-REPLACE	-	MOTOR-EXTRUD1-LN1	7001	MECH	2.0
7/12	BEARING-REPLACE	-	MOTOR-EXTRUD1-LN1	7001	MECH	2.0
Total						4.0

Week 3

Date	Task No.	WO No.	Equipment No.	Cost Center	Expense Class	Hours
7/19	BEARING-REPLACE	-	MOTOR-EXTRUD1-LN1	7001	MECH	2.0
7/19	BEARING-REPLACE	-	MOTOR-EXTRUD1-LN1	7001	MECH	2.0
7/19	EXTRD-MTR-BELT-3M	-	MOTOR-EXTRD2-LN2	7001	MECH	2.0
7/19	EXTRDSCREW-BRNG-6M	-	SCREW-EXTRD1-LN2	7001	MECH	4.0
7/19	EXTRDSCREW-BRNG-6M	-	SCREW-EXTRD1-LN2	7001	MECH	4.0
Total						14.0

Week 4

Date	Task No.	WO No.	Equipment No.	Cost Center	Expense Class	Hours
7/26	BEARING-REPLACE	-	MOTOR-EXTRUD1-LN1	7001	MECH	2.0
7/26	FKLFT-PM-1M	-	-MULTITASK-		COMB	6.0
7/26	BEARING-REPLACE	-	MOTOR-EXTRUD1-LN1	7001	MECH	2.0
7/26	DIE CLEAN	-	DIE-LN2	7001	MECH	2.0
Total						12.0

Datastream Systems, Inc.

Figure 10-14. *Work orders can be grouped together easily according to a specific time period, such as weekly, using CMMS.*

When using a manual system, personnel should use a binder (or binders) that is divided into monthly sections. The PM work orders are inserted into the proper months to create a master schedule. The master schedule is the working copy, from which PM work orders are issued each month. The master schedule can be even more detailed so that PM work orders can be issued by the week or by the day. Choose the scheduling system that best meets the demands of the facility.

PM SYSTEM OPERATION

The most valuable tools supplied by a PM system are the details about how a facility's equipment operates and information on problems experienced with each piece of equipment. This information can be used to decide how much maintenance each piece of equipment needs, what type of equipment is involved, the proper procedures to follow, etc.

All PM work orders are assigned at a specific time, usually the first of each month. Each worker is responsible for completing and returning all completed work orders to the person responsible for filing and logging them or entering them into the CMMS. For example, in a PM program, one copy of the completed PM work order would be placed in the equipment history record file and one copy would be turned in to the chief engineer. Paper systems like this are most common, but paperless systems are available as well.

Data from equipment instruments can be logged electronically in a paperless CMMS.

Paperless Systems

Paperless systems consist of a CMMS loaded with work orders that are filled in and checked off as completed. These records are then transferred directly from these on-site computer terminals to the main computer system. A CMMS permits easy cross-referencing of information. For example, a list of the PM tasks and unscheduled repairs that were performed on a particular piece of equipment can be requested quickly.

CONSTRUCTION AND RENOVATION

Today, renovating an older facility is more cost effective than constructing a new building. This often means that more repair and renovation of commercial properties is required. Along with this increase in renovations comes a need for greater awareness and control of the impact on indoor air quality.

IAQ Fact

EPA and OSHA regulations apply to new construction and renovation projects.

Safe Practices

Routine construction, repair, and renovation can produce high levels of dust and vapors that can affect nearby office employees. To maintain good indoor air quality during any of these procedures, it is essential to isolate the construction area. This does not mean enclosing workers in their own environment of dust and debris.

Isolating the area requires a method of supplying and exhausting air in the construction space. Air can be supplied to a construction, repair, or renovation space by using building system air or 100% outdoor air, depending on the HVAC system and outdoor conditions. The amount of exhaust air varies, depending upon whether the exhaust is a temporary arrangement or part of an existing system. Suppliers can be consulted in order to select green products, which will have the least impact on IAQ.

Barriers can be used to contain the work area and block off or filter return air grills to isolate the construction area. In some cases, it may be acceptable to filter return air back into the building air. However, it is extremely important that tight controls are imposed for maintaining and changing filter media when filtering return air.

There are basic control measures that should be used to control dust and vapor emissions in the following three stages: preliminary steps, duration of the work, and follow-up steps. These practices, however, do not cover regulated work such as asbestos or lead abatement.

Preliminary Steps. Maintenance and renovation work should be scheduled during off-hours whenever possible. Contractors should provide material safety data sheets or manufacturer's specifications for review prior to job authorization. When selecting or specifying products such as new furnishings, carpets, paint, caulk, and adhesives, IAQ workers should compare the emission rates and chemical contents in the manufacturer's specifications. Suppliers can be consulted in order to select green products, which will have the least impact on IAQ.

Duration of Work. For the duration of the construction or renovation work, the construction or renovation area must be isolated (physically and mechanically) from the rest of the facility. Work areas can be sealed with plastic sheeting. Work areas should be under negative pressure. It is possible to shut off or redirect HVAC. Movable air handlers may be necessary to keep area under negative pressure.

Return air grills that could lead to cross-contamination of other occupied spaces must be isolated or filtered. All vapors and dust must be exhausted to the outdoors. It is important to provide dust suppression and ensure daily cleaning of work area. Supplying doormats provides a buffer area for boot and equipment cleaning. Building materials should be stored indoors or covered to prevent wetting, which could cause future IAQ problems once installed.

Follow-Up Steps. After the completion of the project, follow-up steps must be implemented. Personnel must inspect areas for dust or odors prior to reoccupancy and/or prior to releasing contractors. If odors are noted or large quantities of solvents, paints, or adhesives were used, the HVAC system must be operated during off-hours to purge the area of volatile organic compounds (VOCs).

Sink Effect

A *sink effect* is the emission of volatile organic compounds (VOCs) from furniture and other material that have absorbed the VOCs from indoor air. In other words, VOCs and other contaminants are absorbed in carpets, ceiling tiles, partitions, and fabrics and released later, especially at high temperatures and humidity. Also, dust particles can settle to the ground and other horizontal surfaces. People walking by can release these dust particles back into the air. Therefore, smooth rather than fuzzy or textured surfaces for furniture and floors should be chosen. Smooth surfaces are much easier to clean, thus limiting the sink effect. During renovations, rugs and furnishings can be covered with plastic or removed to minimize the sink effect or dust accumulation.

CHEMICAL EMISSIONS

Modern office equipment requires advance planning to ensure good indoor air quality. Office equipment such as copy machines, fax machines, laser printers, blueprint machines, etc. increase the chemical contamination of the indoor environment.

One way to reduce emissions is to educate building occupants. For example, laser printers generate small quantities of ozone and should be turned OFF when they are not needed for extended periods of time. Personal computers operate independent of the printer and can be ON and functional with the printer OFF. Copy machines should be turned OFF during nonworking hours for the same reason.

Limiting the number of ozone-generating office machines within occupied offices is another way to reduce emissions. Placing a fax, a copier, and two laser printers within a single office may be convenient but, depending on the use, may increase the ozone buildup in that space beyond what the normal ventilation is equipped to handle. When placing multiple office machines within one office is unavoidable, testing must be conducted and ventilation rates adjusted accordingly.

In instances where multiple pieces of equipment are congregated (e.g., multiple copy machines, blueprint machines, fax machines, and laser printers), supplemental HVAC equipment may have to be installed to provide the increased exhaust and supply air. When converting office space into special-application equipment rooms that could increase emission levels, managers can add supplemental equipment to accommodate this change. If these changes are made during the design phase of the project, altering the original intent of the base building equipment can be avoided. When possible, this equipment should be located in rooms with increased outdoor airflow.

Case Study

Case Study – Using IAQ Records

Complaint History

The CEO of a large company complained to the management office regarding health symptoms he was experiencing shortly after arriving at work each day, including watery eyes, a stuffy nose, and headaches. The engineering staff responded immediately by visiting the CEO to discuss the issue and gather some preliminary information.

The CEO explained that his discomfort began after space renovations were completed. He believed that the glues or solvents used to lay the carpet may have been the cause. Another theory was that food was left behind the walls during the construction buildout.

When asked, the CEO confirmed that his symptoms continued while in his workspace but subsided when he left. He had no known allergies, and no one else in the building had any similar symptoms.

IAQ Investigation and Resolution

The chief engineer performed an inspection of the building to assess the condition of the damper, filters, and condensate pans. She also checked all IAQ recordkeeping for the last dates on ductwork cleaning, rug cleaning, and air balance reports, as well as the last recorded PM notes. None of the IAQ logs showed anything out of the ordinary or anything that would be expected to cause an IAQ issue.

Near the end of the inspection process, a plant and flower maintenance company made a visit to maintain the plants in the building. After the company watered the plants, the plant fertilizer left a slight odor in the area. Therefore, after noting the odor, the chief engineer removed all the plants and put them in a storage space on another floor. When the CEO returned to work the next day, he had no further symptoms of discomfort. The new chemical fertilizer that the plant vendor was using was the source of the IAQ problems.

Preventive Maintenance Definitions

- *Enterprise asset management (EAM)* is the whole-life, optimal management of the physical assets of an organization to maximize value.
- *Maintenance* is the repair of any sort of mechanical or electrical device that is broken, as well as the implementation of routine actions which keep the device in working order or prevent trouble from arising.
- A *PM work order* is a form or ticket that indicates specific preventive maintenance (PM) tasks to perform at regularly scheduled intervals.
- *Predictive maintenance (PdM)* is maintenance that involves evaluating the condition of equipment by performing periodic or continuous equipment monitoring.
- *Preventive maintenance (PM)* is the periodic inspection and servicing of a building and equipment for the purpose of preventing failures.
- *Proactive maintenance* is a maintenance strategy for stabilizing the reliability of equipment using specialized maintenance services.
- *Reliability-centered maintenance (RCM)* is a maintenance plan used to create a cost-effective maintenance strategy to address dominant causes of equipment failure.
- *Run-to-fail maintenance* is corrective, reactive, and unscheduled maintenance.

**For more information please refer to ATPeResources.com
or http://www.epa.gov.**

1. Name three advantages of a CMMS.

2. List three inspections included on PM work orders.

3. Describe the purpose of the master schedule.

4. Name three items that should be included in the equipment history record file.

5. Name three safe practices that should be followed during construction and renovation work in a facility.

HVAC AIR SYSTEMS

Class Objective:
Define common HVAC systems and describe how HVAC problems can contribute to IAQ issues.

On-the-Job Objective:
Identify the HVAC system in your facility. Discuss the most common sources of IAQ problems.

Learning Activities:
1. Describe the HVAC system in your facility; list ways in which it may be contaminated and contribute to poor IAQ.
2. List the types of filters used in your facility, as well as their efficiency levels.
3. Review the equipment-cleaning procedures in your facility.
4. **On-the-Job:** Examine the HVAC air-handling systems for water inside the unit. If moisture is present, determine its source.

Introduction

Heating, ventilating, and air conditioning (HVAC) systems are often the source of and the pathway for IAQ contaminants. Poor design and maintenance of HVAC systems contribute to the problem. An HVAC system must be properly operated to maintain acceptable IAQ. Preventing IAQ problems is far cheaper than investigating and fixing existing problems. In addition to the importance of implementing a preventive maintenance program, proper operation and maintenance of all facility systems is the easiest way to prevent most IAQ problems.

HVAC SYSTEMS

HVAC systems must be geared to meet both specific and varying application requirements. A commercial building usually contains public space, office space, computer space, retail space, cafeteria space, and/or restaurant space. The air distribution systems of commercial buildings usually operate from a minimum of 12 hr a day, 5 days per week, up to 24 hr per day, 7 days a week. To conserve energy, some buildings can use the main HVAC system at partial load and meet specific occupant needs through smaller supplemental systems. **See Figure 11-1.**

Despite the type of HVAC system used, the system will serve the same general purpose as all HVAC systems—to distribute conditioned air to building spaces for occupant comfort.

In the process of providing environmental comfort, air-handling units (AHUs) circulate all airborne matter, which may include internal odors, outdoor irritants, biological contaminants, gases, and toxic substances. A thorough understanding of the equipment, functions, and controls of the air-handling system in a building is imperative in order to avoid the spread of airborne contaminants.

HVAC Systems

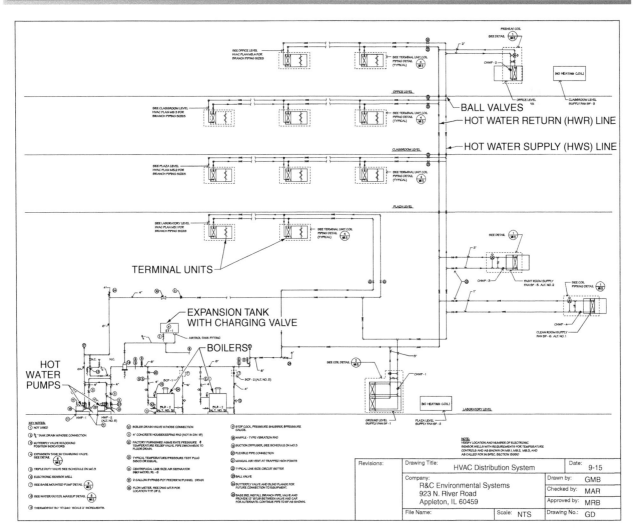

Figure 11-1. Most air distribution systems of commercial buildings operate for occupant needs 24 hr a day, 7 days a week.

HVAC System Characteristics

HVAC systems must maintain building comfort. An obvious characteristic required in an HVAC system is reliability. HVAC equipment must operate at peak performance to maintain good IAQ. Other desired characteristics include maintaining space temperature, relative humidity, and cleanliness.

Temperature. Temperatures in occupied building spaces should be as close to the design setpoints as possible. When people feel uncomfortably hot or cold, they are more likely to complain about IAQ problems. Many building leases now specify required ranges for temperature and relative humidity. **See Figure 11-2.**

RECOMMENDED BUILDING TEMPERATURES FOR RELATIVE HUMIDITY		
RELATIVE HUMIDITY*	**WINTER TEMP†**	**SUMMER TEMP†**
30	68.5–76.0	74.0–80.0
40	68.5–75.5	73.5–79.5
50	68.5–74.5	73.0–79.0
60	68.0–74.0	72.5–78.0

* in percentages
† in °F

Figure 11-2. Recommended temperature ranges differ between the summer and winter seasons.

Excessively high temperatures can increase emission rates and promote the off-gassing of contaminants found in upholstery and fabrics. Off-gassing happens during hot summer days when a building is not being properly cooled. When an HVAC system cannot maintain comfortable temperatures within building spaces, occupants can lose confidence in the system and the building's maintenance staff.

Areas within a building space may have varying demand for heating or cooling. Radiant heating may provide comfort for one occupant but not another. Cold exterior walls may cause convection currents that create drafts. By ensuring proper airflow, warm and cool air go where they are required.

Relative Humidity. Relative humidity (RH) is as critical as temperature for occupant comfort. High relative humidity (above 60%) promotes microbial growth and can promote off-gassing and volatile organic compound (VOC) emissions. Low relative humidity (generally below 30%) can cause respiratory problems and can lead to irritated mucous membranes in humans. All humidification and dehumidification equipment in an HVAC system must operate correctly. Specific spaces within a building that have high or low relative humidity must be conditioned with separate humidity control equipment, but only when the equipment does not conflict with the RH in other parts of the facility.

Cleanliness. The inside and outside of all HVAC equipment must be kept clean and dry. Cleaning minimizes dirt and dust buildup. Dust buildup can become airborne and be distributed by the HVAC system. In the presence of water, even a small collection of dirt and dust provides a breeding ground for microbial contaminants. Drip pans, humidifiers, duct, and coils must be periodically checked for microbial growth. **See Figure 11-3.**

When using cleaning substances and equipment, personnel must follow all manufacturer instructions. Cleaning and other maintenance work must be scheduled when the building or specific space is not occupied. A stationary engineer's job is to control the introduction of contaminants into the air and minimize building occupants' exposure. When a large cleaning operation is performed that results in fumes or objectionable odors, purging the space or the building is a way to remove the contaminants.

IAQ Fact

Environmentally Preferable Purchasing (EPP) *is an on-line resource tool of the EPA that is utilized when developing a cleaning program as part of an effort to improve overall IAQ. EPP emphasizes four product phases: manufacturing, transport and packaging, use, and disposal. Each is an important factor to consider when making cleaning product-purchasing decisions.*

Breeding Ground for Microbial Contaminants

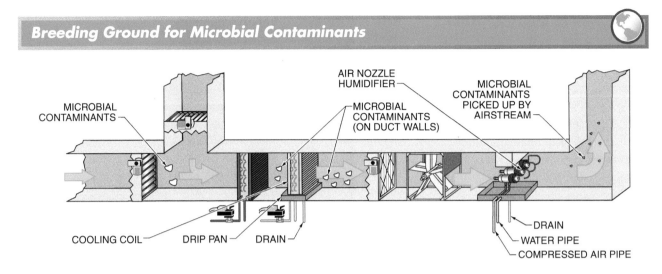

Figure 11-3. When water is present, even a small collection of dirt and dust provides a breeding ground for microbial contaminants in HVAC equipment such as drip pans, humidifiers, ducts, and coils.

COMMON HVAC SYSTEM DESIGNS

HVAC systems are designed to condition and deliver air to building spaces. HVAC systems use automatic controls. These controls can be pneumatic, electric, or electronic. HVAC systems and sequences are similar within buildings, regardless of the type or brand of control system. The three general types of HVAC systems are (all-air) systems, all-water systems, and combined air-water systems. Most HVAC systems include outdoor, mixed, and exhaust-air dampers.

Outside-Air Systems

One way to categorize air-handling systems is to describe the system as either a 100% outside-air (OA) system or as a mixed-air system. A *100% outside-air (OA) system* is a system that does not recirculate any return air from building spaces. Instead, 100% outside-air systems use outdoor air from outside the building for heating and cooling purposes. **See Figure 11-4.** Using 100% outdoor air is a way to prevent any contaminants already in the indoor air from being recirculated back into building spaces. Places where 100% outside-air systems may be used include cafeterias, laboratories, operating rooms in hospitals, and others.

With few exceptions, 100% outside-air systems do not suffer as much from IAQ problems because they continuously use large amounts of outdoor air. The downside of 100% outside-air systems is that there is a heavy energy penalty to pay due to the need to heat, cool, dehumidify, and filter the air at possible extreme conditions. Usually, 100% outside-air systems are required to have both preheat and precool coils (units) that condition the incoming air from extreme conditions.

Mixed-Air Systems

A *mixed-air system* is a unit or section of an HVAC system that brings a specified amount of return air into contact with the outdoor air entering the HVAC system for space use. **See Figure 11-5.** Mixed-air systems assume that the return air is relatively free of contaminants. The amount of outdoor air mixed with return air is determined by the control system. Outside-air dampers and return-air dampers are positioned to obtain the proper amount of mixed airflow. A minimum open setting for the outside-air damper is used to ensure that a specified amount of outdoor air is introduced into the system for ventilation purposes. Some mixed-air systems can also be used as economizers.

Outside-Air Systems

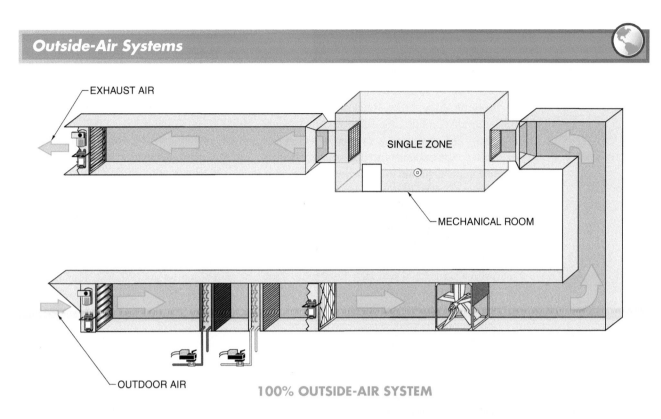

Figure 11-4. *HVAC systems that use 100% outdoor air do not use any return air because the return air may contain contaminants.*

Mixed-Air Systems

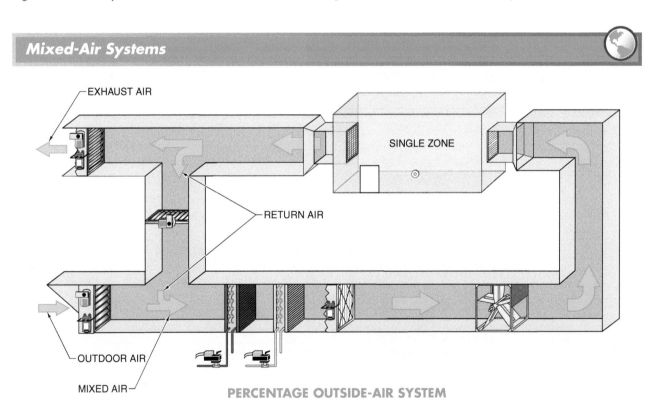

Figure 11-5. *Mixed-air systems allow a percentage of outdoor air to combine with return air for space use.*

Economizer Systems

An *air-handler economizer* is an HVAC unit that uses outdoor air for free cooling. The outdoor air is used instead of mechanical cooling from chilled water or direct expansion air conditioning coils. A *direct expansion (DX) system* is a system in which the refrigerant expands directly inside a coil in the main airstream itself to affect the cooling of the air. Buildings with high internal heat gain require cooling even when it is cold outdoors. An economizer system is a control arrangement that allows cool outdoor air to be used to provide free cooling for buildings with high internal heat gain. For economizer systems to operate effectively, special control schemes must be used that allow economizer operation only when outdoor air is at a proper temperature. Building-economizer systems fall into two groups: dry bulb economizers and enthalpy economizers.

Dry Bulb Economizers. A *dry bulb economizer* is a type of economizer that operates strictly in proportion to the outdoor-air temperature, with no reference to humidity values. As long as the outdoor-air temperature is below a specific value, the air can be used for economizer cooling. **See Figure 11-6.**

Usually, the temperature of the outdoor air must be below 65°F. The exact setting is dependent upon the prevailing climate in specific geographical regions. When the outdoor-air temperature rises and is too warm for free economizer cooling, the economizer cycle is ended. Ending the economizer cycle forces the outside-air damper to a minimum value, which allows enough outside airflow for ventilation purposes only.

Enthalpy Economizers. An *enthalpy economizer* is a type of economizer that uses temperature and humidity levels of the outdoor air to control operation. *Enthalpy* is the total heat content of a substance. For enthalpy control to work, both temperature and humidity sensors are mounted in the outdoor air. However, due to the problems associated with humidity sensors in the outdoor environment, humidity readings are usually taken in the mixed-air airstream using new algorithms for the controller. **See Figure 11-7.** A controller uses the temperature and humidity values to calculate the current enthalpy of the outdoor air in British thermal units per pound (Btu/lb). When the enthalpy of the outdoor air is low enough, the economizer cycle of the HVAC system is used.

Dry Bulb Economizer

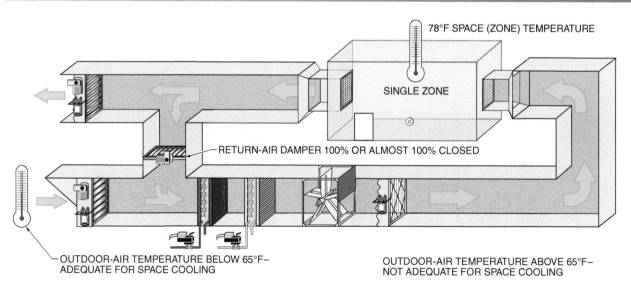

Figure 11-6. A dry bulb economizer is an economizer that operates in proportion to the outdoor-air temperature, with no reference to the humidity values of the air.

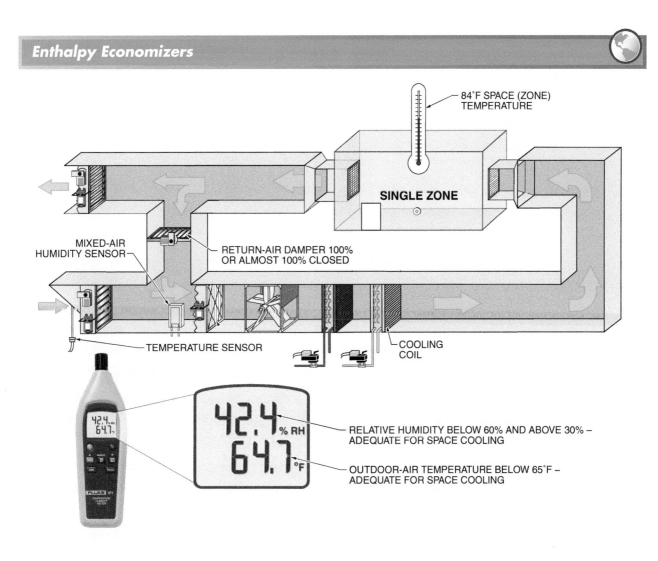

84°F SPACE (ZONE) TEMPERATURE

SINGLE ZONE

MIXED-AIR HUMIDITY SENSOR

RETURN-AIR DAMPER 100% OR ALMOST 100% CLOSED

TEMPERATURE SENSOR

COOLING COIL

42.4 % RH

64.7 °F

RELATIVE HUMIDITY BELOW 60% AND ABOVE 30% – ADEQUATE FOR SPACE COOLING

OUTDOOR-AIR TEMPERATURE BELOW 65°F – ADEQUATE FOR SPACE COOLING

Figure 11-7. Enthalpy economizers use temperature and humidity levels of the outdoor air to control the operation of the HVAC system.

Enthalpy economizer systems allow for variations in the humidity of the air as well as the temperature. In general, enthalpy-controlled systems allow air to be used under more conditions, allowing increased use of the economizer cycle, which promotes energy efficiency. When the outdoor-air enthalpy value is above the setpoint, the economizer cycle is stopped and the outside-air damper is forced to a minimum open value for ventilation purposes only.

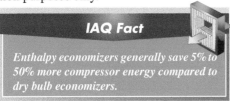

IAQ Fact

Enthalpy economizers generally save 5% to 50% more compressor energy compared to dry bulb economizers.

AIR-HANDLING UNIT COMPONENTS

An *air-handling unit* is the part of an HVAC system that operates to distribute air throughout a building. An air-handling unit consists of several sections including a fan (blower), filter racks, a plenum, heating coils and/or cooling coils, and damper sections. Air handlers are found as small units (terminal units) that usually include a filter, coil, and blower, or large units (makeup air units) that condition 100% of outdoor air for use. **See Figure 11-8.** Air-handling units designed for outdoor use are usually installed on roofs and are therefore known as rooftop units (RTUs).

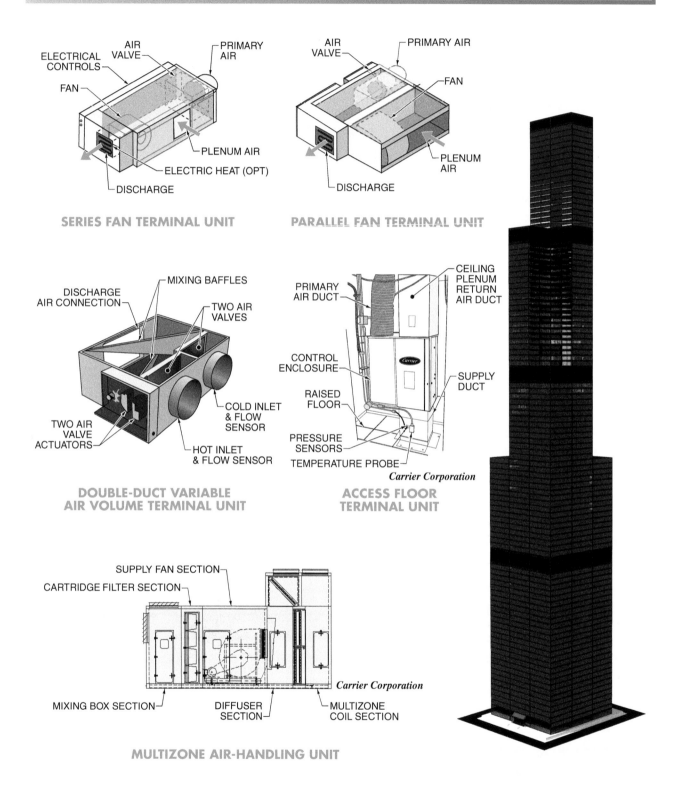

SERIES FAN TERMINAL UNIT

PARALLEL FAN TERMINAL UNIT

DOUBLE-DUCT VARIABLE AIR VOLUME TERMINAL UNIT

Carrier Corporation

ACCESS FLOOR TERMINAL UNIT

Carrier Corporation

MULTIZONE AIR-HANDLING UNIT

Figure 11-8. *Air-handling units are found in many designs, but small units are usually known as terminal units, and large units are usually known as makeup air units.*

Fans

A *fan* is a mechanical device that is part of an air-handling unit, which forces conditioned air throughout a building. Fans do not discriminate as to the quality of the air being distributed. Therefore, other preventive measures are used to maintain the quality of indoor air in occupied spaces. When conditioned air is moving, fans provide the necessary energy (flow) and pressure rating to overcome the resistances to flow found in HVAC systems. Resistances to flow come from terminal equipment, heating and cooling coils, louvers, grills, diffusers, and ductwork.

Fans impose a negative pressure in the duct on the suction side of the fan and a positive pressure in the duct on the outlet side of the fan. All air-handling unit sections must be tightly sealed to prevent air from leaking between sections and in or out of the unit.

Fans that are designed to operate at a constant speed are usually belt driven because the standard motor speed (with pulley ratios) is enough to overcome system static pressures (resistances to flow). Motors and belts that are designed to be inside the duct or fan housing can emit odors or belt material into the airstream. The bearings, shaft, and sheave wear out. Also, excessive operating temperatures can generate airborne contaminants as well. **See Figure 11-9.**

During the cooling season, a fan delivers cold saturated air. The saturated air can combine with airborne dirt and collect on dampers, fan blades, and heating and cooling coils. The corrosion of metal parts can also pollute the airstream. A fan can also be the source of and/or the transmitter of noise pollution to occupied spaces. Acoustical duct lining (usually fiberglass) can become saturated with moisture, causing mold or mildew to form. The mold and mildew can become dislodged and travel with the airstream.

Most spaces in a building are maintained at a slightly positive pressure to prevent the infiltration of air from outside the building from entering occupied conditioned areas. Some spaces, such as restrooms, are maintained under a negative pressure to ensure that contaminants are exhausted. The negative pressure is maintained by means of localized exhaust systems.

Maintenance personnel must check that exhaust fans are not found in building spaces that house gas-fired appliances, such as boilers or water heaters. Exhaust fans could place a boiler room, or the room where a water heater is located, under negative pressure, resulting in drawing flue gas backward, down the flue, and into the room. Negative room pressure has the obvious effect of interfering with the proper gas flue draw and spilling poisonous carbon monoxide into the room. When an exhaust fan is suspected of causing a reverse flow of flue gas, or another exhaust fan in the building is making its negative-pressure effect felt in occupied spaces by causing flue gas to spill into unwanted spaces, management must be consulted immediately. The reverse flow of flue gases is a very dangerous condition.

Fan-Generated Odors and Contaminants

COUPLING WEAR (METAL AND FIBER PARTICLES)

MOTOR HEATS UP (PAINT, LUBRICANT, AND RUBBER ODORS WITH SMOKE FROM WIRE INSULATION)

FAN BEARINGS WEAR (LUBRICANT ODOR WITH LUBRICANT, PLASTIC, AND METAL PARTICLES) INCLUDING VIBRATION SOUNDS

DIRECT DRIVEN AXIAL FAN

SHEAVE WEAR (RUBBER ODOR WITH METAL PARTICLES INCLUDING VIBRATION SOUNDS)

BELT WEAR (RUBBER ODOR WITH RUBBER AND METAL PARTICLES INCLUDING VIBRATION SOUNDS)

FAN BEARINGS WEAR (LUBRICANT ODOR WITH LUBRICANT, PLASTIC, AND METAL PARTICLES INCLUDING VIBRATION SOUNDS)

BELT DRIVEN AXIAL FAN

Figure 11-9. Fans provide the mechanical energy to move air through an HVAC system and to building spaces.

Mixing Plenum and Dampers

The American National Standards Institute (ANSI) and the American Society of Heating, Refrigerating and Air-Conditioning Engineers (ASHRAE) have set ventilation standards as a guide for maintaining proper indoor air quality. The idea behind the standards is to dilute any contaminants by introducing the proper amount of outdoor air into a space. Outdoor air can never be assumed to be free of pollutants because outdoor air may contain vehicle exhaust fumes, boiler or building exhaust gases, acid-laden air, construction dust, trash compactor odors, and a host of other contaminants and irritants. **See Figure 11-10.** It is important to perform air-sample testing of the outside-air intake grills of the system before evaluating the potential problems with the system itself.

According to ANSI/ASHRAE standards, when outdoor air is relatively free of contaminants, providing proper ventilation air is a function of the mixing box (plenum) and its interrelated dampers. Because of the close proximity of the return-air damper to the outside-air damper, it is possible that the return air is short cycling into the outside-air damper, defeating the purpose for which it is intended. A smoke test will verify if this is the case.

The function of outdoor-air (OA), return-air (RA), and exhaust-air (spill) dampers is an interrelated operation requiring periodic inspection, maintenance, and testing for proper operation. **See Figure 11-11.** A partial list of checks that should be performed includes the following:

- Remove the debris restricting airflow.
- Repair any damage to bird and insect screens.
- Clean the damper blades.
- Replace the deteriorated damper blade seals (gaskets).
- Clean and lubricate all moving parts.
- Observe the damper operation through a complete cycle.
- Verify the normally open and normally closed dampers.
- Check for air leakage through the damper when the damper is closed.
- Verify the operation of all controls, operators, and linkages.
- Verify the operation of the gravity-relief damper, if installed.

The mixing box (plenum) must properly blend the correct percentage of outdoor air with return air before the air moves to the next section of the air-handling unit. It is not unusual for a poor mix to occur, especially when parallel-type dampers are installed. Parallel blade dampers may need to be replaced with opposing blade dampers for better control of airflow.

However, instead of replacing the dampers, it may be sufficient to add a baffle screen to the mixing box section to aid in mixing the air. The proper percentage of blended outdoor air and return air can be verified

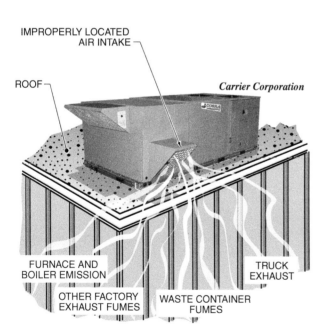

Outdoor-Air Pollutants

IMPROPERLY LOCATED
AIR INTAKE

ROOF

Carrier Corporation

FURNACE AND
BOILER EMISSION

OTHER FACTORY
EXHAUST FUMES

WASTE CONTAINER
FUMES

TRUCK
EXHAUST

Figure 11-10. Due to the amount of contaminants that can be present in inside air, the American National Standards Institute (ANSI) and the American Society of Heating, Refrigerating and Air-Conditioning Engineers (ASHRAE) have set ventilation standards as a guide for maintaining proper indoor air quality.

by taking direct temperature readings. The mixed-air temperature on the downstream side of the filter section, which aids in the mixing process, should also be taken. The percentage of outdoor air for each season can be determined by using the following formulas for summer and winter:

Summer Formula

$$A_S = \frac{T_M - T_R}{T_O - T_R} \times 100$$

Winter Formula

$$A_W = \frac{T_R - T_M}{T_R - T_O} \times 100$$

where:

A_S = summer outdoor air (in %)

T_M = average mixed-air temperature (in °F)

T_R = return-air temperature (in °F)

T_O = outdoor-air temperature (in °F)

100 = constant

A_W = winter outdoor air (in %)

IAQ Fact

While operating a damper through its full cycle, it is important to check that the blades open and close properly, and to check for loose linkages.

Checking Air-Damper Operation

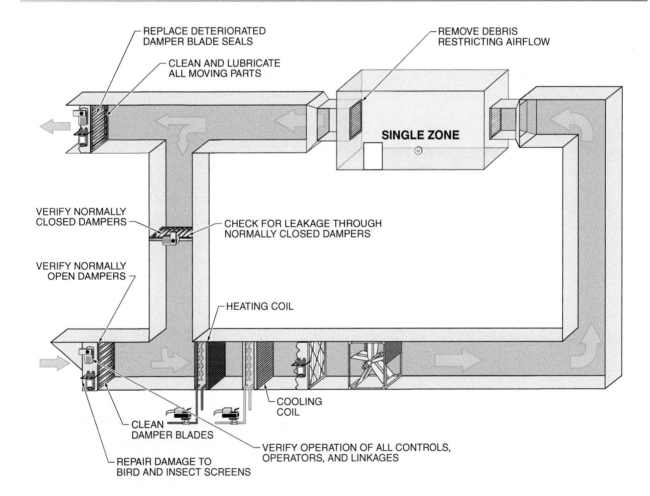

Figure 11-11. *Proper operation of all HVAC system dampers must be periodically checked to ensure proper indoor air quality and energy efficiency.*

Example: Summer Operation

A building operating in the summer has a mixed-air temperature of 85°F, a return-air temperature of 75°F, and an outdoor-air temperature of 100°F. What is the percentage of summer outdoor air being distributed to building spaces? **See Figure 11-12.**

$$A_S = \frac{T_M - T_R}{T_O - T_R} \times 100$$

$$A_S = \frac{85 - 75}{100 - 75} \times 100$$

$$A_S = \frac{10}{25} \times 100$$

$$A_S = \mathbf{40\%}$$

Example: Winter Operation

A building operating in the summer has a mixed-air temperature of 65°F, a return-air temperature of 75°F, and an outdoor-air temperature of 30°F. What is the percentage of winter outdoor air being distributed to building spaces?

$$A_W = \frac{T_R - T_M}{T_R - T_O} \times 100$$

$$A_W = \frac{75 - 65}{75 - 30} \times 100$$

$$A_W = \frac{10}{45 \times 100}$$

$$A_W = \mathbf{22\%}$$

In many cases, it is desirable to determine the percentage of outdoor air from the percentage of carbon dioxide present. This can be accomplished by using the following formula:

$$O_A = \frac{C_R - C_S}{C_R - C_O} \times 100$$

where

O_A = percentage of outdoor air (in %)

C_R = amount of CO_2 in return air (in ppm)

C_S = amount of CO_2 in supply air or mixed air (in ppm)

C_O = amount of CO_2 in outdoor air (in ppm)

Example: CO₂ Operation

A building operating in the summer has a mixed-air amount of CO_2 at 845 ppm, a return-air amount of CO_2 at 1100 ppm, and an outdoor-air amount of CO_2 at 530 ppm. What is the percentage of outdoor air being distributed to building spaces? **See Figure 11-13.**

$$O_A = \frac{C_R - C_S}{C_R - C_O} \times 100$$

$$O_A = \frac{1100 - 845}{1100 - 530} \times 100$$

$$O_A = \frac{255}{570} \times 100$$

$$O_A = 0.447 \times 100$$

$$O_A = \mathbf{44.7\%}$$

AIR-CLEANING METHODS

The three basic strategies used to reduce pollutant concentrations in indoor air are source control, ventilation, and air cleaning. *Source control* is a method of reducing air pollution by identifying strategies to reduce the origin of pollutants. Reducing pollutant origins may involve the use of exhaust systems at specific points where pollutants are present. The source control air-cleaning method requires the correct handling and storage of all chemicals and other pollutants found in a building. *Ventilation* is a method of reducing air pollution by introducing various amounts of outdoor air in order to dilute indoor pollutants. The third method is air cleaning. *Air cleaning* is a method of reducing air pollution that includes electronic air cleaning, ultraviolet (UV) air cleaning, and the use of filters, which includes carbon filters.

IAQ Fact

HEPA filters are high-efficiency particulate air filters used to remove 99.97% of airborne particles that are 0.3 micrometers (μm) in diameter. Due to design, most 0.1 μm sized particles and smaller are trapped.

Outdoor Air Percentage-Temperature Applications

75°F RETURN-
AIR TEMPERATURE

SINGLE ZONE

85°F MIXED-
AIR TEMPERATURE

100°F OUTSIDE-
AIR TEMPERATURE

SUMMER

$$A_S = \frac{T_M - T_R}{T_O - T_R} \times 100$$

$$A_S = \frac{85 - 75}{100 - 75} \times 100$$

$$A_S = \frac{10}{25} \times 100$$

A_s = 40% OF OUTDOOR AIR

75°F RETURN-AIR TEMPERATURE

SINGLE ZONE

65°F MIXED-AIR TEMPERATURE

30°F OUTSIDE-
AIR TEMPERATURE

WINTER

$$A_W = \frac{T_M - T_R}{T_O - T_R} \times 100$$

$$A_W = \frac{75 - 65}{75 - 30} \times 100$$

$$A_W = \frac{10}{45} \times 100$$

A_w = 22% OF OUTDOOR AIR

Figure 11-12. The formula used to determine the percentage of outdoor air being used by air temperature varies between summer and winter.

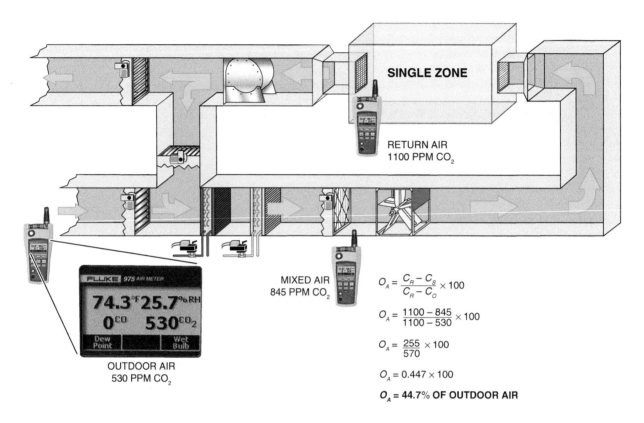

SINGLE ZONE

RETURN AIR
1100 PPM CO_2

FLUKE *975* AIR METER

74.3°F 25.7 %RH
0 CO   530 CO_2

Dew Point   Wet Bulb

OUTDOOR AIR
530 PPM CO_2

MIXED AIR
845 PPM CO_2

$$O_A = \frac{C_R - C_S}{C_R - C_O} \times 100$$

$$O_A = \frac{1100 - 845}{1100 - 530} \times 100$$

$$O_A = \frac{255}{570} \times 100$$

$$O_A = 0.447 \times 100$$

O_A = **44.7% OF OUTDOOR AIR**

Figure 11-13. When the temperature of the outdoor air and indoor air are too close to truly determine, measuring CO_2 becomes the preferred method for determining the percentage of outdoor air being introduced into a system.

Electronic Air Cleaners

An *electronic air cleaner* is a type of air cleaner in which particles in the air are positively charged in an ionizing section and then the air moves to a negatively charged collecting area (i.e., plates) to separate out the particles. As the electrostatically charged particles pass through the negatively charged collecting area, the positively charged dirt particles are attracted to the negatively charged collector plates. **See Figure 11-14.** Depending on the system, the particles clinging to the collector plates are washed off with water or blown off with compressed air.

The efficiency of electronic air cleaners is determined by the velocity and volume of air that passes through the plates and the size of the dirt particles collected. Electronic air cleaners can only remove dust particles effectively at certain uniform air velocities. Baffles or low-efficiency prefilters are used to ensure a uniform flow of air through electronic air cleaners.

Prefilters have the added benefit of eliminating larger particles that otherwise might not be captured and that contribute to plate arcing. Larger particles do not always have enough time to be properly charged and collected by the collector plates. Likewise, smaller particles do not have enough surface area to cling to the plates in the normal operation of an electronic air cleaner.

Proper care is also a major factor in electronic air-cleaner efficiency. It is important to schedule unit inspections and equipment maintenance as part of the PM program. The manufacturer's specifications usually serve as a good baseline. Visual inspections can be used to adjust the interval of service as needed.

Electronic Air Cleaners

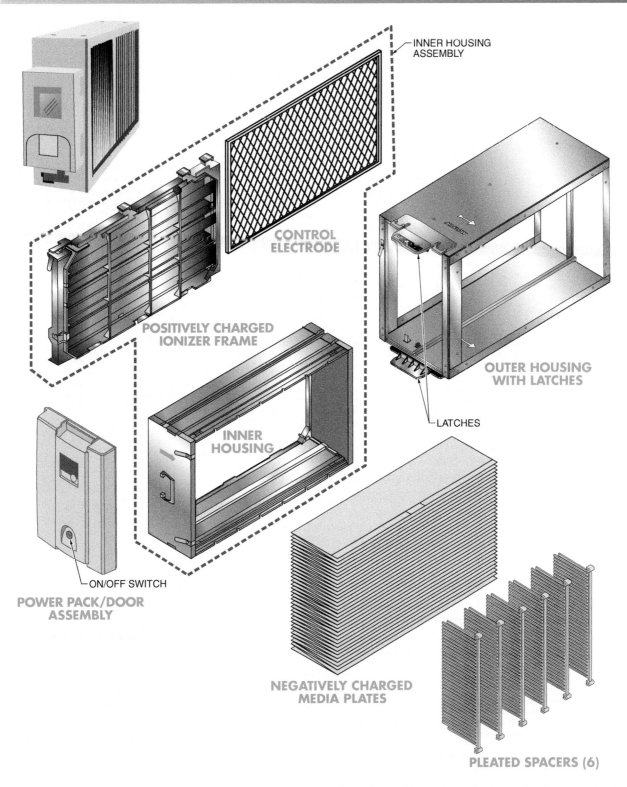

INNER HOUSING ASSEMBLY

CONTROL ELECTRODE

POSITIVELY CHARGED IONIZER FRAME

OUTER HOUSING WITH LATCHES

LATCHES

INNER HOUSING

ON/OFF SWITCH

POWER PACK/DOOR ASSEMBLY

NEGATIVELY CHARGED MEDIA PLATES

PLEATED SPACERS (6)

Figure 11-14. *Electronic air cleaners effectively separate out dirt from the air for small air volumes, but large air volumes require that multiple units be used, which consumes large amounts of energy.*

When cleaning electronic air cleaners, water or compressed air are used to clean all of the plate area. When not properly cleaned, dirt buildup creates a path between the negative and positive elements, which causes arcing. Arcing lowers efficiency and can take some units off-line.

The proper relative humidity should be maintained since excessive moisture can also cause arcing. As with any electronic equipment, it is important to use extreme caution when inspecting or maintaining electronic air cleaners.

Volatile organic compounds (VOCs) and any nonparticulate pollutants are not removed by this method. Electronic air cleaners are used widely in residential systems but not often in large commercial systems due to their energy consumption and the difficulty of constructing plates large enough for the much higher air volumes.

Ultraviolet Air-Cleaners

An *ultraviolet (UV) air-cleaner system* is an air-cleaning system that kills biological contaminants using a specific light wavelength. Ultraviolet systems utilize UV light-generating equipment. Nonbiological pollutants such as particulates and VOCs are not affected by UV systems. Again, ultraviolet systems are better suited for smaller systems or in spot use, such as in a condensate pan or condenser water system where biological growth commonly occurs. **See Figure 11-15.**

Carbon Filters

A *carbon filter* is an air-cleaning method where an activated carbon-filtering medium is used to remove most odors, gases, smoke, and smog from the air by means of an adsorption process. **See Figure 11-16.** Using carbon filters, however, usually results in modifications to the filter section, higher filter costs, increased maintenance time, high-pressure drops, and increased energy costs. When an indoor-air source is the problem, the contaminants must be removed at the source using localized exhaust. By using exhaust fans, the problem

can be solved without having to redesign the air-handling system.

Solving IAQ problems may require a combination of techniques. For example, when construction is taking place on a particular floor, localized exhaust prevents the contaminants from entering the main air distribution system and spreading to other floors.

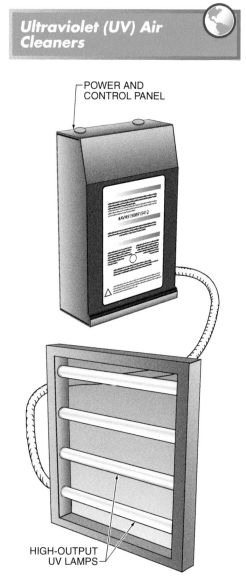

Ultraviolet (UV) Air Cleaners

POWER AND CONTROL PANEL

#AVH51508915412

HIGH-OUTPUT UV LAMPS

Figure 11-15. Ultraviolet (UV) air-cleaner systems kill biological contaminants using a specific light wavelength (approximately 450 nanometers) not visible to the human eye.

Carbon Filters

WHITE SYNTHETIC
FIBER FIRST STAGE

BLACK CARBON
IMPREGNATED FIBER
SECOND STAGE

Figure 11-16. A carbon filter uses an activated carbon-filtering medium to remove most odors, gases, smoke, and smog from the air by means of an adsorption process.

Filters

Many types of filters are used in commercial building applications. A *filter* is an air-cleaning method in which a mechanical device is used to remove particles from the air.

Mechanical Filters. Mechanical filters contain a fiber that has two purposes. The first purpose is to trap any particulate matter, while the second purpose is for the particulates to remain attached to the fibers of the filter medium. The filter medium traps particulate matter in several different ways. The fibers may be arranged in layers, or interwoven using alternating thicknesses of fiber. Sometimes adhesives are applied during the manufacture of the filter or applied as a spray product at the time of installation. **See Figure 11-17.**

Proper care of mechanical filters is important in preventing potential IAQ problems. Guidelines in caring for filters include the following:

- Ensure that filters meet the specifications of the air-handler manufacturer for size and thickness. A filter that is too thick impairs the functionality of the mechanical equipment. A filter that is too thin for the application will not stop enough particles to effectively clean the air.
- Include filter inspections and maintenance in the PM program.
- Change filters according to the manufacturer's specifications. Always replace filters immediately when damaged in any way.
- Keep filters dry by controlling water contamination and relative humidity.
- Filters should be installed so that they fit snugly with no leaks around the edges. The air-handler seams must be properly sealed. Most filters have arrows stamped on them that indicate how the filter must be placed in relation to the direction of airflow.
- Use caution when handling dirty filters. Wear gloves and a respirator to protect against particles released from the filters. Try not to disturb the dirt when moving the filters. Dispose of the filters as prescribed by maintenance procedures or local regulations.

Mechanical Filters

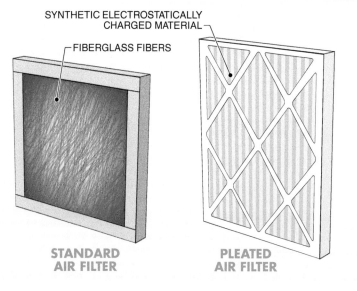

SYNTHETIC ELECTROSTATICALLY
CHARGED MATERIAL

FIBERGLASS FIBERS

STANDARD
AIR FILTER

PLEATED
AIR FILTER

Figure 11-17. In order to not contaminate building spaces, filters must be inspected periodically, of the proper thickness, stored and used properly, correctly installed, and handled correctly when replacing.

Various grades of filters are available. Equipment manufacturer specifications must be checked for the recommended size, thickness, and flammability classification. Filters are classified as class 1 (do not burn with negligible amounts of smoke) and class 2 (burn moderately with some smoke). Some systems require a low flammability classification.

Changing to a thicker filter results in higher energy consumption. Stationary engineers must consult with the equipment manufacturer before changing the thickness or type of filter being used to ensure that the new filter will perform in the equipment as needed.

Antimicrobial filters are appropriate in some cases where outdoor mold and mildew are a problem. When indoor mold and mildew are a problem, filtration is not the answer. The contamination must be identified and eliminated.

When the outdoor-air or indoor-air source is of good quality and the filters are filtering air properly (e.g., there is no bypassing of the filter medium), then the filtration system does not require modification. However, when the mixed-air source contains dust, soot, and/or other particulate matter, a high-efficiency filter may be required.

Different filter designs are used in HVAC systems and equipment simultaneously, such as pleated mechanical filters, bag filters, and cartridge filters.

There are some drawbacks associated with replacing a filter with a high-efficiency filter. A high-efficiency filter may not fit into the filter section, the increased pressure drop caused by the high-efficiency filter may reduce airflow, and the cost may be higher than the building owner wishes to spend. When any of these problems prevent use of a high-efficiency filter, more frequent filter changes may be the only solution.

Filter Selection and Ratings. The quality of filtered air can directly affect IAQ, depending on the type and efficiency of the filtering medium. Filters are rated for efficiency, airflow resistance, and dust-holding capacity. Filter efficiency (i.e., MERV rating) indicates the ability of a filter to remove particulate matter from air. **See Figure 11-18.**

Filter airflow resistance is the pressure drop across a filter at a given velocity. The *dust-holding capacity* is a filter's ability to hold dust without seriously reducing the filter's efficiency. The type and design of a filter determines its efficiency at removing particles of a given size and the amount of energy required to pull or push air through the filter. Filter media can be dry, viscous (sticky), cleanable, activated carbon, high-efficiency, or electronic.

Filters must be selected for their ability to protect both the HVAC system components and general air quality. In many buildings, the best choice is a medium-efficiency, pleated filter with a prefilter. Medium-efficiency, pleated filters have a higher level of removal efficiency than low-efficiency filters. However, medium-efficiency filters last without clogging for longer periods of time than high-efficiency filters.

Filter Media Rating. *Dust spot efficiency* is a filter rating that measures a filter's ability to remove large particles from the air that tend to soil building interiors. Dust-holding capacity is a measure of the total amount of dust that a filter is able to hold during a dust-loading test. In the past, filters were rated by the method known as ASHRAE dust spot rating.

FILTER (MERV) RATINGS

MERV	ASHRAE 52.2 3–10*	ASHRAE 52.2 1–3*	ASHRAE 52.2 0.3–1*	Arrestance*	Dust Spot*	ASHRAE 52.1 Dust Spot‡
1	<20	—	—	<65	<20	>10
2	<20	—	—	65–70	<20	>10
3	<20	—	—	70–75	<20	>10
4	<20	—	—	>75	<20	>10
5	20–35	—	—	80–85	<20	3.0–10
6	35–50	—	—	>90	<20	3.0–10
7	50–70	—	—	>90	20–25	3.0–10
8	>70	—	—	>95	25–30	3.0–10
9	>85	<50	—	>95	40–45	1.0–3.0
10	>85	50–65	—	>95	50v55	1.0–3.0
11	>85	65–80	—	>98	60–65	1.0–3.0
12	>90	>80	—	>98	70–75	1.0–3.0
13	>90	>90	<75	>98	80–90	0.3–1.0
14	>90	>90	75–85	>98	90–95	0.3–1.0
15	>90	>90	85–95	>98	>95	0.3–1.0
16	>95	>95	>95	>98	>95	0.3–1.0
17†	>99	>99	>99	—	>99	0.3–1.0
18†	>99	>99	>99	—	>99	0.3–1.0
19†	>99	>99	>99	—	>99	0.3–1.0
20	>99	>99	>99	—	>99	0.3–1.0

MERV	Application
1–4	Minimum residential, minimum commercial, equipment protection
5–8	Commercial, industrial, superior residential
9–12	Superior commercial, superior industrial, best residential
13–16	Medical facilities, surgical rooms, best commercial, smoke protection
17–20	Clean rooms, hazardous materials

* in microns in %
† future use
‡ in microns

Figure 11-18. *Filters are rated for efficiency (MERV rating), airflow resistance, and dust-holding capacity.*

Filter Replacement. Filters are usually scheduled to be replaced when the filter has performed a specific amount of work (i.e., time in use). To determine whether a filter must be replaced, an electronic manometer is used to measure the increase in pressure drop across the filter and/or a visual inspection is performed to assess the amount of dirt clogging the filter. **See Figure 11-19.**

Periodic visual inspections help to establish a baseline PM program so that filters are changed at preset intervals. The intervals can be set up by month, quarter, or total hours of operation. The interval serves as a way to approximate how often filters should be changed. The intervals can be lengthened or shortened as required to conform to varying conditions of operation and outdoor air quality.

For example, a PM program may call for a filter to be changed when a manometer indicates that the pressure drop across the filter reads 0.9 inches of water column (in. WC). Other pressure drop values are used based on the filter design, airflow resistance, and dust-holding capacity of the filter. This may make economic sense, but other factors must be considered. In some areas, excessive pollen accumulation in the spring season will require that the filter be replaced more frequently. Other conditions may require a filter to be changed outside of the regular time interval. For instance, water from a nearby humidifier might accidentally dampen the filter. Adjustments to the filter time interval will help prevent IAQ problems such as excessive particles in the air, reduced airflow to occupants, or microbial growth from molds or mildew.

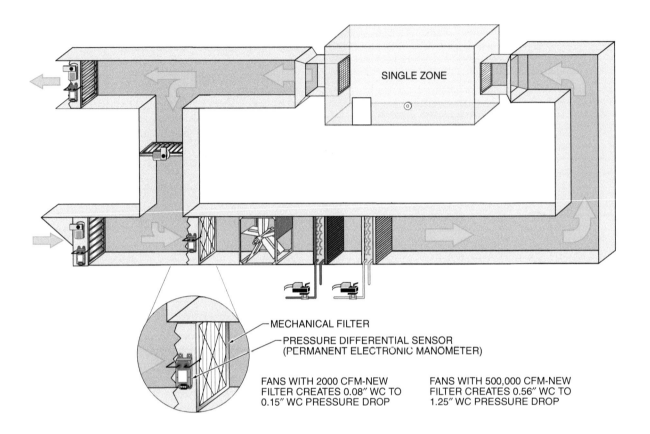

Figure 11-19. The pressure drop across a filter is checked by an electronic manometer to determine if a filter is clogged, which would result in filter replacement.

VENTILATION AIR

Stationary engineers must ensure that a correct percentage of outdoor air is brought into an HVAC system for occupied spaces. Variable-air-volume (VAV) boxes, economizers, and outdoor-air dampers must not close 100% or to a percentage that allows outdoor airflow to fall below minimum standards for ventilation.

The system balance should be checked against specifications. Diffusers and return grills must not become blocked or obstructed. Room partitions should not limit air circulation inside a space. Room partitions must not extend above a height that would interfere with the proper circulation of conditioned air in a space.

HVAC System Ductwork

An air-handling unit must be connected to a good distribution system (ducts) in order to perform properly. Space limitations have been known to compromise the design of air distribution ducts. Control devices, such as main and branch volume dampers or blowers controlled by variable frequency drives, help distribute the air according to the design engineer's load calculations. The proper sizing, location, and aerodynamic design of supply diffusers help to distribute air evenly throughout spaces. The proper location of return-air grills ensures a sweeping effect of the air, negating short cycling of the conditioned supply air. **See Figure 11-20.**

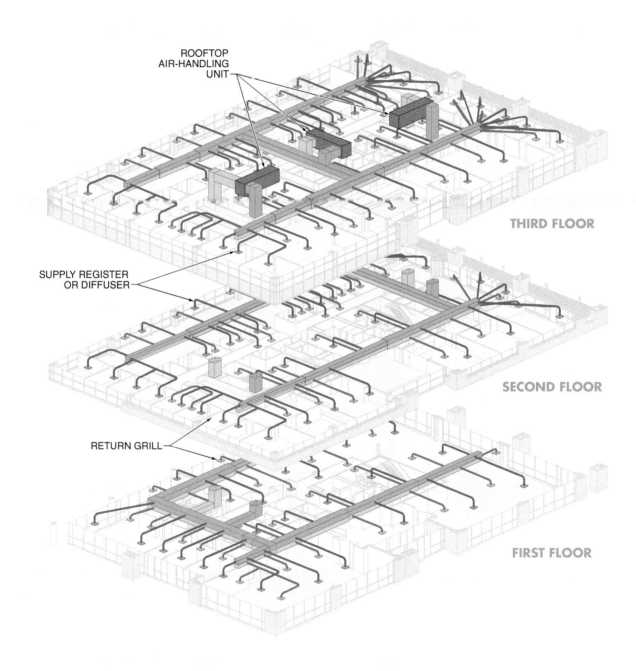

ROOFTOP
AIR-HANDLING
UNIT

THIRD FLOOR

SUPPLY REGISTER
OR DIFFUSER

SECOND FLOOR

RETURN GRILL

FIRST FLOOR

Figure 11-20. The proper sizing of ducts and the location and design of diffusers, registers, and grills affect the amount of air movement in building spaces.

Duct systems can leak conditioned air to unwanted areas, such as ceiling plenums, wall cavities, and crawl spaces. Leaks not only rob a distribution system of expensive conditioned air, but also deprive occupants of the precise amount and quality of air intended for their comfort. An air-balancing report provides an overview of the intended supply air and the resulting distributed air. The difference between the intended supply air and resulting distributed air is the air lost in the system to unintended areas. In some buildings, finding system leaks can be a difficult if not an impossible job.

Comfort has a direct effect on how occupants perceive the quality of air being supplied to their spaces. Temperature and humidity are key factors to comfort. However, the velocity of the air moving through the space is important in comfort as well. ASHRAE recommends room air movement to be around 50 fpm. Too high a velocity can cause drafts, but too low a velocity can be perceived as an IAQ problem.

Work performed on a duct system can also generate indoor air quality problems. Any work that disturbs system ducts can introduce or loosen dirt particles and convey the particles to occupied spaces via the airflow. Extreme care must be used when performing inspections or traverses within the duct system. Replacing an air-handling unit, which involves disconnection and reconnection of ductwork, can easily release contaminants into the HVAC system.

It is important to limit the jostling or movement of ducts as much as possible. When ducts are exposed, the area should be cleaned and the inside of the duct should be vacuumed to the greatest depth possible. It is necessary to install temporary filters at the air diffusers and run the system to catch any remaining loose dirt and particles. During this temporary filtering procedure, a reduction of air volume to affected spaces is likely. Temporary filtering should be kept in mind when scheduling work and/or cleaning ducts.

Cooling Coils

The *cooling coil* of an HVAC system is a heat exchanger that allows chilled water to be used for cooling air. Cooling coils usually consist of a finned coil, copper tubes, and aluminum fins. **See Figure 11-21.** The air is cooled using chilled water or a refrigerant. While the cooling coil is cooling the air, it is also removing moisture from the air (condensate) at the same time. When the air leaves the cooling coil, it is at a very high relative humidity.

It is easy to recognize the condensation process that a cooling coil produces, but most people do not realize that the entire duct distribution system contains the high relative humidity air. The cold, humid air traveling through the duct does not attain a lower humidity level until the air leaves the air diffuser to offset the heat load in a room. In some extreme cases, the air passing over the drain pan can lift moisture out of the pan and spray it over the surfaces of the air handler.

The cooling coil section and the distribution duct can have a significant effect on indoor air quality. Dirt that attaches to the wet cooling coil and/or stagnant water in the drain pan of the coil section can lead to bacteria growth, mold formation, and mildew. Mold and mildew can become airborne and thrive in the cool, damp surroundings of the duct system. The airborne mold and mildew eventually enter the conditioned spaces and become absorbed in the lungs of the occupants. Wet insulation in the cooling coil section does not dry out during the cooling season and is known to breed bacteria.

Periodic inspection and cleaning of the interior parts of the cooling coil section, including the main drain pan, intermediate coil drain pans, water traps, and cooling coils are essential. **See Figure 11-22.** Drain pans must be pitched properly to allow for positive drainage. Even when drain pans are properly positioned, the condensate water trap must prevent air from entering a coil section that is under vacuum and from escaping a coil section that is under pressure.

Cooling Coils

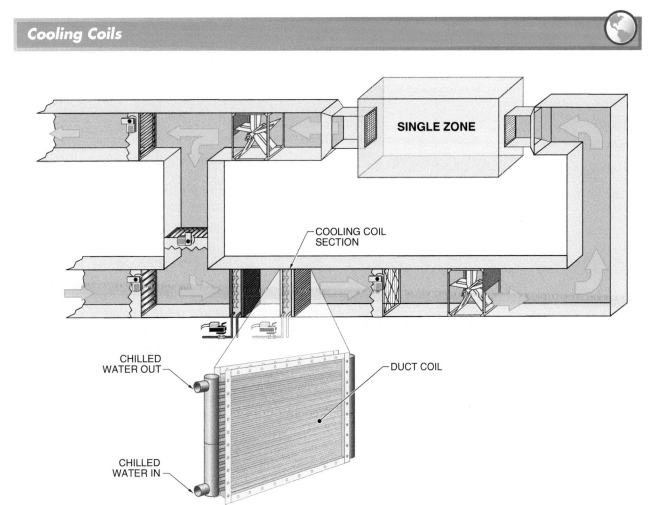

Figure 11-21. *Cooling coils usually consist of a finned coil, copper tubes, and aluminum fins that allow chilled water to be used for cooling air.*

Cooling Coil Drainage

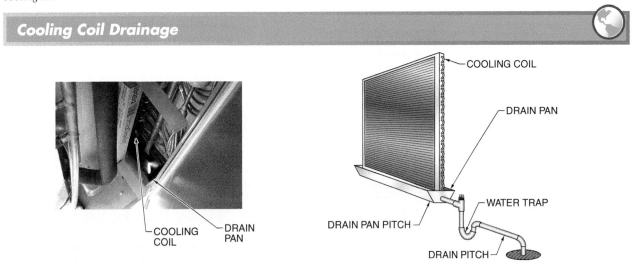

Figure 11-22. *Drain pans must be pitched to allow drainage and must be periodically checked for blockages in the pan and/or drain pipes.*

Cooling Coil/Dehumidifier Section. Spray coil dehumidifiers are rarely designed for new commercial buildings, but many exist in older buildings. Any system that has standing water in its sump may be a haven for bacteria growth. The evaporative cooling effect with recirculated sprayed water assists in the cooling of air. However, bacteria in the water ends up being directly injected into the airstream.

Eliminating the spray on the chilled water coil may mean sacrificing cooling capacity. Care must be taken when monitoring the condition of the water in these sumps. Water treatment can also complicate matters.

Raising chilled water temperatures is an energy-conservation measure often used to save money. As the chilled water temperature increases, the amount of moisture in the supply air also increases. Care must be taken not to increase the relative humidity in the occupied spaces to the point where mold, mildew, and bacteria can thrive.

In an effort to provide enough conditioned air to occupants, fan speeds are sometimes increased without consideration of the carryover effect of water particles into the duct system. According to ANSI/ASHRAE guidelines, an air-handling unit mounted horizontally should not have a coil face velocity of more than 500 fpm. A vertically mounted unit can have a velocity up to 600 fpm without undue carryover of moisture into the duct system.

Coil fin spacing can range from 2 fins per inch to 14 fins per inch. The closer the fin spacing, the higher the air velocity across the coil and the greater the carryover of water into the duct system. Since, by design, most cooling coils produce water condensate and distributes moist air into the duct system, special attention must be paid to the cooling coil section of an air distribution system.

Heating Coils

A *heating coil* lowers the relative humidity of the air as the coil increases air temperature (i.e., sensible heat). When a conditioned space has enough of a latent heat load, there may be no need to add humidity to the air. The conditioned space should have a relative humidity of between 30% to 60%, which minimizes static electricity, bacteria, viral, and fungal growth. Humidity levels below 30% tend to rob moisture from occupants, especially from the lungs and mucous membranes, which can result in respiratory infections, dry skin, and general discomfort.

Heating Coil/Humidifier Section. Most air-handling systems use some form of humidity control during the winter heating periods. This is generally accomplished with the introduction of water into the airstream on the output side of the heating coil. **See Figure 11-23.** Steam from boilers must not be introduced into the HVAC airstream under any circumstances.

The amount of moisture introduced into the airstream is controlled by a control system and by humidity sensors in the ducts. The humidity-control loop must be set up properly to ensure that the system does not add too much or too little moisture to the supply air. The airstream must absorb 100% of the moisture so that there is no precipitation accumulating in the duct system.

Water pan humidifiers pose the same problem as any standing water reservoir by providing the ideal growth environment for bacteria, viruses, and fungi. Furthermore, warm or hot standing water provides an ideal environment for Legionella bacteria to grow.

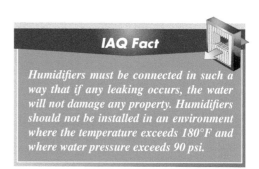

IAQ Fact

Humidifiers must be connected in such a way that if any leaking occurs, the water will not damage any property. Humidifiers should not be installed in an environment where the temperature exceeds 180°F and where water pressure exceeds 90 psi.

Humidifiers

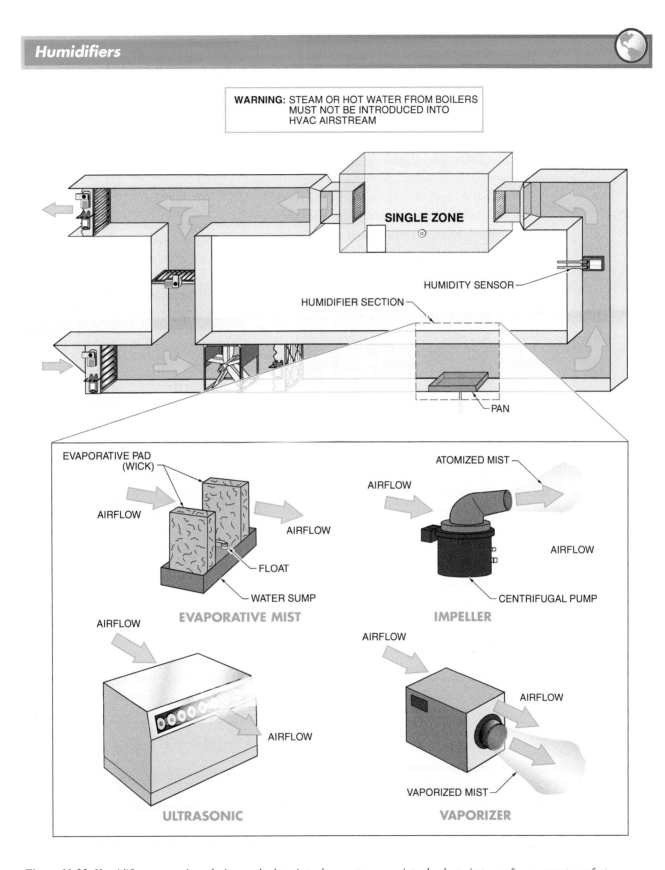

Figure 11-23. Humidifiers use various design methods to introduce water vapor into the duct airstream for occupant comfort.

COMMON AIR-HANDLING SYSTEMS

There are several types of air distribution systems that will meet the environmental requirements of a building. Common air-handling units are usually centrally located in a mechanical equipment room (MER) and serve large areas or zones of a building. Common air-handling systems include:

• single-zone systems (constant volume)

• multizone systems (constant volume)

• dual-duct systems (constant volume)

• reheat systems (constant volume)

• variable-air-volume (VAV) systems

Single-Zone Systems

A *single-zone system* is a type of air system that serves a single temperature control zone and is the simplest form of air system. **See Figure 11-24.** Single-zone systems are completely responsive to the requirements of the space. Well-designed air systems maintain temperature and humidity efficiently and can be shut down without affecting the operation of adjacent areas.

Economizer control arrangements can be used with unit ventilators and central station systems. A single-zone system supplies a constant volume of air to the conditioned space. The supply-air temperature can be held constant and the fan cycled to control space temperature, or the fan can be run continuously and the air temperature reset to maintain conditions. There are a variety of single-zone systems, including room fan coil units, unit ventilators (with a special sequence control), induction units, and central station fan systems.

IAQ Fact

Setpoint range used in HVAC systems indicates how much an actual temperature is allowed to vary around the setpoint temperature. A thermostat with a 2°F setpoint range, set at 72°F for cooling, calls for cooling when the temperature is 73°F and stops calling for cooling when the temperature is 71°F.

Single-Zone Air-Handling Units

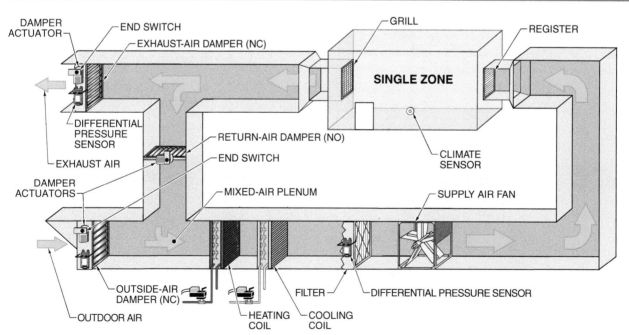

Figure 11-24. Single-zone air-handling units provide primary HVAC capability to one zone or area.

Multizone Systems

A *multizone system* is a type of air system that serves a relatively small number of zones from a single, central air-handling unit. **See Figure 11-25.** The system delivers a constant volume of air. In response to zone thermostats, the requirements of the various zones are met by mixing cold and warm air through zone dampers at the central air handler. The cold deck and hot deck temperatures may be varied in response to zone loads. The mixed conditioned air is distributed throughout the building by a system of single-zone ducts. All of the various economizer control arrangements can be used with a multizone system.

The fan of a multizone system runs continuously during the time that building spaces are occupied. An averaging thermostat in the hot deck modulates the heating valve to maintain the setpoint temperature (which may be reset according to the outdoor temperature). An averaging thermostat in the cold deck modulates the cooling coil valve to maintain the setpoint temperature. A space thermostat modulates the mixing dampers at the central fan to supply a mix of hot and cold air that satisfies the space load.

Dual-Duct Systems

A *dual-duct system* is a type of air system in which all the air is conditioned in a central fan system and distributed to conditioned spaces through two parallel main ducts. One duct carries cold air and the other carries warm air, thus providing air sources for both heating and cooling at all times. Dual-duct systems are used in buildings that also have multiple zones and can serve a wide variety of loads, but are considered to be energy inefficient and are very expensive to install. **See Figure 11-26.**

Up-to-date system pressure measurements are required to maintain the energy efficiency of a multizone system at a high level.

Multizone Air-Handling Units

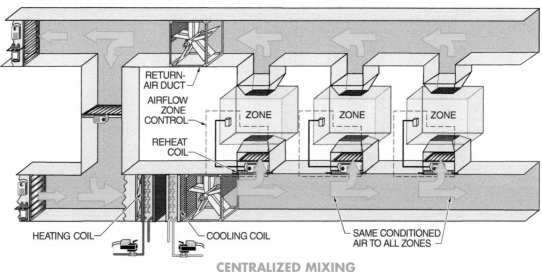

RETURN-AIR DUCT
AIRFLOW ZONE CONTROL
REHEAT COIL
ZONE
ZONE
ZONE
HEATING COIL
COOLING COIL
SAME CONDITIONED AIR TO ALL ZONES
CENTRALIZED MIXING

Figure 11-25. Multizone air-handling units use mixing dampers in the air handler to provide HVAC to a small number of zones.

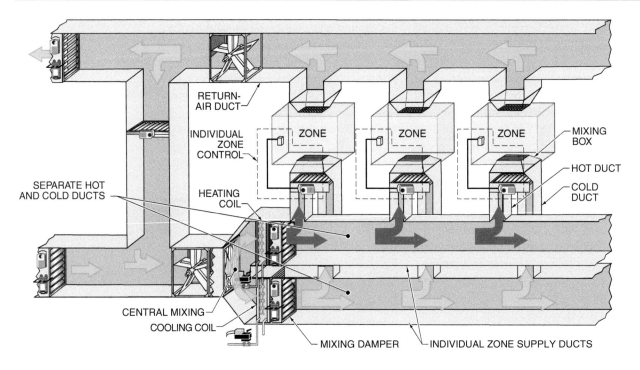

CENTRALIZED MIXING

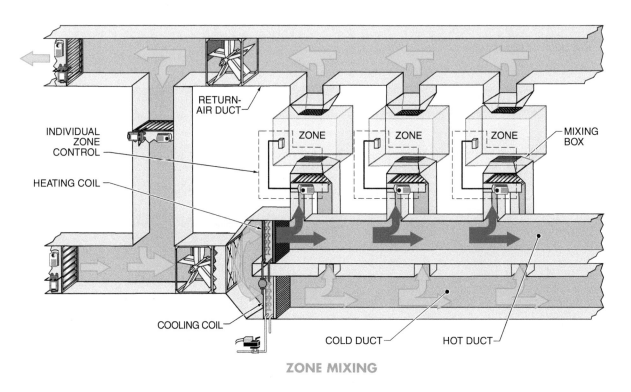

ZONE MIXING

Figure 11-26. *Dual-duct systems use hot ducts and cold ducts to carry conditioned air to mixing boxes in several zones to provide whatever comfort is required in a specific zone.*

In each conditioned space or zone, a mixing valve that responds to a room thermostat mixes the warm and cold airstreams in proper proportions to satisfy the space load. The proportion of cold or warm air varies, but the total amount of air is always the same. For this reason, dual-duct mixing systems are generally considered to be superior for indoor air quality purposes because the total amount of air does not decrease according to temperature demand as it does with VAV systems. However, in application, mixing boxes actually deliver a constant volume to building zones.

Reheat Systems

A *reheat system* is a type of air system that permits zone or space control for areas of different exposure, provides heating or cooling of perimeter areas of unequal exposure, or provides process or comfort control where close temperature control is required. **See Figure 11-27.** Unfortunately, reheat systems are also known for being energy inefficient.

A reheat system delivers a constant volume of air at a temperature that is cold enough to satisfy the maximum cooling load. Terminal reheat coils under the control of a space thermostat heat the air to a temperature that satisfies the zone load. Reheat systems are used in hospitals, laboratories, clean rooms, solid-state electronic component manufacturing facilities, and other such spaces. All of the various economizer control arrangements can be used with reheat systems.

Variable-Air-Volume Systems

A *variable-air-volume (VAV) system* is an air-handling unit that provides air at a constant temperature but varies the amount delivered to various loaded zones. **See Figure 11-28.** VAV systems usually use a variable-frequency drive or variable-pitch fan blades to vary the amount of air. A central fan system delivers air at a constant temperature.

The volume of air delivered to a space is reduced as the space temperature drops. Volume regulation is controlled by a variable-volume air handler. When the space temperature equals the setpoint temperature, either the supply air must be shut OFF completely (which can impact indoor air quality) or reheat must be added to avoid overcooling the space.

Terminal Reheat Air-Handling Units

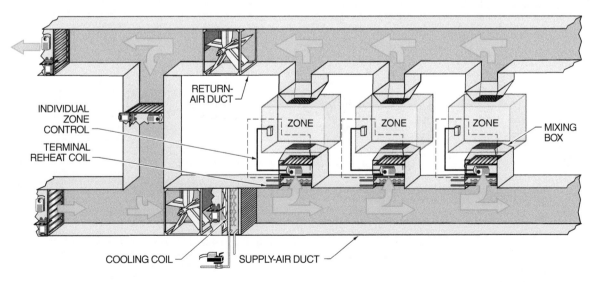

Figure 11-27. Reheat systems utilize independent heating coils in the duct of each zone controlled by a thermostat in the zone.

Variable-Air-Volume (VAV) Air-Handling Units

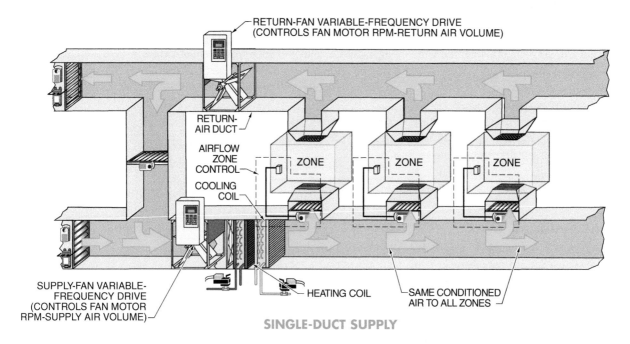

RETURN-FAN VARIABLE-FREQUENCY DRIVE
(CONTROLS FAN MOTOR RPM-RETURN AIR VOLUME)

RETURN-
AIR DUCT

AIRFLOW
ZONE
CONTROL

COOLING
COIL

ZONE

ZONE

ZONE

SUPPLY-FAN VARIABLE-
FREQUENCY DRIVE
(CONTROLS FAN MOTOR
RPM-SUPPLY AIR VOLUME)

HEATING COIL

SAME CONDITIONED
AIR TO ALL ZONES

SINGLE-DUCT SUPPLY

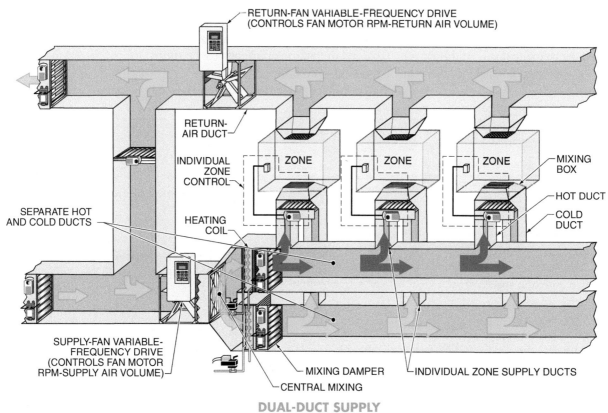

RETURN-FAN VARIABLE-FREQUENCY DRIVE
(CONTROLS FAN MOTOR RPM-RETURN AIR VOLUME)

RETURN-
AIR DUCT

INDIVIDUAL
ZONE
CONTROL

HEATING
COIL

ZONE

ZONE

ZONE

MIXING
BOX

HOT DUCT

COLD
DUCT

SEPARATE HOT
AND COLD DUCTS

SUPPLY-FAN VARIABLE-
FREQUENCY DRIVE
(CONTROLS FAN MOTOR
RPM-SUPPLY AIR VOLUME)

MIXING DAMPER

CENTRAL MIXING

INDIVIDUAL ZONE SUPPLY DUCTS

DUAL-DUCT SUPPLY

Figure 11-28. VAV air-handling units usually use a variable-frequency drive-controlled fan to vary air volumes in a duct in order to maintain the desired static pressure.

The volume of air per space variation is difficult to predict and depends on the varying climatic conditions in the area. Consequently, in a VAV system, the amount of outdoor air delivered to each zone changes continuously. The ANSI/ASHRAE Standard 62.1-2007, *Ventilation for Acceptable Indoor Air Quality,* specifies the minimum outdoor-air requirements for various facilities. Keeping the ANSI/ASHRAE standard in mind, the VAV system may have to be modified to provide a minimum primary air supply for the climatic conditions found in a space.

A VAV system is relatively energy efficient when compared with reheat, multizone, or dual-duct systems. All of the various economizer control arrangements are used with VAV systems. Several variations of VAV systems exist. Each variation has a different impact on indoor air quality.

A VAV terminal box can throttle airflow from 100% to 0% of the design flow. A reduction in airflow can reduce ventilation airflow to a space below an acceptable quantity when the space temperature is satisfied and the damper is fully closed. Having the dampers 100% closed on a variable-air-volume terminal box does cause many air quality complaints. **See Figure 11-29.**

A variable-air-volume terminal box will always have a minimum damper setting determined by building occupancy requirements. Airflow is throttled from 100% to the minimum flow required to satisfy ventilation requirements and to maintain IAQ. This arrangement can cause overcooling in some or all zones when the space temperature is satisfied but the minimum amount of air continues to enter the space.

A variable-air-volume terminal box may have both a minimum setting and a reheat coil. Even though the minimum amount of air for the space may be more than the amount required for cooling, the reheat of that air accommodates the desired space temperature. This reheating has the advantage of satisfying ventilation requirements and contributing to acceptable IAQ standards.

Variable-Air-Volume (VAV) Terminal Boxes

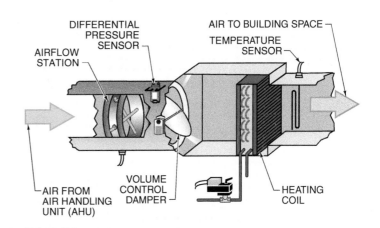

Figure 11-29. Even though variable-air-volume terminal boxes can throttle airflow from 100% to 0% of design flow, all VAV terminal boxes have a minimum flow setting to maintain ventilation standards.

FACILITY PRESSURES

Facilities are usually operated at a slightly positive pressure so that contaminants do not leak into the building. A positive pressure is especially important on lower floors due to the stack effect of the building's design. The stack effect occurs when hot air rises through the facility elevator shafts, stairwells, and service columns. **See Figure 11-30.**

Stack Effect

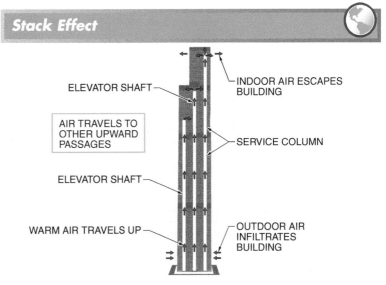

Figure 11-30. The stack effect in a building occurs when hot air rises up through the building elevator shafts, stairwells, and service columns, similar to the operation of a boiler chimney.

Outdoor air is drawn into lower floors, bringing in contaminants from the street level. The stack effect becomes larger as the temperature difference between the indoor and outdoor environment increases. This stack effect can be minimized by keeping the lower floors at a positive pressure in order to limit the inflow of unfiltered, possibly contaminated air.

Positive pressure is created by supplying enough air to a space so that the space has a higher pressure in relation to the atmospheric pressure. Pressure monitors measure the differential pressure between the indoor space and the outdoor or atmospheric pressure. Pressure monitors are used to maintain positive pressure in a space by controlling the fan speed or volume control dampers.

Odors

Odors by themselves are indicators of the presence of an odor-emitting substance. Odors may be pleasant or not, or harmful or not, depending upon the source and the sensitivity of each individual. Some people suffer symptoms when exposed to common chemicals, such as perfumes or air fresheners.

To prevent odors, all pan drain traps must be filled with water to prevent sewer gases from entering the building or HVAC systems. Vacant floor plumbing traps that may have lost their water seal due to evaporation over a period of time must be inspected and filled. **See Figure 11-31.** All drain and sewer line leaks must be repaired immediately.

Energy Conservation Techniques

Per ANSI/ASHRAE Standard 62.1-2007, *Ventilation for Acceptable Indoor Air Quality,* the use of outdoor air should be increasingly used to provide a dilution effect to indoor airborne contaminants. Applying ANSI/ASHRAE Standard 62.1-2007 may increase energy consumption. The increased fan static pressure required for high-efficiency filters can also increase energy costs.

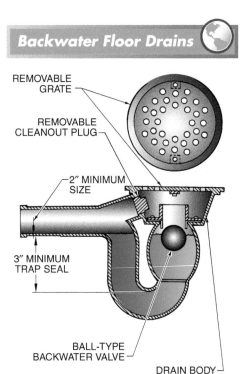

Backwater Floor Drains

REMOVABLE GRATE

REMOVABLE CLEANOUT PLUG

2" MINIMUM SIZE

3" MINIMUM TRAP SEAL

BALL-TYPE BACKWATER VALVE

DRAIN BODY

Figure 11-31. All pan and floor drain traps must be filled with water to prevent sewer gases from entering a building or HVAC system.

To offset any rise in energy costs caused by applying ANSI/ASHRAE Standard 62.1-2007, introducing heat recovery equipment into the air distribution system should be considered. The following is a partial list of air-to-air or air-to-water heat exchangers that may be integrated into the HVAC system of a building:

- rotary wheel heat exchanger
- fixed-plate counterflow heat exchangers
- crossflow plate heat exchangers
- heat pipe heat exchanger
- run-around coil-to-coil heat exchanger
- boiler flue gas heat exchanger
- steam condensate heat exchanger
- wastewater heat recovery

Retrofit projects, such as insulated windows, thermal window films, improved building insulation, and electric utility rebate programs for energy-efficient improvements made to the HVAC or lighting systems, can all help offset any increases in energy costs that result from an increase in ventilation air use.

All-Water Systems

All-water (hydronic) systems provide heating or both heating and cooling to terminal units serving an occupied space. Terminal units include baseboard radiation, freestanding radiators, wall and floor radiant panels, fan coil units, and other configurations.

All-water systems may be either one-pipe, two-pipe, three-pipe, or four pipe. A one-pipe system transmits hot water to terminal units during the heating season and transmits chilled water during the cooling season when cooling is provided. A two-pipe system can transmit both hot water and chilled water simultaneously so that a terminal unit, serving an area, can provide either heating or cooling.

One-Pipe Hydronic Systems. A *one-pipe hydronic system* is a type of hydronic piping system that uses a single pipe as both a supply pipe and return pipe. **See Figure 11-32.** One-pipe systems are limited to residential and small commercial systems. A major disadvantage of the one-pipe system is that each terminal unit uses the water discharged from the previous terminal unit. Thus, the system rapidly loses efficiency.

Two-Pipe Hydronic Systems. A *two-pipe hydronic system* is a type of hydronic piping system that has a separate supply pipe and return pipe. **See Figure 11-33.** Each coil has a separate take off from the supply line, ensuring that adequate water flow is available. The hot and chilled water sources are connected to a common return pipe and valves. Manual valves, automatic valves, and balancing valves are used to ensure adequate flow and control.

Three-Pipe Hydronic Systems. A *three-pipe hydronic system* is a type of hydronic piping system where hot or cold water can be introduced to a terminal unit (e.g., coil). Some three-pipe systems switch between hot water in the winter and chilled water in the summer. During a change in seasons, a three-pipe system is utilized so that either water source can be selected for the specific situation. **See Figure 11-34.** This may be accomplished by using a manual system, outside-air thermostat, mixing valves, or other methods. Care must

be used with changeover systems because introducing warm water to the evaporator of a chiller will cause high-head pressures and mechanical problems in the chiller.

One-Pipe Hydronic Systems

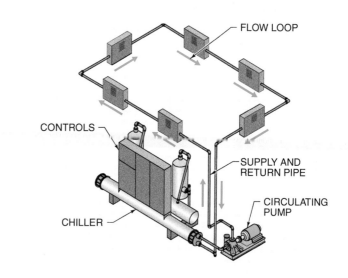

FLOW LOOP

CONTROLS

SUPPLY AND RETURN PIPE

CIRCULATING PUMP

CHILLER

Figure 11-32. One-pipe hydronic systems have a single pipe that acts as the supply pipe and return pipe for the flow loop, connecting one terminal unit to the next terminal unit.

Two-Pipe Hydronic Systems

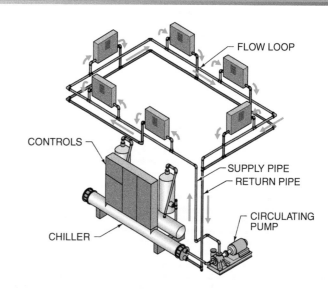

FLOW LOOP

CONTROLS

SUPPLY PIPE
RETURN PIPE

CIRCULATING PUMP

CHILLER

Figure 11-33. Two-pipe hydronic systems have a separate supply pipe and return pipe at each terminal unit.

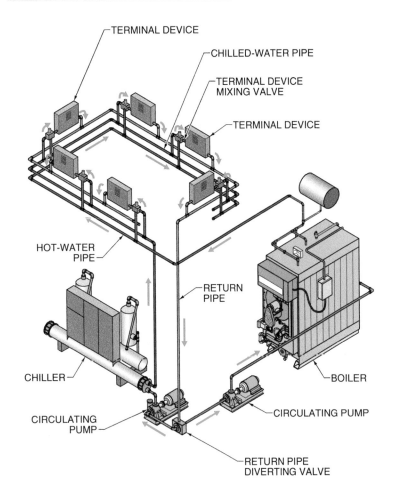

TERMINAL DEVICE

CHILLED-WATER PIPE

TERMINAL DEVICE
MIXING VALVE

TERMINAL DEVICE

HOT-WATER
PIPE

RETURN
PIPE

CHILLER

BOILER

CIRCULATING
PUMP

CIRCULATING PUMP

RETURN PIPE
DIVERTING VALVE

Figure 11-34. Three-pipe hydronic systems have a hot-water loop and a cold-water loop so that hot or cold water can be introduced to any terminal unit at any time.

Four-Pipe Hydronic System. A four-pipe hydronic system uses separate piping for heating and cooling. The terminal units are connected to both heating and cooling pipes. Water flow to the terminal units is controlled by mixing and diverting valves that allow either hot or cold water to flow through any terminal unit. Four-pipe hydronic systems are expensive to install but provide excellent control of air temperature. Four-pipe hydronic systems are more economical to operate than three-pipe hydronic systems. **See Figure 11-35.**

A *steam heating system* is a heating system that uses steam to carry heat from the point of generation to the point of use. Steam heating systems are used because steam carries more British thermal units per pound of steam than water, allowing steam to be used where there is a limited amount of room for the distribution system. Another advantage to steam heating systems is that the steam does not need to be pumped to the location of use. Steam is a vapor that flows any time there is a difference in pressure from one area (boiler) to another area (terminal unit). The boiler is located in a central plant and the steam is distributed to individual terminal units or buildings. **See Figure 11-36.**

Direct-Return Push and Reverse-Return Systems. Another design consideration for water distribution systems is the routing of the supply and return piping. A *direct-return system* is a system that is piped so that the terminal unit with the shortest supply pipe also has the shortest return pipe. **See Figure 11-37.** This means that the total piping is less, which means that the total resistance is lowest and the flow the greatest. A *reverse-return system* is a system consisting of pipes in which a coil with the shortest supply pipe also has the longest return pipe. This means that the relative pipe runs to each coil are about the same. This helps with water-balancing issues.

Primary/Secondary Systems. In large multi-building facilities, primary/secondary systems are often used. **See Figure 11-38.** In primary/secondary systems, the hot or chilled water source has the supply connected to the return with its own pumping system. This is the primary loop. A separate supply take off and return pipe is available at each building with a separate pump. This is known as the secondary loop. There may be multiple secondary loops used, allowing each zone the capability of using as much water as required.

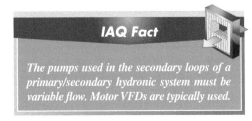

IAQ Fact

The pumps used in the secondary loops of a primary/secondary hydronic system must be variable flow. Motor VFDs are typically used.

Four-Pipe Hydronic Systems

TERMINAL DEVICE — SOLENOID — CHILLED-WATER SUPPLY PIPE

— TERMINAL DEVICE MIXING VALVE

— TERMINAL DEVICE

— THREE-WAY VALVE

HOT-WATER RETURN PIPE

HOT-WATER SUPPLY PIPE

CHILLER

CIRCULATING PUMP

CHILLED-WATER RETURN PIPE

BOILER

CIRCULATING PUMP

HOT-WATER RETURN PIPE

Figure 11-35. A four-pipe hydronic system uses supply and return heating piping and supply and return cooling piping.

Steam Heating Systems

STEAM TRAP

STEAM TERMINAL UNIT

STEAM STRAINER

FLOW VALVE

STEAM CONDENSATE (RETURN PIPE)

STEAM HEADER (SUPPLY PIPE)

CITY MAKEUP WATER

CONDENSATE TANK

FEEDWATER PUMP

STEAM BOILER

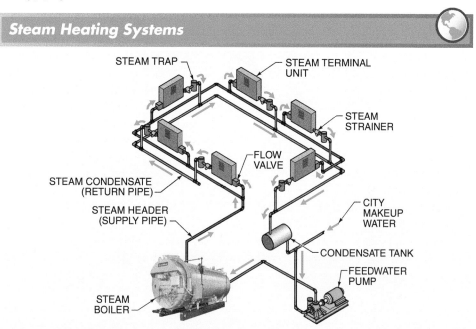

Figure 11-36. Steam heating systems consist of a high-pressure boiler, fittings, accessories, steam supply piping, terminal units, a condensate return system, and controls.

Direct-Return and Reverse-Return Systems

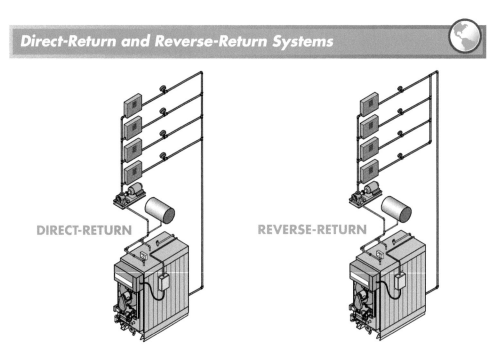

Figure 11-37. In direct-return systems, the shortest supply line has the shortest return. In reverse-return systems, the shortest supply line has the longest return.

Primary/Secondary Systems

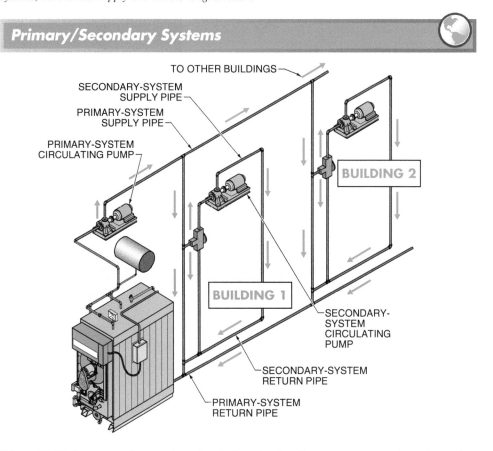

Figure 11-38. Large complexes such as hospitals and university campuses use primary/secondary systems to distribute water to buildings.

Air and Water Systems

An *air and water system* provides both air and water to terminal units serving an occupied space. Terminal units can include hot- and cold-water piping or refrigerant piping, a heat exchanger, a fan, drain pain, sensors, and controls. Room terminal units usually include induction units, fan coil units, or air-supply diffusers with a heating coil.

A water system may be either a one-pipe or two-pipe system. A one-pipe system transmits hot water to terminal units during the heating season and transmits chilled water during the cooling season when cooling is provided. A two-pipe system can transmit both hot water and chilled water simultaneously so that a terminal unit serving a space can provide either heating or cooling at any time.

Electric Radiant Heating Systems

An *electric radiant heating system* is a heating system that transfers heat from an electric resistance heating element to air using radiant energy waves. *Radiant heating* is heating that occurs when a surface (resistance heating element) is heated and the surface gives off heat in the form of radiant energy waves. As the heating element heats up, the wire gives up heat to the air passing over the element, which raises the temperature of the air. **See Figure 11-39.** Different versions of electric radiant heat panels are made for indoor use or outdoor use, with designs available for terminal unit use.

Electric Baseboard Heating Systems. An *electric baseboard heating system* is an electric heater that is enclosed in a low cabinet that fits along the baseboard of a space. Electric baseboard heaters are smaller than typical electric space heaters and use a section of fins to transfer heat.

The fins provide a large heat transfer surface area. Electric baseboard heaters can be 12″ high and are available in any length. Electric baseboard heaters are installed at floor level along the outside walls of building spaces. **See Figure 11-40.**

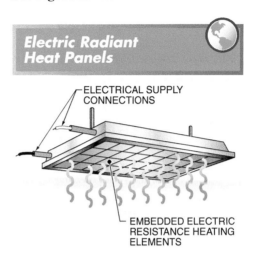

Figure 11-39. Radiant heat panels have resistance heating elements and radiate heat directly to the area below.

Figure 11-40. Electric baseboard heaters are smaller than electric space heaters, and are less expensive to install but more costly to operate than other heating systems.

Case Study

Case Study – Office Building

IAQ Issues

A 1,600,000 sq ft, 52-story commercial office building has commercial office tenants on all floors except the ground floor, which is 100% occupied by retail tenants. The facility is 90% occupied and has been operating as a commercial office building for 10 years.

The stationary engineer started receiving complaints from the third floor of the facility. Complaints started almost immediately and were spread out intermittently across the floor with the exception of the northwest corner of the building, where the temperature and airflow problems seemed to be universal.

Investigation

An investigation revealed that new tenants had moved into the entire third floor after having new carpeting installed and furniture delivered. The tenants on the third floor complained about a lack of airflow, too high a temperature, and miscellaneous odors.

Airflow measurements showed approximately a 60% reduction in designed airflow to the diffusers across the floor with the exception of the problem corner, which showed zero airflow. After inspecting all of the air-handling system's related controls and components, it was discovered that the third-floor damper linkage was loose and had slipped to a position that only allowed about 30% of the total designed airflow. Further examination of the ventilation system revealed a broken fire-damper link on a branch duct that served the problem corner of the building.

Resolution

Results from temperature and relative humidity readings confirmed an inability to maintain temperature and relative humidity setpoints. Correcting the damper linkage problem restored full airflow to the floor with the exception of the problem corner.

A previous tenant had a fire damper installed in the branch duct to comply with the fire code for a special application room in that corner of the building. The repair of the fire damper resolved the remaining airflow problem.

HVAC Air Systems Definitions . . .

- An *air and water system* provides both air and water to terminal units serving an occupied space.

- *Air cleaning* is a method of reducing air pollution that includes electronic air cleaning, ultraviolet (UV) air cleaning, and the use of filters, which includes carbon filters.

- An *air-handler economizer* is an HVAC unit that uses outdoor air for free cooling.

- An *air-handling unit* is the part of an HVAC system that operates to distribute air throughout a building.

- A *carbon filter* is an air-cleaning method where an activated carbon-filtering medium is used to remove most odors, gases, smoke, and smog from the air by means of an adsorption process.

- The *cooling coil* of an HVAC system is a heat exchanger that allows chilled water to be used for cooling air.

- A *direct-return system* is a system that is piped so that the terminal unit with the shortest supply pipe also has the shortest return pipe.

- A *dry bulb economizer* is a type of economizer that operates strictly in connection to the outside-air temperature, with no reference to humidity values.

- A *dual-duct system* is a type of air system in which all the air is conditioned in a central fan system and distributed to conditioned spaces through two parallel main ducts.

- The *dust-holding capacity* is a filter's ability to hold dust without seriously reducing the filter's efficiency.

- *Dust spot efficiency* is a filter rating that measures a filter's ability to remove large particles from the air that tend to soil building interiors.

- A *DX system* is a system in which the refrigerant expands directly inside a coil in the main airstream itself to affect the cooling of the air.

- An *electric baseboard heating system* is an electric heater that is enclosed in a low cabinet that fits along the baseboard of a space.

- An *electric radiant heating system* is a heating system that transfers heat from an electric resistance heating element to air using radiant energy waves.

- An *electronic air cleaner* is a type of air cleaner in which particles in the air are positively charged in an ionizing section and then the air moves to a negatively charged collecting area (i.e., plates) to separate out the particles.

- *Enthalpy* is the total heat content of a substance.

- An *enthalpy economizer* is a type of economizer that uses temperature and humidity levels of the outside air to control operation.

- A *filter* is an air-cleaning method in which a mechanical device is used to remove particles from the air.

- *Filter airflow resistance* is the pressure drop across a filter at a given velocity.

- A *heating coil* lowers the relative humidity of the air as the coil increases air temperature (i.e., sensible heat).

- A *mixed-air system* is a unit or section of an HVAC system that brings a specific amount of return air into contact with the outside air entering the HVAC system for space use.

Definitions

. . . HVAC Air Systems Definitions

Definitions

- A *multizone system* is a type of air system that serves a relatively small number of zones from a single, central air-handling unit.
- A *100% outside-air (OA)* system is a system that does not recirculate any return air from building spaces.
- A *one-pipe hydronic system* is a type of hydronic piping system that uses a single pipe as both a supply pipe and return pipe.
- *Radiant heating* is heating that occurs when a surface (resistance heating element) is heated and the surface gives off heat in the form of radiant energy waves.
- A *reheat system* is a type of air system that permits zone or space control for areas of different exposure, provides heating or cooling of perimeter areas of unequal exposure, or provides process or comfort control where close temperature control is required.
- A *reverse-return system* is a system consisting of pipes in which coil with the shortest supply pipe also has the longest return pipe.
- A *single-zone system* is a type of air system that serves a single temperature control zone and is the simplest form of air system.
- *Source control* is a method of reducing air pollution by identifying strategies to reduce the origin of pollutants.
- A *steam heating system* is a heating system that uses steam to carry heat from the point of generation to the point of use.
- A *three-pipe hydronic system* is a type of hydronic piping system where hot or cold water can be introduced to a terminal unit (e.g., coil).
- A *two-pipe hydronic system* is a type of hydronic piping system that has a separate supply pipe and return pipe.
- An *ultraviolet (UV) air-cleaner system* is an air-cleaning system that kills biological contaminants using a specific light wavelength.
- A *variable-air-volume (VAV) system* is an air-handling unit that provides air at a constant temperature but varies the amount delivered to various loaded zones.
- *Ventilation* is a method of reducing air pollution by introducing various amounts of outdoor air in order to dilute indoor pollutants.

**For more information please refer to ATPeResources.com
or http://www.epa.gov.**

Review Questions

1. Explain the purpose of HVAC systems.

2. List three types of air-handling systems.

3. What are some common control strategies for IAQ?

4. What device is used to measure the pressure drop across the filters?

5. Name two types of common water distribution systems.

HVAC CONTROLS

Saylor-Beall Manufacturing Company

Class Objective

Define the major aspects of HVAC control systems and their role in the IAQ process.

On-the-Job Objective

Ensure that HVAC control systems are working properly to maintain ideal IAQ conditions.

Learning Activities:

1. How do HVAC controls impact IAQ?
2. Describe the four main groups of components of a pneumatic control system.
3. List the kinds of IAQ problems that may be the result of the malfunction of transmitters or controllers.
4. **On-the-Job:** Review and examine any HVAC control systems in the facility.

Introduction

The quality of indoor air does deteriorate when HVAC controls are not functioning properly. The purpose of an environmental control system is to ensure a proper level of comfort within a facility. When the control system is not functioning properly, the safety and health of facility occupants could be jeopardized. HVAC control systems that are not working properly lead to a variety of IAQ issues.

PNEUMATIC CONTROL SYSTEMS

A *pneumatic control system* is a control system in which compressed air is used to provide power for the system. Pneumatic control systems were developed in the early 1900s as a primary method of controlling the environment in commercial buildings. **See Figure 12-1.**

Pneumatic control systems were developed because the existing electric control systems were not as safe as pneumatic systems, and pneumatic systems offered more precise control. With the advent of building automation systems, electronic (low-voltage) control systems have become a popular substitute for pneumatic controls systems. However, building automation systems can and do operate successfully with pneumatic systems, and such control schemes are commonly found in commercial buildings.

Pneumatic control systems can be separated into four main groups of components based on their function. The four groups are the air compressor station, transmitters and controllers, additional components, and controlled devices. **See Figure 12-2.**

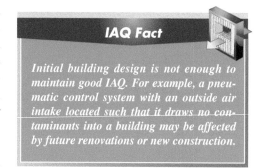

IAQ Fact

Initial building design is not enough to maintain good IAQ. For example, a pneumatic control system with an outside air intake located such that it draws no contaminants into a building may be affected by future renovations or new construction.

Effect of Pneumatic Controls on IAQ

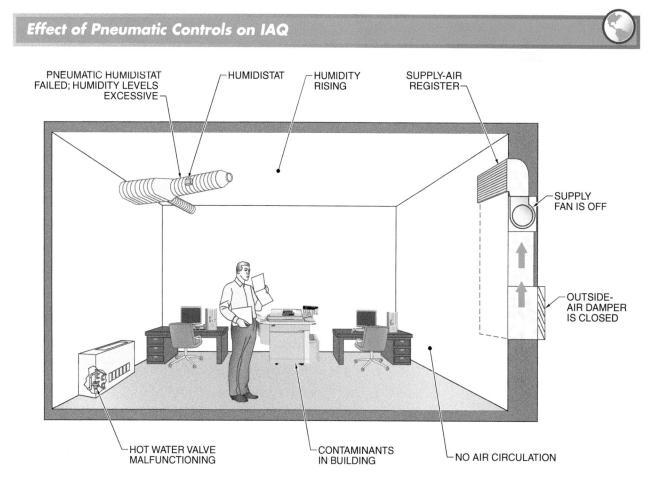

PNEUMATIC HUMIDISTAT FAILED; HUMIDITY LEVELS EXCESSIVE

HUMIDISTAT

HUMIDITY RISING

SUPPLY-AIR REGISTER

SUPPLY FAN IS OFF

OUTSIDE-AIR DAMPER IS CLOSED

HOT WATER VALVE MALFUNCTIONING

CONTAMINANTS IN BUILDING

NO AIR CIRCULATION

Figure 12-1. Pneumatic controls must operate properly to control the environment.

Pneumatic Control Systems

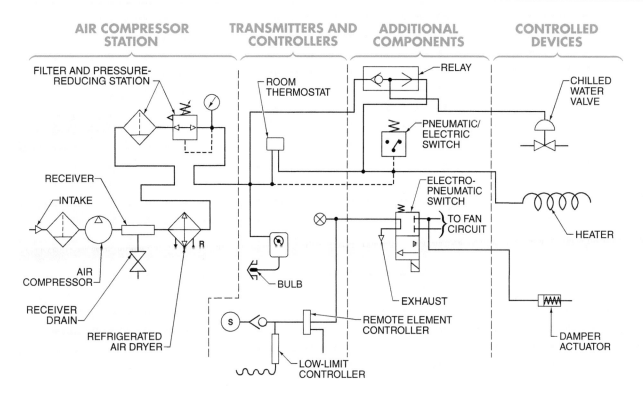

Figure 12-2. Pneumatic control systems include an air compressor station, transmitters and controllers, additional components, and controlled devices.

AIR COMPRESSOR STATIONS

The air compressor station is the heart of a pneumatic system. Air provided by an air compressor station must be clean, dry, and oil free, or the HVAC control system may operate improperly or fail. When the air compressor station is not operating properly, the lack of a correct air supply to controllers and actuators may cause outside air dampers and chilled water valves to close. This would cause a loss of ventilation air and a lack of dehumidification.

When ventilation air is lost, indoor air contaminants are not removed and levels may increase. When the cooling valve is closed, dehumidification is lost and the indoor humidity will rise. Rising humidity can contribute to the growth and increase of mold and other harmful biological organisms.

These are just two examples of how problems in the pneumatic control system impact HVAC components that can lead to IAQ problems. A malfunction in the HVAC control system can cause a chain reaction, which may cause serious implications for the building's indoor air quality. It is therefore critical for stationary engineers to operate and maintain these control systems in a manner that prevents malfunctions and protects the quality of air, providing a safe and healthy working environment for workers.

An air compressor station consists of an air compressor and its auxiliary components. **See Figure 12-3.** An air compressor takes in air from the atmosphere. The auxiliary components remove particulate matter, moisture, and oil and control the air supply pressure and volume.

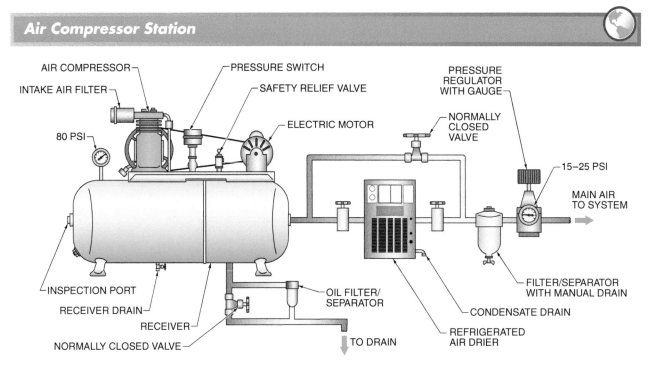

Figure 12-3. An air compressor station includes the air compressor and auxiliary components such as the air drier, drains, filters, and pressure-regulating valve stations.

Air Compressors

An *air compressor* is a component that takes air from the atmosphere and compresses it to increase its pressure. Air compressors convert the mechanical energy provided by an electric motor into the potential energy of compressed air. The vast majority of air compressors used in HVAC control systems are positive displacement compressors.

Additional Components

The additional components of an air compressor station must provide clean air that is free of moisture to the HVAC controls in a building, and they must regulate the pressure and volume of the system. Particulates are removed from the air supplied to a compressor station by the outside air intake air filters.

Most air compressor systems use the ambient air of the mechanical room as intake air for the compressor. The intake filter is located within a couple of inches of the compressor pump. Some compressors use air taken from an outside air intake to reduce the amount of dirt entering the system by obtaining cleaner, cooler air from outside the building. Oiled air intake filters are also used to further clean intake air entering a compressor. Oil in pneumatic control systems originates at the air compressor, which is usually oil-lubricated.

Oil removal filters are used to remove oil droplets from the air in pneumatic systems. Air drawn into the air compressor inlet contains moisture that must be removed. One method of removing moisture is through the use of a refrigerated air dryer. Refrigerated air driers use refrigeration to lower the temperature of compressed air. This process cools the compressed air and allows the moisture in the air to condense out. The condensate is then directed to a drain.

The correct pressure and volume of compressed air is required for proper operation of a pneumatic control system. When a controller receives air at an incorrect pressure, the result is the development of erratic system performance and excessive air compressor wear. Devices used in an air compressor station to correct the pressure and volume of air include pressure switches, pressure regulators, and safety relief valves.

Pressure switches start and stop an air compressor motor based on the pressure in the receiver. The pressure switch allows the air compressor to maintain correct air pressure within the receiver. Pressure regulators provide the proper air pressure to the controllers. A *pressure regulator* is a modulating valve that provides a constant outlet pressure to downstream components. A *safety valve* is a device designed to automatically open and release pressure from the receiver if and when pressure rises to an unsafe level. An air receiver must contain a safety relief valve to protect individuals and the system components against overpressurization.

TRANSMITTERS AND CONTROLLERS

Pneumatic transmitters sense the temperature, pressure, or humidity in a building space or system and send a signal to controllers. Controllers in turn send a pneumatic signal to controlled devices that regulate heating, cooling, or humidification. Pneumatic transmitters may be one-pipe or two-pipe devices.

A *one-pipe (low-volume) device* is a device that uses a small amount of the compressed air supply (restricted main air) to operate. The compressed air passes through an orifice (restrictor) that meters the airflow to the transmitter. Pneumatic transmitters used in building HVAC applications are one-pipe (low-volume) devices. One-pipe devices have only one air connection to the transmitter. A *two-pipe (high-volume) device* is a device that uses the full volume of compressed air available to operate. Two-pipe devices have separate main air and output air line connections to the transmitter.

When transmitters and controllers are out of calibration or are operating improperly, they may contribute significantly to IAQ problems. For example, controllers operating improperly may cause outside air dampers to be closed instead of open, reducing the volume of incoming ventilation air. Good indoor air quality is contingent on HVAC systems maintaining minimum ventilation air to prevent indoor air contaminants from reaching unhealthy levels. Water or steam valves may be in the wrong position, causing improper humidity levels and allowing growth of biological organisms. Preventing excess moisture from being introduced into HVAC systems and subsequently the building space is an effective way to eliminate an environment favorable to mold growth. Transmitters and controllers include pneumatic thermostats, limit thermostats, pneumatic humidistats, pneumatic pressurestats, and receiver controllers.

Pneumatic Thermostats

A *pneumatic thermostat* is a device that senses room temperature and alters branch line pressure to controlled devices to maintain a space's temperature. Pneumatic thermostats are available in many shapes and sizes and with many options. All pneumatic thermostats operate on the same basic principles. In a pneumatic thermostat, a sensing device (bimetallic element) is mounted under a cover and exposed to the air in the room. A *bimetallic element* is a temperature sensing device that consists of two different metals joined together at one end. **See Figure 12-4.**

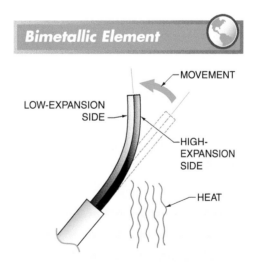

Figure 12-4. A bimetallic element is a sensing device that consists of two metals with different expansion and contraction rates that respond to temperature increase or decrease.

The two different metals within a bimetallic element have different rates of expansion and contraction. One of the metals has a high rate of expansion and contraction while the other has a low (almost zero) rate of expansion and contraction. When the room temperature increases, the bimetallic element bends toward the metal that has the low rate of expansion and contraction.

The amount of bending is proportional to the change in temperature at the bimetallic element. The bimetallic element bends toward the metal that has the high rate of expansion and contraction if the room temperature decreases. The bimetallic element is attached to the thermostat on one end, while the other end is free to bend.

The free end is used to cover or uncover a bleedport, depending on whether the room temperature increases or decreases. A *bleedport* is an orifice that allows a small volume of air to be expelled into the atmosphere. The bleedport is connected to the main (supply) air inlet.

The bleedport has a small diameter and meters the airflow through it. When room temperature is high, the bimetallic element bends toward the bleedport, preventing air from escaping. This increases the branch line pressure, which alters the controlled device.

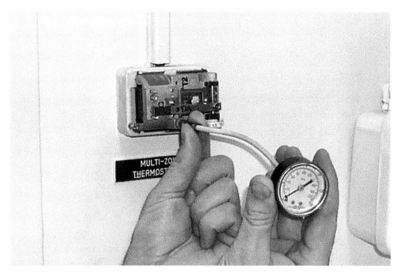

Pneumatic thermostats use changes in compressed air pressure to control the temperature of a room.

Branch line pressure is the pressure in the air line that is piped from the thermostat to the controlled device. Normally, an increase in branch line pressure causes the temperature in the room to decrease. When room temperature is too low, the bimetallic element bends away from the bleedport, allowing more air to escape and decreasing the branch line pressure. This action causes the controlled devices to increase the building space temperature to its setpoint. *Setpoint* is the setting of the desired temperature in the room or space.

The room temperature changes, the bends, the actuator strokes, and the amount of controlled medium added to the building space are proportional to one another. A thermostat may be direct-acting or reverse-acting. In a direct-acting thermostat, as the temperature increases, the branch line pressure increases and vice versa. In a reverse-acting thermostat, as the temperature increases, the branch line pressure decreases and vice versa.

Limit Thermostats

A *limit thermostat* is a pneumatic thermostat that causes a change in a controlled device whenever the set temperature or pressure is reached. Many HVAC systems, such as air-handling units and unit ventilators, require the addition of a limit thermostat to provide HVAC system component safety to the control system. **See Figure 12-5.** When some type of limit protection is not provided, the system may suffer a freeze-up or some other problem. Limit thermostats are not used for primary building space temperature control but are used only as protection against failure.

Limit thermostats are usually one-pipe devices located in series with a primary controller such as a standard two-pipe thermostat. Limit thermostats are commonly located in the discharge of a heating coil and are set at a minimum temperature that protects the coil or system. In a normally operating control system, limit thermostats do nothing.

Limit Thermostat

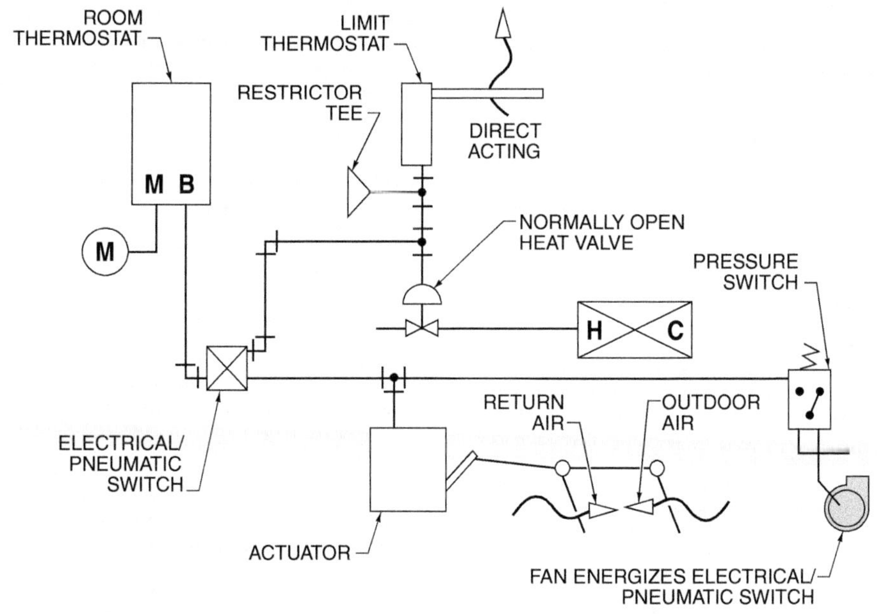

Figure 12-5. *Limit thermostats cause a change in a controlled device whenever the set temperature or pressure is reached.*

A limit thermostat causes the valves or dampers to go to a maximum condition (open or closed) to try to correct the situation, depending on the application. In the case of a low limit application, the normally open (NO) heating valve is forced wide open in an attempt to bring the low temperature up. A discharge limit application is often referred to as a freezestat in the field. The setpoint in this application is usually 35°F to 40°F.

A tremendous amount of damage can result from the failure of a freezestat designed to protect a fan coil from freezing. Water damage as a result of a coil freezing can produce water and moisture in places where it could be hard to clean such as in the air-handler room, inside the air handler, or in the ductwork. Water damage from a coil freezing can extend into occupied spaces and poses more complicated clean-up problems.

Pneumatic Humidistats

A *pneumatic humidistat* is a controller that uses compressed air to open or close a device that maintains a certain humidity level inside a duct or area. Pneumatic humidistats are used to sense and control the humidity in a building space or duct. Humidity, along with temperature, affects human comfort in a building. As humidity increases, mold becomes a concern. Mold grows rapidly at a humidity of 50% or higher. Proper inspection, cleaning, and calibration of a humidistat are required to prevent mold growth. When humidifiers are in use, using a sling psychrometer on a daily basis to record and track the humidity levels in occupied spaces helps avoid IAQ problems caused by high humidity.

The construction of a pneumatic humidistat is similar to that of a thermostat, with the exception that the bimetallic element is replaced by a humidity-sensing (hygroscopic) element. **See Figure 12-6.** The hygroscopic element accepts moisture from, or rejects moisture to, the surrounding air. Pneumatic humidistats are available in both one-pipe and two-pipe versions as well as direct-acting or reverse-acting.

Pneumatic Humidistats

Figure 12-6. *Pneumatic humidistats are used to sense and control the humidity in a building space or duct.*

Humidity sensing and control is not the same as temperature sensing and control. In general, all humidity controllers and sensors have potential problems that temperature devices do not. These problems affect both pneumatic and electronic control systems and include dirt, corrosive gases, and drift.

Dirt in duct systems and rooms can coat the hygroscopic element and reduces the ability of the material to accept moisture from, or reject moisture to, the ambient air. Some air-handling units in applications such as hospitals and manufacturing plants circulate gases along with the air that can corrode and damage the sensing element. Virtually all humidity sensors change their operating characteristics over time, a condition known as drift.

Pneumatic Pressurestats

Pressure inside a duct or building is a major variable that must be controlled in HVAC control systems. A *pneumatic pressurestat* is a controller that maintains a constant air pressure in a duct or area. Pneumatic pressurestats are designed to sense and control the pressure inside a duct or building space, which is measured in inches of water column (in.WC). **See Figure 12-7.**

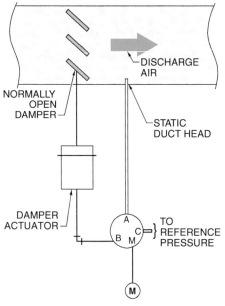

Pneumatic Pressurestats

Figure 12-7. *Pneumatic pressurestats are designed to sense and control the pressure inside a duct or building space.*

Many modern mechanical systems, such as variable-air-volume systems, rely on a specific air pressure inside a duct. The specific air pressure is used to supply terminal units, which open and close to maintain the temperature in the building space. Pressurestats are used to maintain duct air balance parameters and ensure that the internal building space pressure is maintained at an acceptable level.

In an air distribution system, it is essential to have the correct static pressure to ensure balanced airflow. A lower than necessary static pressure can cause too little airflow and air can become stagnant. Too high a static pressure can cause problems by delivering too much airflow, reducing the heat transfer necessary to maintain the correct temperature and pressure in the space. There can be many negative results from over-pressurization of a facility.

One of the most common indicators of an over-pressurized building is that the exterior doors do not close on their own because of the air pressure against them. Pressurestats have two connections that enable the pressurestat to measure a single pressure or the difference in pressure between ducts or areas. However, sophisticated pressure control is normally accomplished with receiver controllers and transmitters, not with simple pressurestats.

Receiver Controllers

Thermostats, humidistats, and pressurestats control the environment inside building spaces. Additional controls are required to achieve precise control of complex HVAC

systems. HVAC systems include air-handling units, boilers, heat exchangers, chillers, and cooling towers. Receiver controllers provide the flexibility and accuracy needed to control these systems.

A *receiver controller* is a device that accepts one or more input signals from pneumatic transmitters and produces an output signal based on the setup of the controller. **See Figure 12-8.** Receiver controllers are used for HVAC system control. Receiver controllers are not normally used to control the temperature, humidity, and pressure in one room of a commercial building.

Receiver controllers are available in a variety of designs. Most receiver controllers use a force-balance design. A *force-balance design* is a design in which the controller output is determined by the relationship among mechanical pressures. The force-balance design is similar to a lever, with calibration and adjustments serving to change the fulcrum point at the middle of the lever.

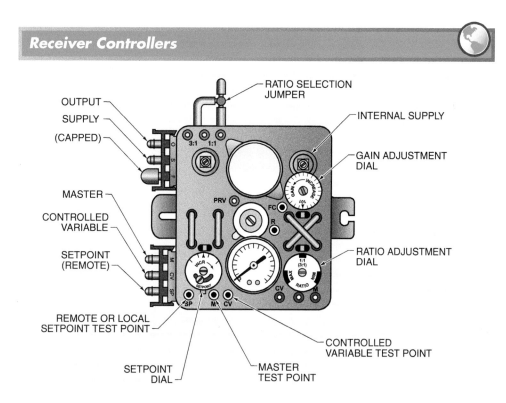

Figure 12-8. Receiver controllers are used for HVAC system control. A receiver controller accepts one or more input signals from pneumatic transmitters and produces an output signal based on the setup of the controller.

Some receiver controllers use stacked diaphragms in place of the lever arrangement. The stacked diaphragm arrangement is considered to be the most reliable and durable design. The disadvantage of the stacked diaphragm design is that the moving parts may deteriorate and have increased friction over time, causing the controller to drift or go out of calibration. Receiver controllers are designed as single-input or dual-input.

Single-Input Receiver Controllers. A *single-input receiver controller* is a receiver controller that is designed to be connected to only one transmitter and to maintain only one temperature, pressure, or humidity setpoint. **See Figure 12-9.** Many air-handling units and central plant systems use single-input receiver controllers to maintain specific desired setpoints in building mechanical systems.

Receiver controllers help to maintain indoor air quality because they have the ability to impact IAQ on a larger scale than most other pneumatic components. Receiver controllers have to be calibrated in an exact manner to achieve the control necessary in a HVAC pneumatic control system. Incorrect operation of a receiver controller can cause a wide range of temperature, humidity, and airflow problems.

Dual-Input Receiver Controllers. A *dual-input (reset) receiver controller* is a receiver controller in which the change of one variable, commonly the outdoor air-temperature, causes the setpoint of the controller to automatically change (reset) to match the changing condition. **See Figure 12-10.** One transmitter measures the variable that is being controlled, such as hot water temperature. The other transmitter measures the variable that causes the controlled variable to change, such as outdoor air temperature. The controller setpoint change occurs automatically as the outdoor air temperature changes.

One example of the importance of receiver controllers is the outdoor air reset schedule for a hot water system. When the receiver controller is out of calibration, the hot water temperature in the system is not sufficient to provide adequate heating within the facility. Many air-handling units and central plant fan systems use dual-input receiver controllers to automatically change setpoints in the building mechanical systems.

Figure 12-9. *A single-input receiver controller is designed to be connected to only one transmitter and maintain only one temperature, pressure, or humidity setpoint.*

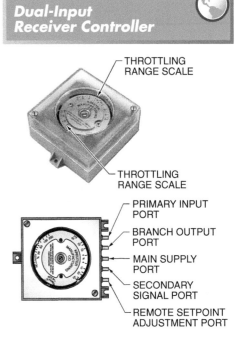

KMC Controls

Figure 12-10. *A dual-input receiver controller is designed to take the change of one variable and reset the setpoint of the controller to match the changing condition.*

AUXILIARY COMPONENTS

An *auxiliary component* is an object used in a pneumatic control system that produces a desired function when actuated by the output signal from a pneumatic controller. Auxiliary components are used to change the direction or amount of airflow, or are used as interfaces. An *interface* is a device that allows two different types of components or systems to interact with each other. An example of an interface is the way an automation system interacts with a transducer to alter the pneumatic signal to an end device.

Auxiliary components are commonly located between the controller and controlled device. When auxiliary components are not operating properly, they cause dampers to be closed or devices such as fans and cooling compressors to be turned off. These situations cause the loss of proper ventilation air, which allows the buildup of harmful IAQ contaminants in a space.

Auxiliary components are categorized based on their function. An auxiliary component may perform more than one function, depending on how it is piped into a space. Auxiliary component functions include changing flow direction, changing pressure, and interfacing between two devices.

An example of a pneumatic auxiliary component that changes the direction of airflow in a control system is a switching relay. Auxiliary components that change air pressure include minimum position relays and positioners. Interface auxiliary components include electric/pneumatic (EP) switches and pneumatic/electric (PE) switches. Pneumatic control systems require interface auxiliary components to enable the control of electric and electronic devices and systems.

IAQ Fact

Receiver controllers can minimize contaminants within a room by changing room pressure. A facility usually requires the pressure of a clean room to be greater than surrounding rooms.

Switching Relays

A *switching relay* is a component that switches airflow from one circuit to another. **See Figure 12-11.** Switching relays are used to change systems from summer to winter operation and create pneumatic logic circuits to enable and disable control system functions.

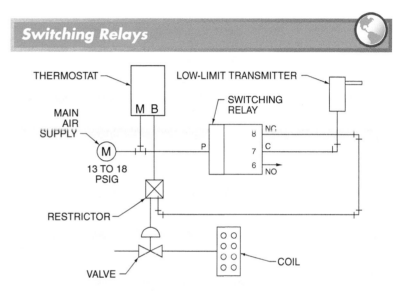

Switching Relays

Figure 12-11. *Switching relays are used to change airflow from one circuit to another.*

Switching relay ports (air connections) include a pilot (P) port, common (C) port, normally open (NO) port, and normally closed (NC) port. The pilot pressure is the input pressure from a controller that causes the relay output to switch from one port to another. The pilot port is a dead-ended port, meaning that the air at the pilot port does not go through the relay.

When pilot pressure is above the relay switch point, the normally closed and common ports are connected. When the pilot pressure is below the relay switch point, the normally open and common ports are connected. The relay switch point may be adjustable or fixed at the factory. Manufacturers use port designations such as P, NO, NC, and C, or combinations such as A, B, C, and D, or 5, 6, 7, and 8. Manufacturer literature is used to determine the port nomenclature of a particular switching relay.

Minimum-Position Relays

A *minimum-position relay* is a relay that prevents outside air dampers from completely closing. Outside-air dampers cannot be permitted to close completely in some applications due to ventilation air requirements. Minimum-position relays allow a minimum amount of outdoor air to flow through the air-handling unit and into the building space. Minimum-position relays are available with a decal that indicates the percentage of outside airflow. The output pressure of a minimum-position relay should be checked to ensure that the outside-air dampers are open sufficiently to allow the proper amounts of outdoor air.

Minimum outside air-damper settings are critical to good indoor air quality. It is difficult to correctly set their minimum position by visual inspection. The damper linkage usually has a percentage indicator, but it has to be set after determining the minimum and maximum travel of the damper blades. Airflow measurements should be used to ensure accuracy.

Minimum-position relays are low-pressure-limit devices in that the device does not allow the output pressure to fall below a predefined, adjustable low limit. For example, in looking at part of a control system, a pneumatic low-limit thermostat acts as a mixed-air controller with an averaging element in the mixed-air plenum. **See Figure 12-12.**

The output of the pneumatic low-limit thermostat is piped to the input of the minimum-position relay. The output of the relay is piped to the outside-air damper. An outside air lockout causes the input signal at the minimum-position relay to fall to zero, causing the outside-air damper to move to its minimum position.

Positioners

A *pneumatic positioner (pilot positioner)* is an auxiliary component mounted to a damper or valve actuator that ensures that the damper or actuator moves to a given position. **See Figure 12-13.** Sending a certain air pressure to an actuator does not guarantee that an actuator extends a certain length. The extension of an actuator may be opposed by rust, corrosion, or external forces acting on the damper or valve, such as airflow striking the damper blade.

IAQ Fact

Electric/pneumatic positioners operate by comparing two signals: the analog input signal from the controller and the actuator position signal from the feedback assembly. The two signals are fed into a comparator which adds air or exhausts air.

Minimum-Position Relays

WARM RETURN AIR
OUTSIDE AIR LOCKOUT
COOL OUTDOOR AIR
MINIMUM-POSITION RELAY
PNEUMATIC LOW-LIMIT THERMOSTAT
AVERAGING ELEMENT
MIXED-AIR PLENUM
OUTSIDE-AIR DAMPER

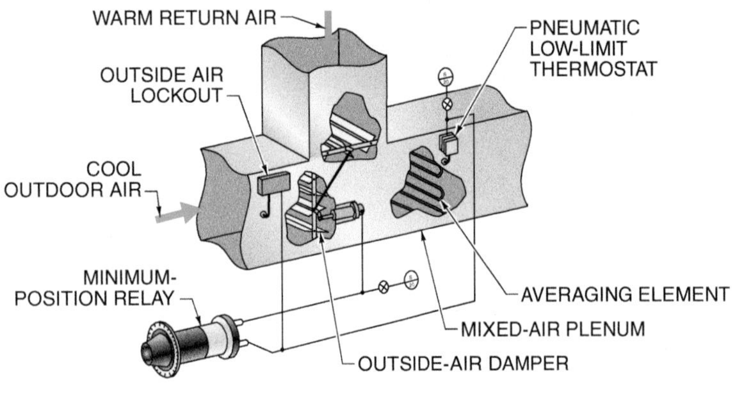

Figure 12-12. Minimum-position relays are used to provide fresh air ventilation and prevent outside air dampers from completely closing.

Positioners

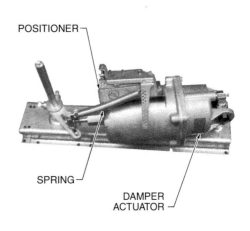

POSITIONER
SPRING
DAMPER ACTUATOR

Figure 12-13. Pneumatic positioners ensure damper or valve actuators move to a given position.

A positioner overcomes external forces and causes a damper to move a specific amount. The operation of the positioner should be checked to ensure that the outside-air dampers are open sufficiently to allow proper amounts of outdoor airflow. Failure to check the operation of the positioner may cause the outside-air dampers to remain closed, thus allowing indoor-air contaminants to build up.

Positioners are usually attached on the side of a damper or valve actuator and have an arm that connects to the actuator shaft through a spring assembly. Pneumatic positioners are used to alter the effective spring range of an actuator. In some applications, it is necessary to closely sequence two or more damper or valve actuators. The actuator spring range must be closely adjusted to obtain the sequencing. Normally, the spring ranges of actuators are nonadjustable. The spring ranges can only be changed if the actuator is replaced.

A positioner moves an actuator through its spring range when it receives a specific, adjustable pressure input. For example, an outside-air damper actuator has a spring range of 3 psi to 11 psi. The actuator is required to move from open to closed when the air pressure input changes from 3 psi to 13 psi. When a positioner is calibrated properly and when the input pressure to the positioner changes from 3 psi to 13 psi, the pressure to the actuator changes from 3 psi to 11 psi. Thus, the actuator appears to have a spring range of 3 psi to 13 psi.

Electric/Pneumatic Switches

An *electric/pneumatic (EP) switch* is a component that enables an electric control system to interface with pneumatic HVAC system components. In an electric/pneumatic switch, an electrical signal operates a pneumatic device. Electric/pneumatic switches are often referred to as three-way, solenoid-operated air valves (SAVs).

The solenoid has two wires that are connected to the appropriate point in the system. Solenoids of different voltages are available, with 24 VAC, 120 VAC, and 240 VAC being

the most common. Standard 60 Hz and 50 Hz models are available.

Electric/pneumatic switches have a NO port, NC port, and a common port. When no electrical power is applied to the solenoid, the NO and common ports are connected internally. The NC port is disconnected from the other ports. When electric power is applied to the solenoid, the NC port and the common port are connected internally. The NO port is disconnected from the other ports. Manufacturer's designations include 1, 2, 3, and A, B, and C.

One application for electric/pneumatic switches is to control a damper by the operation of a fan. **See Figure 12-14.** In this application, the electric/pneumatic switch solenoid is wired in parallel with the fan or fan starter. When the fan is ON, the electric/pneumatic switch is energized. When the fan is OFF, the electric/pneumatic switch is de-energized.

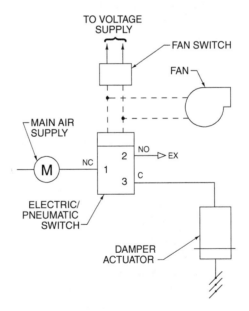

Figure 12-14. Electric/pneumatic switches enable a pneumatic device such as a damper to respond to the operation of an electrical device such as a fan.

Main (supply) air is connected to the normally closed (1) port. The common (3) port is connected to the controller (damper actuator) at the air-handling unit control system. The normally open (2) port is vented to the atmosphere. When the fan starts, the electric/pneumatic switch is energized, the normally closed and common ports are connected, and the main air supply flows to the damper, causing the damper actuator to extend (open damper). When the fan stops, the electric/pneumatic switch is de-energized, the normally open and common ports are connected, and the actuator air is vented to the atmosphere, causing the damper actuator to retract (close damper).

Electric/pneumatic switch failure may result in outside-air dampers failing to open upon fan start. Failure of the outside-air damper EP switch can cause persistent indoor air quality problems by depriving the system of ventilation air. This condition could exist for a long period of time if the outside-air damper operation is not visually inspected on a daily basis. Since electric/pneumatic switches use an electromagnetic field, some technicians remove the top cover of the solenoid and use a small screwdriver to detect the presence of a magnetic field. Digital multimeters may also be used to check voltage and/or resistance at the solenoid coil.

IAQ Fact

EP and PE switch failures frequently cause significant indoor environmental problems. In some instances, switch failure can lead to additional equipment damage. For example, if a PE switch fails to turn on a hot water pump, the result could be a frozen coil.

Pneumatic/Electric Switches

A *pneumatic/electric (PE) switch* is a component that allows an air pressure signal to energize or de-energize an electrical device such as a fan, pump, compressor, or electric heating element. Pneumatic/electric switches perform a task that is the opposite of that of electric/pneumatic switches and cannot be interchanged with them. Pneumatic/electric switches have one inlet air port that activates an electric switch through a bellows mechanism. Pneumatic/electric switches have an NO electrical terminal, an NC electrical terminal, and a common electrical terminal.

The NC and common terminals are connected electrically when the incoming air pressure is lower than the setpoint. In this case, the NO terminal is disconnected. The NO and common terminals are connected electrically when the incoming air pressure is higher than the setpoint. In this case, the NC terminal is disconnected. A voltmeter may be used to check whether the pneumatic/electric contacts are open or closed.

A pneumatic/electric switch has an adjustable pressure setpoint and differential. These adjustments allow for the electrical device to be turned on or off at different controller setpoints. The differential prevents short cycling due to small changes in the incoming air pressure. A differential of 1 psi means that the pressure must change a minimum of 1 psi from the setpoint before the pneumatic/electric switch changes state. The adjustable setpoint allows staging or sequencing of different electrical devices from the same input signal.

Pneumatic/electric switches are also used to allow the signal from an outside air transmitter to energize and de-energize a hot water pump. **See Figure 12-15.** In this application, a pneumatic/electric switch is piped to an outside air transmitter. The pneumatic/electric switch energizes a hot water pump when the outdoor air temperature is below 55°F.

The NC and common terminals disconnect when there is a rise in pressure. Any failure causes the hot water pump to continue running. The setpoint of the pneumatic/electric switch must correspond to the pressure output of the transmitter at 55°F. The pressure corresponding to 55°F, in this application, is 9.9 psi for a −40°F to 160°F transmitter.

Pneumatic/Electric Switches

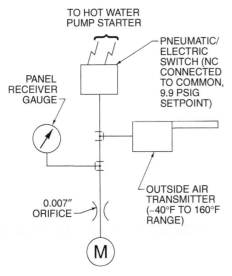

TO HOT WATER PUMP STARTER

PNEUMATIC/ ELECTRIC SWITCH (NC CONNECTED TO COMMON, 9.9 PSIG SETPOINT)

PANEL RECEIVER GAUGE

0.007" ORIFICE

OUTSIDE AIR TRANSMITTER (–40°F TO 160°F RANGE)

M

Figure 12-15. Pneumatic/electric switches allow the signal from an outdoor-air transmitter to be used to energize and de-energize a hot water pump.

CONTROLLED DEVICES

A *controlled device* is the object that regulates the flow of liquid or air in a system to provide a heating, air conditioning, or ventilation effect. Controlled devices are driven by the compressed air that is supplied from a controller, which causes the controlled device to open or close properly. Common controlled devices include actuators that deliver the heating or cooling into the building spaces or system, such as dampers for regulating airflow, and valves for regulating water or steam flow.

Actuators

An *actuator* is a device that accepts an air signal from a controller, which causes mechanical motion to occur. In HVAC control systems, pneumatic actuators accept airflow from the controller and cause the actuator shaft or valve stem to move. The movement of the actuator shaft or valve stem regulates the flow of air through a damper and the flow of water or steam

through a valve. **See Figure 12-16**. If for any reason the actuators are damaged or are operating incorrectly, valves and/or dampers may not open or close properly. A common damper actuator problem occurs when the linkage slips and the damper does not fully open or close. Improper valve or damper operation may cause indoor air quality problems such as inadequate airflow, temperature, or humidity conditions. Any actuators found to be damaged must be repaired promptly.

Actuators

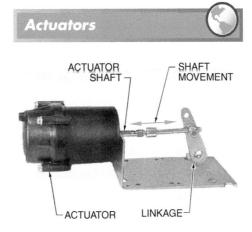

ACTUATOR SHAFT

SHAFT MOVEMENT

ACTUATOR

LINKAGE

Figure 12-16. An actuator and linkage control damper regulate the temperature of a building space.

Dampers control the amount of outdoor air that enters a building. While outdoor air is needed for proper ventilation, too much or too little air can lead to IAQ issues.

Pneumatic Actuators. The components of a pneumatic actuator include the end cap with air fitting, diaphragm, piston cup, and spring and shaft assembly. The end cap with air fitting provides a connection for the air line from the controller to the actuator. Any time the actuator is disassembled or repaired, the end cap bolts and air fitting should be checked for leaks. The actuator body provides a housing and support for the other parts of the actuator. The actuator body protects the other components from damage due to impact.

The diaphragm transmits the force of the incoming air pressure to the piston cup and then to the spring and shaft assembly. The piston cup transmits the force generated by the air pressure against the diaphragm to the spring and shaft assembly. Normally, the piston cup requires no maintenance or replacement. The spring and shaft assembly transforms the air pressure into mechanical movement. In a damper actuator, the shaft is connected through a linkage to a damper. In a valve actuator, the shaft is connected to the stem of the valve.

Actuators may be damper or valve actuators. **See Figure 12-17.** The basic components of damper and valve actuators are the same, but the physical configuration is different. Damper actuator components are normally enclosed in the actuator body and have a longer stroke than most valve actuators.

Direct-Coupled Actuators. Direct-coupled actuators are high quality, inexpensive, and highly reliable. **See Figure 12-18.** *Direct-coupled actuators* are actuators that are directly attached to the damper or valve without the use of a linkage. Direct-coupled actuators are used almost exclusively in most modern control systems. Direct-coupled actuators may use either electro-hydraulic or electromechanical methods to create movement.

Direct-Coupled Actuator

Figure 12-18. Direct-coupled actuators are clamped to the damper shaft.

Electrohydraulic actuators usually consist of a hollow cylinder with a piston inserted in it. The two sides of the piston are alternately pressurized and depressurized by a small hydraulic pump to achieve a controlled, precise movement of the piston and, in turn, of the shaft connected to the piston. The physical linear displacement is only along the axis of the piston/cylinder. Electrohydraulic actuators are no longer installed in new HVAC control systems.

The vast majority of direct-coupled actuators used today are electromechanical actuators. Electromechanical actuators convert rotary motion from an electric motor into linear displacement. There are many designs of electromechanical actuators, but most electromechanical actuators use a bidirectional 24 VAC motor.

Damper and Valve Actuators

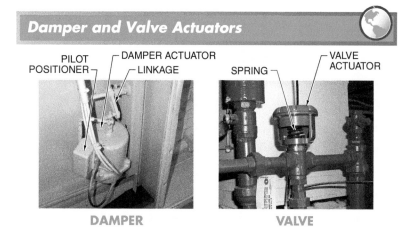

PILOT POSITIONER — DAMPER ACTUATOR — LINKAGE — VALVE ACTUATOR — SPRING

DAMPER **VALVE**

Figure 12-17. Actuators may be damper or valve actuators. Damper actuator components are usually enclosed in the actuator body. Valve actuator components are usually visible from the outside.

Valve actuators often use an adapter kit that has a continuous thread machined on its circumference and running along its length (similar to the threading on a bolt). A nut is threaded onto this shaft and has corresponding threads. To keep the nut from rotatating with the shaft, the nut is interlocked with a stationary part of the actuator. When the shaft is rotated, the nut is driven either up or down the threads, depending on the direction of rotation. When a shaft is connected to the nut by linkages, this shaft converts rotary motion into linear displacement.

Linear displacement is used to cause a valve stem to rise or lower as needed. Most electromechanical actuators are rated by how long it takes to drive from fully open to closed, with 30 sec, 60 sec, 90 sec, and 120 sec being common. The motor can also be reversed or stopped as desired, allowing the valve to open or close to a specific position as well as to remain in place. A return spring is sometimes included on linear pneumatic actuators that will cause the damper or valve to return to a fail-safe position upon power loss. The type of fail-safe control is known as floating or incremental control.

Dampers

A *damper* is a movable piece of metal in a duct used to control the flow of air. Outside air dampers that are operating improperly can cause indoor air problems. Dampers that are stuck may fail to admit sufficient ventilation air to flush out contaminants. Damper seals that are damaged or missing may allow excessive airflow, causing high humidity levels that allow the growth of microorganisms. Dampers are available in different designs and sizes. Damper designs include parallel, opposed, and round blade dampers. **See Figure 12-19.**

A *parallel blade damper* is a damper in which adjacent blades are parallel and move in the same direction with one another. Parallel blade dampers are the most common type of damper used in HVAC systems. An *opposed blade damper* is a damper in which adjacent blades move in opposite directions from one another. Opposed blade dampers are more expensive than parallel blade dampers, but opposed blade dampers provide excellent flow characteristics and precise air control. A *round blade damper* is a damper with a round blade. Round blade dampers are used in systems consisting of round ductwork. Round blade dampers are used primarily with small, variable-air-volume (VAV) terminal boxes. Round blade dampers are simple in operation and provide good air control.

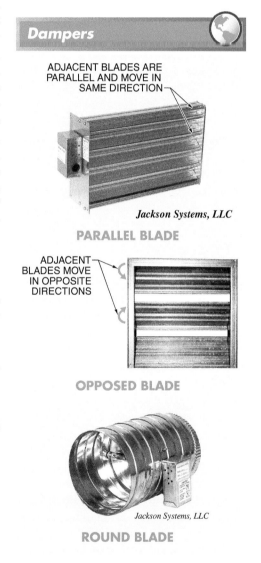

Dampers

ADJACENT BLADES ARE PARALLEL AND MOVE IN SAME DIRECTION

Jackson Systems, LLC

PARALLEL BLADE

ADJACENT BLADES MOVE IN OPPOSITE DIRECTIONS

OPPOSED BLADE

Jackson Systems, LLC

ROUND BLADE

Figure 12-19. Dampers are used to control the flow of air and include parallel, opposed, and round blade dampers.

All dampers list the square feet of the damper blade area, the pressure drop across the damper, and the cubic feet per minute (cfm) of airflow at a given pressure drop. As the area of the damper blade is increased, a larger actuator diaphragm is required to produce the necessary force to open or close the damper. Therefore, a correlation exists between damper size and actuator size. In large applications, high airflow requirements can be met by joining small damper sections together into large sections and connecting the actuators to a common air line, thus driving all damper actuators at the same time.

This is also an application where pilot positioners can be used to drive multiple damper banks. Damper binding and misalignment can lead to ventilation problems. When an outside air damper binds in the open position, the conditioned space can get too much ventilation air. This causes the humidity levels in the space to be too high or too low. In variable-air-volume systems, if the damper binds or is out of adjustment, it can cause too much or too little airflow into the conditioned space.

Valves

A *valve* is a device that controls the flow of fluids in an HVAC system. Hot water is used in heating systems and chilled water is used in cooling systems. Warm water is used in cooling towers or water source heat pumps. Steam is used by coils to provide heat in a building space, in heat exchangers to provide hot water, or in some cases is discharged directly into air handlers to provide humidification. The type of fluid, temperature, and pressure is known collectively as the service of a valve. The service of a valve cannot be changed without changing the capacity or function of the valve. Valves that are sized, operated, or maintained improperly contribute to indoor air quality problems by not allowing proper heat transfer through coils, or proper humidification from humidifiers.

The components of a control valve include the valve body, stem, disc, packing, and

actuator. **See Figure 12-20.** The valve body consists of the outer housing through which steam or water passes. The valve body also contains the means for attaching the valve to the piping. Valves are attached to piping using threads, flared fittings, or flanged fittings. The valve body is usually constructed of cast iron, bronze, steel, or stainless steel.

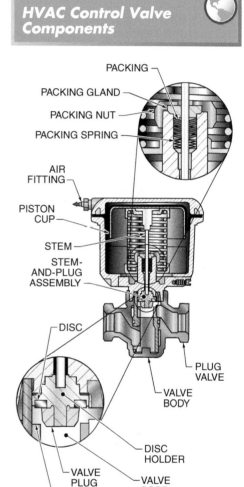

Figure 12-20. Control valves control the flow of fluids in an HVAC system. Control valve components include the valve body, stem, disc, packing, and actuator.

The valve stem transmits the force of the actuator to the valve plug. The valve stem is attached to the actuator piston cup and moves up and down, positioning the valve plug and disc. The valve plug position allows a variable amount of fluid to flow through the valve. The shape of the plug determines the valve flow characteristics. The valve plug has a flat disc that contacts the valve seat. The flow through the valve is shut OFF when the disc contacts the valve seat. The packing and the packing nut prevent leakage of the fluid along the sides of the valve stem. Different types of packing are used for hot water, chilled water, and steam valves. Control valves may be two-way or three-way.

Two-Way Valves. A two-way valve is a valve that has two pipe connections. **See Figure 12-21.** Two-way valves are available in a variety of pipe diameter sizes, from 1/2″ to 27″ or larger. Two-way valves are either normally open or normally closed. A *normally open (NO) valve* is a valve that allows fluid to flow when the valve is in its normal position. A *normally closed (NC) valve* is a valve that does not allow fluid to flow when the valve is in its normal position. Normally closed valves have a plug on the bottom of the valve body, which enables the disassembly and removal of the disc and other internal parts.

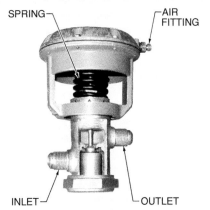

Figure 12-21. Two-way valves have one inlet and one outlet.

Two-way valves are used in steam, hot water, and chilled water applications. Steam valves are always two-way valves because steam does not require a return to the steam header or boiler. The disadvantage of using two-way valves in hot or chilled water applications is that there is no water flow through the valve when closed. A lack of water flow can cause pump wear or damage in large systems because the pump is forced to operate at an unstable load when a number of valves are closed at one time.

Pump or system relief valves are installed to maintain a relatively constant supply pressure, which reduces pump wear. These systems can become complex and costly due to the required piping and control systems. In addition, the sources of the hot or cold water (boiler, converter, or chiller) require a constant water flow in order to operate at their maximum energy efficiency and to prevent damage. Two-way valves often do not meet these needs.

Three-Way Valves. A *three-way valve* is a valve that has three pipe connections. Three-way valves are used in HVAC systems to control the flow of water because they provide a constant system pressure in the supply piping and through the pumps, boilers, heat exchangers, and chillers. Three-way valves may be mixing valves or diverting valves. **See Figure 12-22.**

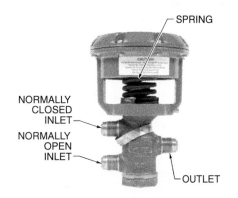

Figure 12-22. Mixing valves and diverting valves have similar designs. Mixing valves have two inlets and one outlet. Diverting valves have one inlet and two outlets.

A *mixing valve* is a three-way valve that has two inlets and one outlet. Mixing valve ports are referred to as common, normally open, and normally closed. Two of the ports, which are the normally open and normally closed ports, are water inlets. The common port is the outlet, and the port designations cannot be switched. A mixing valve disc has an egg-shaped appearance.

When used as a mixing valve, the normally open port is usually piped to the hot water supply, while the normally closed port is piped to the chilled water supply. In this way, the valve fails to the heating position when a problem occurs. The piping may be reversed so that the valve fails to the cooling position instead of the heating position. Also, a mixing valve is often used in a bypass valve application as a means of changing the flow through a system.

A *diverting (bypass) valve* is a three-way valve that has one inlet and two outlets. Diverting valve ports are referred to as common, normally open, and normally closed. One outlet port is normally connected to a heat exchanger or coil (a cooling tower), and the other outlet port is connected to the system bypass.

Diverting valves are used in applications that require a two-position (ON/OFF) flow. Diverting valves cannot be used in mixing applications. Diverting and mixing valves may require balancing valves installed in the piping to balance the flow between supply and return. A diverting valve disc has an hourglass shape.

Case Study – Massachusetts High School

Case Study

Background

The management company of a large office building in the mid-Atlantic region of the United States decided that, rather than replacing aging air compressors on the building's fire-suppression "dry-pipe" sprinkler systems, they would tie them into the very large pneumatic control air compressor. The facility's pneumatic control air compressor stations had more than enough reserve capacity to accommodate the additional load.

The old dry-pipe compressors were removed and new pneumatic piping was tied in from the central plant. There were over 30 dry-pipe-system zones ultimately tied in as described. Each system seemed to work well and the compressors handled the additional load.

Catastrophe Strikes

Several months later, on a summer day, the facility experienced a major catastrophe. On that fateful day, one of the dry-pipe systems tripped and filled the sprinkler pipes with water. Because someone forgot to install a check valve on the air-line feeding, the water from the dry-pipe-system zone back-fed into the rest of the facility's pneumatic system. The water filled every pneumatic main and branch line up to the fourth floor.

Resolution

A check valve was subsequently installed in the dry-pipe system, and all other systems were double-checked. The engineers blew out the air lines for several days to purge all moisture. All of the pneumatic controls in the affected areas had to be replaced, including hundreds of thermostats, as well as other controls. While the moisture was being purged, the VAV boxes went into their fail-safe position—full cooling—which lowered temperatures below setpoint in the occupant space.

HVAC Controls Definitions . . .

- An *actuator* is a device that accepts a signal from a controller, which causes mechanical motion to occur.
- An *air compressor* is a component that takes air from the atmosphere and compresses it to increase its pressure.
- An *auxiliary component* is an object used in a pneumatic control system that produces a desired function when actuated by the output signal from a pneumatic controller.
- A *bimetallic element* is a temperature sensing device that consists of two different metals joined together at one end.
- A *bleedport* is an orifice that allows a small volume of air to be expelled into the atmosphere.
- *Branch line pressure* is the pressure in the air line that is piped from the thermostat to the controlled device.
- A *controlled device* is the object that regulates the flow of liquid or air in a system to provide a heating, air conditioning, or ventilation effect.
- A *damper* is a movable piece of metal in a duct used to control the flow of air.
- *Direct-coupled actuators* are actuators that are directly attached to the damper or valve without the use of a linkage.
- A *diverting (bypass) valve* is a three-way valve that has one inlet and two outlets.
- A *dual-input (reset) receiver controller* is a receiver controller in which the change of one variable, commonly outside air temperature, causes the setpoint of the controller to automatically change (reset) to match the changing condition.
- An *electric/pneumatic (EP) switch* is a component that enables an electric control system to interface with pneumatic HVAC system components.
- A *force-balance design* is a design in which the controller output is determined by the relationship among mechanical pressures.
- An *interface* is a device that allows two different types of components or systems to interact with each other.
- A *limit thermostat* is a pneumatic thermostat that causes a change in a controlled device whenever the set temperature or pressure is reached.
- A *minimum-position relay* is a relay that prevents outside air dampers from completely closing.
- A *mixing valve* is a three-way valve that has two inlets and one outlet.
- A *normally closed (NC) valve* is a valve that does not allow fluid to flow when the valve is in its normal position.
- A *normally open (NO) valve* is a valve that allows fluid to flow when the valve is in its normal position.
- A *one-pipe (low-volume) device* is a device that uses a small amount of the compressed air supply (restricted main air) to operate.
- An *opposed blade damper* is a damper in which adjacent blades move in opposite directions from one another.
- A *parallel blade damper* is a damper in which adjacent blades are parallel and move in the same direction with one another.
- A *pneumatic control system* is a control system in which compressed air is used to provide power for the system.

Definitions

. . . HVAC Controls Definitions

- A *pneumatic/electric (PE) switch* is a component that allows an air pressure signal to energize or de-energize an electrical device such as a fan, pump, compressor, or electric heating element.

- A *pneumatic humidistat* is a controller that uses compressed air to open or close a device that maintains a certain humidity level inside a duct or area.

- A *pneumatic pressurestat* is a controller that maintains a constant air pressure in a duct or area.

- A *pneumatic positioner (pilot positioner)* is an auxiliary component mounted to a damper or valve actuator that ensures that the damper or actuator moves to a given position.

- A *pneumatic thermostat* is a device that senses room temperature and alters branch line pressure to controlled devices to maintain a space's temperature.

- A *pressure regulator* is a modulating valve that provides a constant outlet pressure to downstream components.

- A *receiver controller* is a device that accepts one or more input signals from pneumatic transmitters and produces an output signal based on the setup of the controller.

- A *round blade damper is* a damper with a round blade.

- *Setpoint* is the setting of the desired temperature in the room or space.

- A *single-input receiver controller* is a receiver controller that is designed to be connected to only one transmitter and to maintain only one temperature, pressure, or humidity setpoint.

- A *switching relay* is a component that switches airflow from one circuit to another.

- A *three-way valve* is a valve that has three pipe connections.

- A *two-pipe (high-volume) device* is a device that uses the full volume of compressed air available to operate.

- A *valve* is a device that controls the flow of fluids in an HVAC system.

**For more information please refer to ATPeResources.com
or http://www.epa.gov.**

Review Questions

1. Explain why the proper operation of HVAC control systems is important to IAQ.

2. List the auxiliary components of an air compressor station and the types of contaminants they remove from the system.

3. What are some of the problems that may occur if pneumatic transmitters and controllers malfunction?

4. Describe the difference between electric/pneumatic switches and pneumatic/electric switches.

5. Explain how dampers and valves relate to actuators.

BUILDING AUTOMATION SYSTEMS FOR IAQ

Fluke Corporation

Class Objective

To understand the devices and components of a building automation system, including how these devices and components are used to maintain good IAQ, how they maintain efficient lighting levels, and how they keep the building secure.

On-the-Job Objective

Ensure that a building automation system is working properly to maintain ideal IAQ, lighting, and security conditions.

Learning Activities:

1. Describe the function of distributed direct digital control (DDC) systems.
2. Name the types of building automation controllers used in distributed DDC systems.
3. Explain the difference between analog input devices and digital input devices, and analog output components and digital output components.
4. Describe the two types of control loops used with direct digital control strategies.
5. Explain the use of direct digital control algorithms.
6. **On-the-Job:** List the direct digital control features used in your facility.

Introduction

Building automation systems control HVAC equipment, lighting, security, and other essential functions of a building. These systems also indicate abnormal conditions (alarms) for factors that affect IAQ and provide information regarding HVAC equipment energy consumption and maintenance. How a building automation system is used and updated affects IAQ and occupant health in a building. Routinely testing and calibrating sensors and controllers related to a building automation system are essential to keeping the system operating properly.

261

BUILDING AUTOMATION SYSTEMS

A *building automation system (BAS)* is a system that uses microprocessors (computer chips) to control the energy-using components in a building. Building automation systems are the most common control system installed in commercial buildings today. Most modern building automation systems utilize a distributed direct digital control (DDC) system architecture.

Distributed Direct Digital Control Systems

A *distributed direct digital control (DDC) system* is a building automation control system that has multiple central processing units (CPUs) at the controller level. **See Figure 13-1.** In distributed DDC systems, each controller makes decisions. Distributed DDCs began replacing central-direct digital control systems in the 1980s.

Distributed DDC System Architecture. The difference between older building automation systems and distributed DDC systems is that each controller in a distributed DDC system has the ability to make its own decisions. The decision-making ability is spread out, or distributed, throughout the control system. Distributed direct digital control systems usually have a master controller.

The master controller provides communication support, a convenient place to connect a PC or modem for monitoring purposes, and connections for global data. *Global data* is data needed by all controllers in a network. Global data includes such items as outdoor-air temperature, humidity, and system electrical demand.

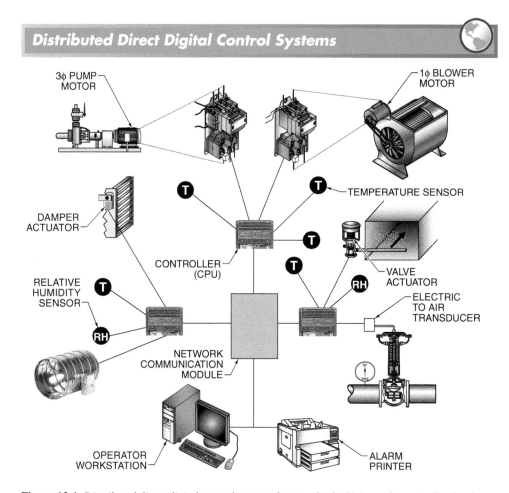

Distributed Direct Digital Control Systems

Figure 13-1. *Distributed direct digital control systems have multiple CPUs at the controller level.*

Stand-alone control is the ability of a controller to function on its own. Stand-alone control is less risky than other types of control because a controller failure is local and has minimal effect on the entire system. Because stand-alone control enables a system to function without the operation of a failed controller, stand-alone control has increased the reliability of building automation systems.

Controllers in a network commonly share information through the communication network that connects them. Controllers that require information from a failed controller in the network are unable to access that information. The operating controllers function at a reduced level of efficiency until the failed controller is repaired or replaced.

Distributed DDC System Functions. Distributed DDC systems have the same control functions as central DDC systems. Functions include duty cycling, electric demand control, and time clocks. Distributed DDC systems can also perform precise closed-loop temperature control, humidity control, and pressure control.

Advantages of distributed DDC systems include improved reliability and increased capacity over previous systems. The stand-alone controller feature makes distributed DDC systems better than any previous building automation control system. Distributed DDC systems are modular and easily expandable. A DDC building automation system can be started with a minimum number of controllers and expanded as future additions are required.

BUILDING AUTOMATION SYSTEM CONTROLLERS

Building automation systems, such as distributed DDC systems, usually require a number of controllers. When connected by a communications network, the controllers provide comprehensive control and monitoring of the HVAC equipment in a building. Controllers are often referred to as modules, control modules, and panels.

The two main types of building automation system (BAS) controllers include application-specific controllers and network communication modules.

Application-Specific Controllers

An *application-specific controller (ASC)* is a controller designed to control only one section or type of HVAC system. **See Figure 13-2.** In most applications, application-specific controllers are more cost-effective than other types of controllers. Application-specific controllers are more cost effective because most of the software programming is done beforehand, reducing setup time.

Most manufacturers install ASCs on the HVAC equipment at the factory. This is referred to as original equipment manufacturer (OEM) installation. Field installation and retrofit problems are minimized by purchasing HVAC equipment with the ASCs installed at the factory. Application-specific controllers include unitary, air-handling unit, and variable-air-volume air-handling unit controllers.

Application-Specific Controllers

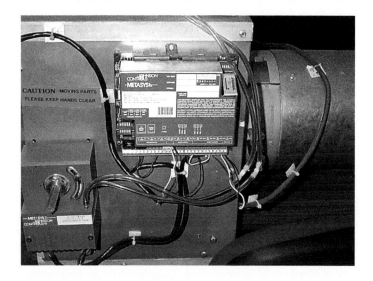

Figure 13-2. *Application-specific controllers are designed to control only one type or part of an HVAC system.*

Unitary Controllers. A *unitary controller* is a type of controller designed for basic zone control using a standard wall-mounted temperature sensor. **See Figure 13-3.** Unitary controllers are designed to control packaged HVAC equipment such as rooftop units, heat pumps, and fan coil units. Unitary controllers are usually compact to allow field mounting at the packaged HVAC equipment. Rain-tight and heated enclosures are often used, which allow operation of multiple stages of heating and cooling, economizer dampers, heat pump reversing valves, and supply fans. Some unitary controllers provide supply-air sensing capabilities, airflow switches, and dirty-filter-condition switches.

Air-Handling Unit Controllers. An *air-handling unit (AHU) controller* is a controller that contains input terminals and output terminals required to operate large central-station air-handling units. **See Figure 13-4.** AHU controllers control humidification, static pressure, and indoor air quality. Many central-station air-handling units have complex control sequences.

Dual-duct and multizone air-handling units have operating requirements that cannot be met by a unitary controller. Air-handling unit controllers are larger and more complex than unitary controllers. AHU controller software is also more complex than unitary controller software, which facilitates managing an air-handling unit's complex control sequences.

Variable-Air-Volume Air Terminal Unit Controllers. A *variable-air-volume (VAV) air terminal unit controller* is a controller that modulates the damper inside a VAV air terminal unit or rpm of the blower motor(s) with signals to variable frequency drives to maintain a specific building space temperature. VAV air terminal unit controllers are similar in appearance to unitary controllers. The difference is in the factory programming of the erasable, programmable read-only memory (EPROM) chip. **See Figure 13-5.**

VAV air terminal unit controllers modulate the primary damper inside the VAV terminal box or rpms of the blower motors. Reheat valves can also be controlled. A flow input signal to the controller can be set up to indicate airflow (in cubic feet per minute). The controller uses the flow rate input signal to control the speed of the blower inside the VAV air-handling unit. The software for most variable-air-volume air-terminal unit controllers allows VAV air-handling units to provide air flow measurement readouts.

Unitary Controllers

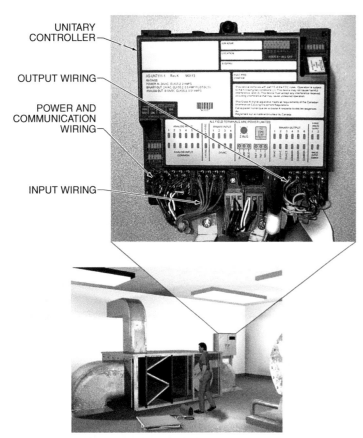

UNITARY CONTROLLER

OUTPUT WIRING

POWER AND COMMUNICATION WIRING

INPUT WIRING

Figure 13-3. *Unitary controllers are designed for simple zone control using a standard wall-mounted temperature sensor.*

IAQ Fact

VAV air-handling units can be performance assessed using AHU Performance Assessment Rules (APAR) software. APAR is applicable to single-duct variable-volume AHUs with air-side economizers.

Air-Handling Unit Controllers

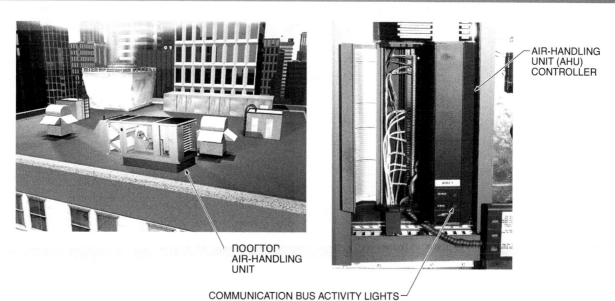

ROOFTOP
AIR-HANDLING
UNIT

AIR-HANDLING
UNIT (AHU)
CONTROLLER

COMMUNICATION BUS ACTIVITY LIGHTS

Figure 13-4. *Air-handling unit controllers are used when the sophisticated control of large central station air-handling units is required.*

VAV Air Terminal Unit Controllers

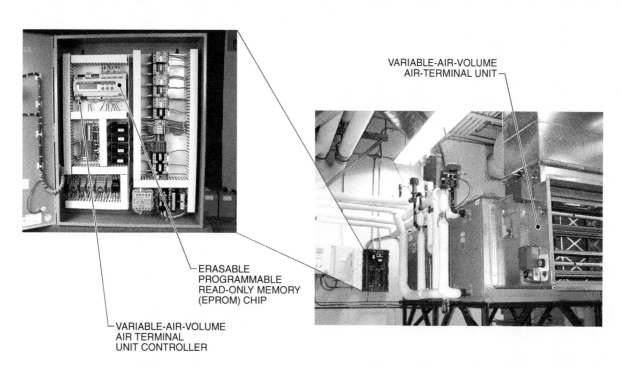

VARIABLE-AIR-VOLUME
AIR-TERMINAL UNIT

ERASABLE
PROGRAMMABLE
READ-ONLY MEMORY
(EPROM) CHIP

VARIABLE-AIR-VOLUME
AIR TERMINAL
UNIT CONTROLLER

Figure 13-5. *Variable-air-volume (VAV) air terminal unit controllers are similar in appearance to unitary controllers. The difference is in the factory programming of the erasable programmable read-only memory (EPROM) chip.*

A common AHU controller application is the control of a variable-air-volume (VAV), air-handling unit (AHU) to maintain a minimum static pressure in the supply duct. **See Figure 13-6.** The VAV AHU heating and cooling controls maintain a constant discharge air temperature of 55°F. An output signal from the air-handling unit controller is provided to control (increase) the volume of air provided by the supply fan. The air volume is increased by controlling the speed of the blower motor using an electric motor drive or by opening dampers to admit a greater amount of air.

Variable-air-volume air-handling unit controllers are also used in pressure-independent VAV terminal boxes with reheat applications. In a pressure-independent application, building space temperature is controlled by the amount of airflow to the space. In addition to controlling building space temperature, a flow sensor is installed at the inlet to the VAV terminal box. **See Figure 13-7.** The flow sensor (pitot tube or differential pressure sensor) is connected to the VAV air-handling unit controller to measure the amount of airflow. An output connection from the controller is provided to the coil of the reheat valve, so the reheat valve can be opened on a call for heat.

Network Communication Modules

A *network communication module (NCM)* is a special controller that coordinates communication from controller to controller on a network and provides a location for operator interface. **See Figure 13-8.** In most building automation systems, individual controllers do not communicate directly with each other. Network communications modules usually have a number of ports (series and parallel) for connecting operator interface devices to the system.

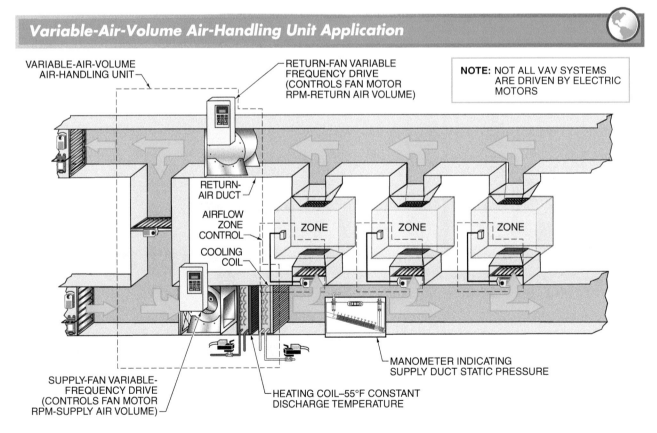

Variable-Air-Volume Air-Handling Unit Application

VARIABLE-AIR-VOLUME AIR-HANDLING UNIT

RETURN-FAN VARIABLE FREQUENCY DRIVE (CONTROLS FAN MOTOR RPM-RETURN AIR VOLUME)

NOTE: NOT ALL VAV SYSTEMS ARE DRIVEN BY ELECTRIC MOTORS

RETURN-AIR DUCT

AIRFLOW ZONE CONTROL

COOLING COIL

ZONE

ZONE

ZONE

SUPPLY-FAN VARIABLE-FREQUENCY DRIVE (CONTROLS FAN MOTOR RPM-SUPPLY AIR VOLUME)

HEATING COIL–55°F CONSTANT DISCHARGE TEMPERATURE

MANOMETER INDICATING SUPPLY DUCT STATIC PRESSURE

Figure 13-6. *Variable-air-volume (VAV) air-handling unit controllers are used to control VAV air-handling units in order to maintain a minimum static pressure in supply ducts.*

Generally, a network communication module obtains information from individual controllers and passes it along to other controllers in the network. One network communication module is normally required in small- to medium-size buildings having up to 200 controllers. Large installations may have several network communication modules that are networked together.

OPERATOR INTERFACE METHODS

A variety of methods are available that allow stationary engineers to access information and use the building's automation system. An *operator interface* is a device that allows a technician to access and respond to building automation system information. Operator interface methods include the use of on-site and off-site devices. Choosing the correct operator interface device(s) and using them properly are required for the proper performance of any system.

On-Site Operator Interface Methods

In medium- and large-sized buildings, such as hospitals, colleges, or office buildings, maintenance or management personnel commonly access and troubleshoot the automation system of a building as part of normal day-to-day responsibilities. On-site methods and tools used in large buildings to access and troubleshoot a building automation system include desktop personal computers (PCs), alarm printers, notebook PCs, portable operator terminals, keypad displays, and dumb terminals. **See Figure 13-9.**

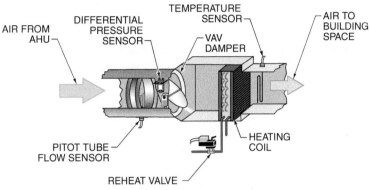

VAV Terminal Box with Reheat

Figure 13-7. *Variable-air-volume (VAV) air-handling controllers are used to control building space temperature and air volume in VAV air-handling systems.*

Controller Network Communication Modules

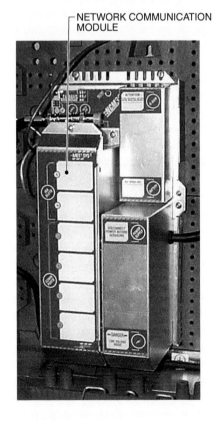

NETWORK COMMUNICATION MODULE

Figure 13-8. *Network communication modules coordinate communication from controller to controller on a network and provide a location for operator interface.*

IAQ Fact

The main reason for dissatisfaction with building automation systems is the lack of effective training. When personnel are unable to properly use a system's operator interface devices, they are not able to perform routine tasks as expected. A training needs analysis survey should be performed on a building's automation system to determine the course content for all personnel.

DESKTOP PCs

ALARM PRINTERS

NOTEBOOK PCs

PORTABLE OPERATOR TERMINALS

KEYPAD DISPLAYS

DUMB TERMINALS

Figure 13-9. *Desktop personal computers (PCs), alarm printers, notebook PCs, portable operator terminals, keypad displays, and dumb terminals are used in large buildings to access and troubleshoot a building's automation system.*

Off-Site Operator Interface Methods

The use of web browser interfaces is an important BAS development. It allows a stationary engineer to access and view the BAS over the Internet using a computer that is running web browser software. Web browser capability is available via a software package, which usually runs on a dedicated web server, or is built into the highest-level controllers provided with the BAS system. Stationary engineers can take advantage of this capability to monitor and control the BAS from any computer equipped with a web browser. Due to security concerns, stringent firewall and password protection considerations must be given to off-site operator interfaces. **See Figure 13-10.**

Also, the use of an Internet communications protocol, called XML, can aid in the use of the Internet for building control.

Server Data

A *server computer* is a large-capacity, hard-drive computer attached to a network. When used with building automation systems, server computers have multiple purposes. The server is often responsible for issuing email alarms and alerts. Server computers also are used to store large amounts of historical data regarding system performance. For IAQ, this might involve the trending of ventilation flow amounts, fan operation, temperature, humidity, and pollutant measurements such as carbon dioxide.

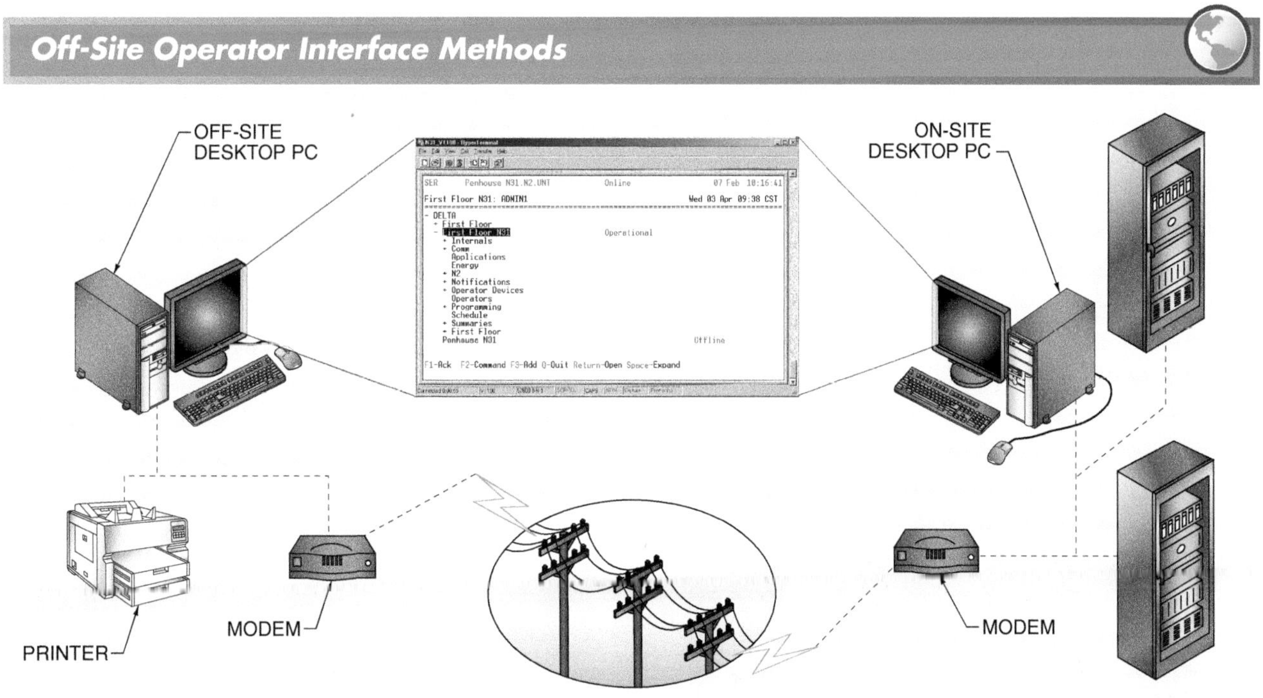

OFF-SITE
DESKTOP PC

ON-SITE
DESKTOP PC

PRINTER

MODEM

MODEM

Figure 13-10. *A building automation system with web browser software packages allows a stationary engineer to communicate and change settings from anywhere in the world.*

Software tools are available on the server computer that permit operating personnel, as well as regulatory agencies, the ability to pull desired data out and then format the data for ease of use. This is especially important when trying to determine causes of IAQ problems. The server may also host needed tools such as a computerized maintenance management system (CMMS). Some CMMSs can transfer data directly from building automation systems.

BUILDING AUTOMATION SYSTEM INPUT DEVICES

A *building automation system input device* is an object such as a sensor that indicates building conditions to a BAS controller. The central processing unit (CPU) of a BAS controller makes decisions based on the information received from input devices. The decisions are used to change the state of output components. BAS input devices send analog signals or digital signals to the BAS controller.

Analog Input Devices

An *analog input device (AI)* is a sensor that indicates a variable such as temperature, pressure, or humidity and causes a proportional electrical signal change at the building automation system controller. **See Figure 13-11.** Analog input devices provide stationary engineers with an actual readout of the variable on a personal computer or other operator interface device. Analog input devices are usually more expensive than digital input devices.

The proper analog input device must be selected for a particular task in a commercial building. Factors to consider when selecting an input device are the price, reliability, its proposed location, environment, and the humidity level in which the input device will operate. Once the input device is selected, proper installation procedures must be followed to ensure proper device operation. Analog input devices include temperature sensors, humidity sensors, pressure sensors, and specialized IAQ devices such as carbon dioxide and formaldehyde sensors.

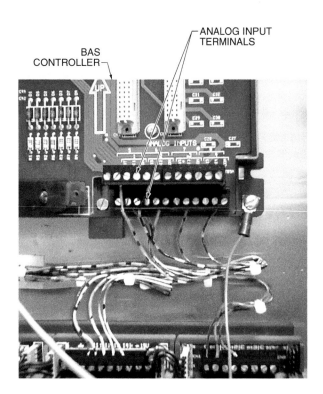

**BAS CONTROLLER—
ANALOG INPUT TERMINAL STRIP**

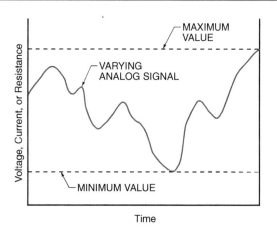

ANALOG SIGNAL

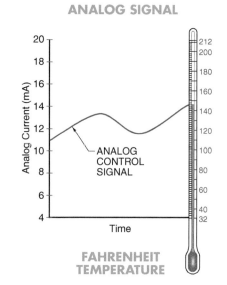

FAHRENHEIT TEMPERATURE

Figure 13-11. *Analog input devices send 4 mA to 20 mA or 0 V to 10 or 20 V signals to a building automation system's controllers, causing a proportional electrical signal change to output components.*

Analog Temperature Sensors. An analog temperature sensor is an analog input device that measures the temperature in a duct, pipe, or room and sends a signal to a controller. Temperature sensors are the most common analog input devices used in building automation systems. Temperature sensors are available in a variety of packages and mounting configurations. Common temperature sensor designs include wall-mounted, duct-mounted, immersion (well-mounted), and averaging. **See Figure 13-12.** Averaging temperature sensors are the most common design used in large buildings.

A *wall-mounted temperature sensor* is the most common temperature sensor used in building automation systems and is used to sense air temperatures in building spaces. A *duct-mounted temperature sensor* is used to sense air temperatures inside ductwork. An *immersion (well-mounted) temperature sensor* is used to sense water temperature in piping systems. An *averaging temperature sensor* is used in large duct systems to obtain an accurate temperature reading. Temperature sensors that are designed to sense outdoor air temperatures are also available.

All temperature sensors must be placed in the proper location to sense the desired temperature. For example, outdoor-air temperature sensors must be located away from exposure to direct sunlight. All manufacturers provide sunshields to prevent direct sunlight from affecting outdoor-air temperature sensor operation. In addition, averaging sensors must be installed properly by stringing the flexible sensing element across the opening of the duct. Standard temperature sensor analog output signal values are 4 mA to 20 mA and 0 VDC to 10 VDC.

Analog Humidity Sensors. Human comfort, product integrity, and corrosion prevention require that the humidity level in a commercial building be controlled. An *analog humidity sensor* is an analog input device that measures the amount of moisture in the air and sends a proportional signal to a controller. The most common humidity sensors measure the percent of relative humidity (%RH), while other humidity sensors measure dew point or absolute humidity. Most humidity sensors use a hygroscopic element. A *hygroscopic element* is a sensor mechanism that changes its characteristics as the humidity level in the air changes. **See Figure 13-13.**

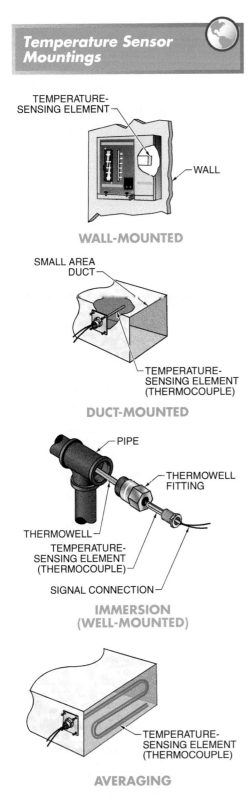

Temperature Sensor Mountings

TEMPERATURE-SENSING ELEMENT

WALL

WALL-MOUNTED

SMALL AREA DUCT

TEMPERATURE-SENSING ELEMENT (THERMOCOUPLE)

DUCT-MOUNTED

PIPE

THERMOWELL FITTING

THERMOWELL

TEMPERATURE-SENSING ELEMENT (THERMOCOUPLE)

SIGNAL CONNECTION

IMMERSION (WELL-MOUNTED)

TEMPERATURE-SENSING ELEMENT (THERMOCOUPLE)

AVERAGING

Figure 13-12. *Temperature sensors are available in a variety of packages and mounting configurations including wall-mounted, duct-mounted, immersion (well-mounted), and averaging.*

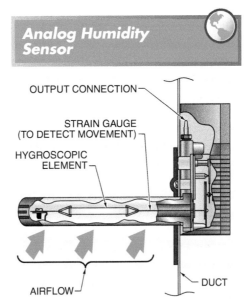

Analog Humidity Sensor

OUTPUT CONNECTION

STRAIN GAUGE (TO DETECT MOVEMENT)

HYGROSCOPIC ELEMENT

DUCT

AIRFLOW

Figure 13-13. *Humidity sensors contain a hygroscopic element that changes physical characteristics as the humidity changes.*

Accurate humidity sensing and control is more difficult than temperature sensing and control. Humidity sensors are susceptible to dirt on the hygroscopic element. Many humidity sensors drift over time, with inaccuracies of 1% per year being common. Also, when certain gases or solvents are present in the air, the gases will damage the sensing element or prevent the element from functioning properly.

Proper sensor function is essential in process or product storage environments where humidity accuracies must be maintained. Humidity sensors must be checked a minimum of once per year for accuracy because the human body is not an accurate indicator of humidity level. Any test instrument used to check the accuracy of humidity sensors must be regularly checked against a known standard.

Humidity sensors are often combined in a single package with temperature sensors to simplify humidity and temperature sensing. Analog humidity sensors use the same 4 mA to 20 mA and 0 VDC to 10 VDC standard analog signal values as temperature sensors.

Analog Pressure Sensors. An *analog pressure sensor* is an analog input device that measures the pressure in a duct, pipe, or room and sends a signal to a controller. Pressure sensors are used in HVAC applications to provide an indication of the actual pressure inside a duct or pipe to a BAS controller. Piezoelectric elements (pressure-sensitive crystals) are used in some pressure-sensing applications. Bellows elements are used in many piping applications. **See Figure 13-14.** Standard signal values of analog pressure sensors include 4 mA to 20 mA and 0 VDC to 10 VDC.

Care must be taken with the installation and location of pressure sensors. Pressure sensors located in an improper area sense an incorrect pressure. Tubing and accessories must be installed according to manufacturer recommendations. Also, it is necessary to ensure the sensor range is correct for the expected system pressures. For example, a sensor measuring between 0 inches of water column (in. WC) to 2 in. WC must not be used when duct static pressure is higher than 2 in. WC. Some pressure sensors require an additional power supply and may require three or four wires instead of two. Many pressure sensors are polarity-sensitive, requiring that the installation of the pressure sensor be to manufacturer specifications only.

IAQ Analog Input Devices. Specialized analog input devices are available that can measure the system variables needed for effective IAQ measurement and control. For example, gas sensors are used to detect the amount of explosive or dangerous gas in the air. Refrigerant sensors are used to detect the amount of refrigerant in the air due to leaks.

Carbon dioxide sensors are used to determine when indoor air quality is acceptable. Indoor air quality is related to the level of carbon dioxide in the air due to human activity. Carbon monoxide sensors are used to sense the amount of toxic exhaust fumes at ground level entrances to a building and inside parking garages. Analog input signals from devices such as variable-frequency drives (VFDs) are often used to indicate motor speed (in revolutions per minute), motor current (in amperes), and frequency (in hertz) to a building automation system controller.

In IAQ applications, a differential pressure sensor in a duct can act as an airflow measuring station (AMS). An airflow measuring station consists of a number of sensors at regular intervals in a duct. The sensor location is the same as that used in performing a duct traverse. **See Figure 13-15.** There are different types of AMSs available.

A pitot tube-type airflow-measuring device has pressure outlets at the array that are connected to a differential pressure transmitter (DPT). The DPT is then wired to the building automation system controller. After mathematical scaling is performed at the controller, direct airflow readouts in cubic feet per minute (cfm) are available. These readouts are used for fan control, alarming, and for long-term data storage (maintenance information).

Analog Pressure Sensors

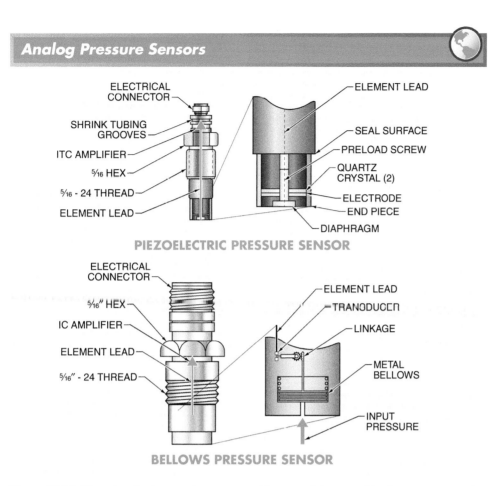

Figure 13-14. *Piezoelectric elements (pressure-sensitive crystals) are used in duct pressure-sensing applications, while bellows elements are used in many piping applications.*

Airflow Measuring Stations

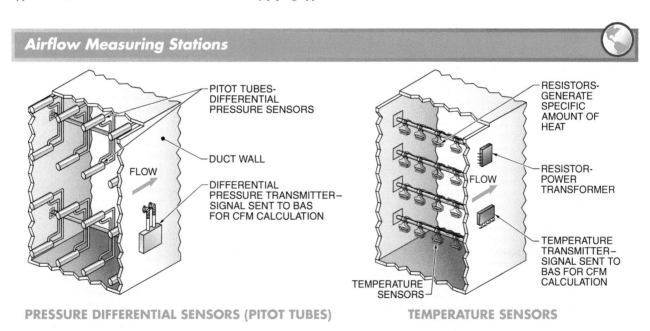

Figure 13-15. *In IAQ applications, differential pressure sensors or pitot tubes and temperature sensors in a duct can act as airflow measuring stations (AMS).*

A second type of AMS uses electronic resistance temperature sensors in a duct. A current is passed through the wires and resistors in the duct. The resistors are placed at intervals that correspond to pitot-tube-traverse measuring points. The temperature at the sensors is related to the airflow across the array.

Digital Input Devices

A *digital (binary) input device* is a sensor that produces only an ON or OFF signal. **See Figure 13-16.** Digital input devices differ from analog input devices in that analog input devices provide a varying readout of the actual value, while digital input devices indicate when a value is above or below a certain setpoint. For example, an analog input device may provide a reading of 74.5°F. A digital input device may indicate that a temperature is above or below 70°F.

Digital input devices do not provide as much information as analog input devices. However, digital input devices are less expensive than analog input devices and are often used as status points in a system. Digital input devices are used to indicate when a fan is ON or OFF, a pump is ON or OFF, a filter is plugged or in good condition, or a room is occupied or unoccupied. Manual switches, start buttons, stop buttons, and selector switches fall into the category of digital input devices.

The input signaling (analog or digital) selected for an application is often based on cost and functionality. For example, an analog input signal is used when a variable (68°F to 72°F) is required. A digital input signal is used when only the status (ON or OFF) of a device is required. Digital input devices are commonly designated as dry contact closure devices.

Digital Input Device Symbols

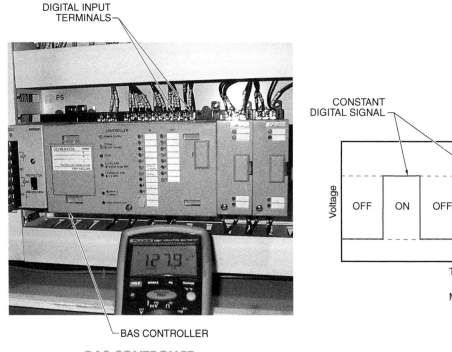

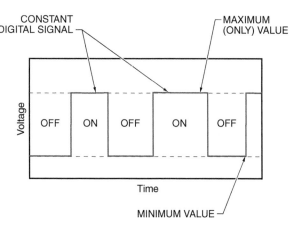

Figure 13-16. *A digital input device signal is a binary (ON or OFF) signal.*

The input terminals of dry contact closure devices require an external power supply. Dry contact closure devices require that the wiring bring voltage to the controller terminals to function. Digital input devices include thermostats, humidistats, and differential pressure switches.

Thermostats. A *thermostat* is a digital input device and controller. Limit thermostats indicate whether a temperature is above or below a certain value. **See Figure 13-17.** Limit thermostats are usually used to indicate an improper temperature level and are not used as primary temperature controllers. Limit thermostats are usually inserted into a duct or strapped to a pipe. Limit thermostats also have a setpoint that determines the temperature at which the thermostat changes state (ON or OFF).

A building automation system controller only reads whether the thermostat is open or closed. The BAS software loaded into the controller indicates whether an open thermostat indicates a high or low temperature. Normally open and normally closed contacts are available for operations such as opening on a rise in temperature or closing on a rise in temperature.

Humidistats. A *humidistat* is a digital input device that indicates that a humidity setpoint has been reached. Humidistats are often used to indicate the humidity level in a duct or building space. Like digital thermostats, digital humidistats have a setpoint with normally open and normally closed contacts. Digital humidistats are usually less expensive than analog humidity sensors. Care must be taken when installing a humidistat and selecting the setpoint. **See Figure 13-18.**

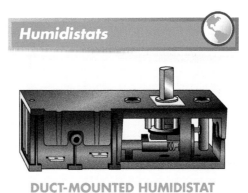

Humidistats

DUCT-MOUNTED HUMIDISTAT

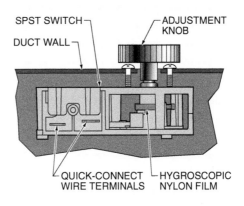

SPST SWITCH

DUCT WALL

ADJUSTMENT KNOB

QUICK-CONNECT WIRE TERMINALS

HYGROSCOPIC NYLON FILM

Figure 13-18. *Humidistats are digital input devices used in ducts and indicate when a humidity setpoint has been reached.*

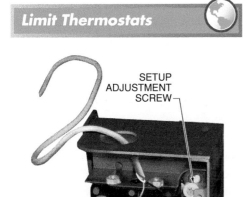

Limit Thermostats

SETUP ADJUSTMENT SCREW

WIRE TERMINATION SCREWS

Figure 13-17. *Limit thermostats maintain a temperature above or below an adjustable setpoint.*

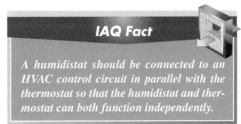

IAQ Fact

A humidistat should be connected to an HVAC control circuit in parallel with the thermostat so that the humidistat and thermostat can both function independently.

Differential Pressure Switches. A *differential pressure switch* is a digital input device that opens or closes contacts because of the difference between two pressures. The setpoint of a differential pressure switch must be adjusted correctly and the correct contacts (normally open or normally closed) selected for a given application. Differential pressure switches are often used to indicate a difference in pressure across a fan or pump or to indicate the filter condition of an air-handling unit.

Differential pressure switches usually have a section of tubing located on the suction (inlet) side of a fan or pump and another section of tubing located on the discharge side. **See Figure 13-19.** Differential pressure switch tubing must be installed to avoid close bending, kinking, or long piping runs. Any air leaks must be located and corrected before differential pressure switches are placed in service.

Differential pressure switches must not be located too close to a fan or pump where high pulsations and vibrations occur. Also, differential pressure switches must not be installed too close to ductwork elbows, tees, or other obstructions that affect the pressure measurement. It is important to always read and follow the manufacturer's installation recommendations.

Differential Pressure Switches

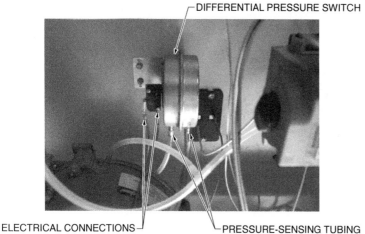

DIFFERENTIAL PRESSURE SWITCH

ELECTRICAL CONNECTIONS

PRESSURE-SENSING TUBING

Figure 13-19. *Differential pressure switches are often used to indicate a difference in pressure across a fan or pump or to indicate the condition of an air-handling unit filter.*

BUILDING AUTOMATION SYSTEM OUTPUT COMPONENTS

A *building automation system output component* is a piece of electrical equipment that changes state (ON or OFF) in response to a command from a building automation system controller. BAS output components may be either directly wired to the controller or wired to an interface such as a transducer or relay, which changes the value of the output signal. Building automation system output components are either analog (positioning between 0% and 100%) or digital (ON or OFF).

Analog Output Components

An *analog output component* is a piece of electrical equipment that receives a continuous signal that varies between two values. Analog output components receive signals in the ranges of 4 mA to 20 mA, 0 VDC to 5 VDC, or 0 VDC to 10 VDC. Analog output components that are proportionally controlled include such items as damper actuators, valve actuators, and variable-frequency drives (VFDs). Analog output components lack standardization. For example, a manufacturer may produce output components with a specific voltage range that are incompatible with other system equipment voltage ratings.

Analog output components can be connected directly to the controller or have their signal changed by a transducer or relay. Variable-frequency drives often receive analog output signals from a BAS controller ranging from 4 mA to 20 mA or 0 VDC to 10 VDC. The analog controller output signal causes the variable-frequency drive to speed up or slow down a system motor. Analog output signals from controllers are also connected directly to motor-actuated valves and dampers for proportional positioning.

Transducers. A *transducer* is an electrical component that changes one type of proportional control signal into another type of signal. Most transducers are connected directly to controlled components such as valve actuators, damper actuators, and variable-frequency drives. These controlled components use common current and voltage ranges such as 4 mA to 20 mA or 0 VDC to 10 VDC.

Some transducers can receive an input current signal or voltage signal and output a variable resistance, such as 0 Ω to 135 Ω. These transducers are used with actuators that require a variable-resistance input. Other transducers are available that use an analog input signal to control multistage heating and cooling units.

Electric/pneumatic transducers usually have an analog input signal, such as 4 mA to 20 mA, from the building automation system controller and output a proportional air pressure signal of 0 psi to 20 psi. **See Figure 13-20.** The transducer output air pressure is often adjustable to match the actual spring range of the actuator. Pneumatic transducers are used in applications that have existing pneumatic valve and damper actuators. Pneumatic transducers enable pneumatic actuators to be used with modern building automation systems, simplifying installation and reducing job costs.

Digital Output Components

A *digital output component* is a piece of electrical equipment that receives an ON or OFF signal. The software of a BAS controller evaluates the information received as input signals, and, depending on the programming, the controller sends a command (ON—voltage signal or OFF—0 V, no signal) changing the state of the controlled digital output component. The change of state is caused by the controller sending a digital output signal from the output section of the controller.

Most digital output terminals of a building automation system controller are triac controlled. Triac-switched controller output terminals require an external power supply that is 12 V or 24 V. A *triac* is a solid-state switching device used to switch alternating current (AC). **See Figure 13-21.** Digital output components include indicator lights, relays, incremental output components (stepper motors), and pulse-width-modulated output components (electric motor drives).

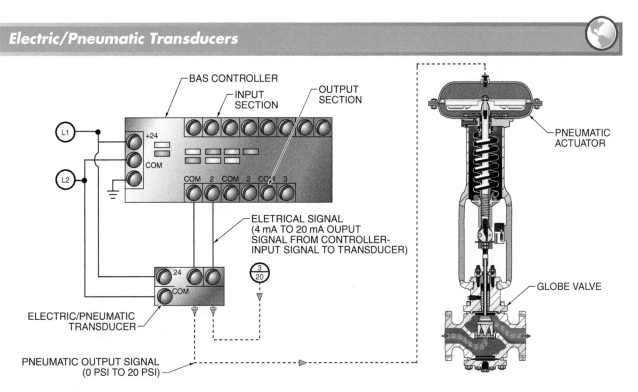

Electric/Pneumatic Transducers

Figure 13-20. *Electric/pneumatic transducers have an analog input signal, such as 4 mA to 20 mA, from the building automation system and an output air pressure signal of 0 psi to 20 psi.*

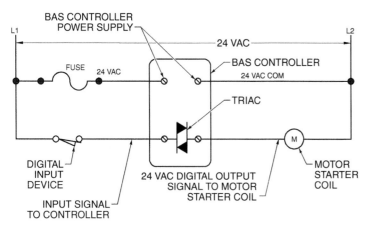

CONTROLLER APPLICATION

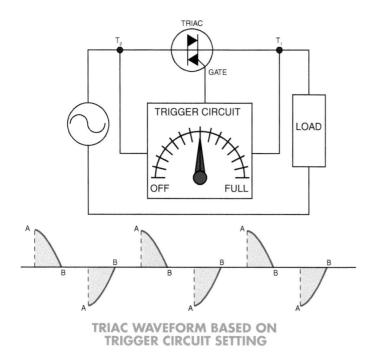

**TRIAC WAVEFORM BASED ON
TRIGGER CIRCUIT SETTING**

Figure 13-21. *Triac-switched controller output terminals require an external power supply and are used to switch alternating current (AC).*

Application-specific controllers are commonly used to control packaged rooftop units and heat pumps. Rooftop units and heat pumps usually contain low-voltage control systems rated at 24 VAC. Most BAS controllers use triacs for output terminal switching, which are rated as Class II (low-voltage) devices. Therefore, application-specific controllers can only directly control low-voltage and low-current output components. The output terminals of an application-specific controller can be wired directly to the input terminals of a rooftop unit or heat pump controller.

In many cases, the existing rooftop unit or heat pump transformer can be used to provide power for the application-specific controller output terminals. Care must be taken to avoid interference problems from spark ignition devices and to ensure the transformer is of the proper size, which is measured in volt-amps (VA). Stationary engineers can often use the standard color-coding of wires to make the proper connections.

Relays. The majority of controlled components, such as motors, are line-voltage components that operate a line voltage of 120 VAC and higher. BAS controller output terminals are generally rated at low voltage. Therefore, a relay must be used to allow the low-voltage-rated triacs in the BAS controller output terminals to switch line-voltage components. **See Figure 13-22.** Triac output terminals are usually not powered, so an external power supply such as one from a 24 VAC transformer must be used. The application-specific relay can be supplied by the manufacturer or purchased separately.

DIRECT DIGITAL CONTROL STRATEGIES

A *control strategy* is a BAS software method used to control the energy-using equipment in a building. A *direct digital control (DDC) strategy* is a control strategy in which a building automation system performs closed-loop temperature, humidity, or pressure control. A *direct digital control (DDC) system* is a building automation system in which controllers are wired directly to controlled components to turn them ON or OFF. **See Figure 13-23.**

Relays

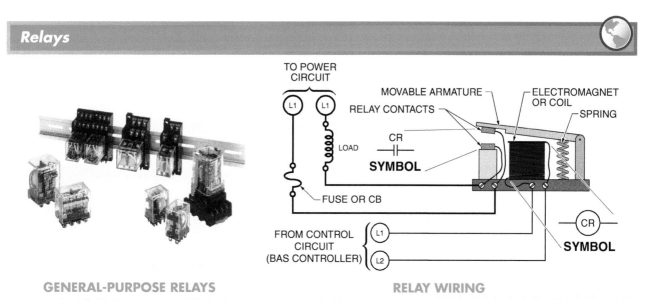

GENERAL-PURPOSE RELAYS

RELAY WIRING

Figure 13-22. *Relays are used to allow low-voltage-rated triacs of building automation system controllers to switch line-voltage components ON and OFF.*

Direct Digital Control Systems

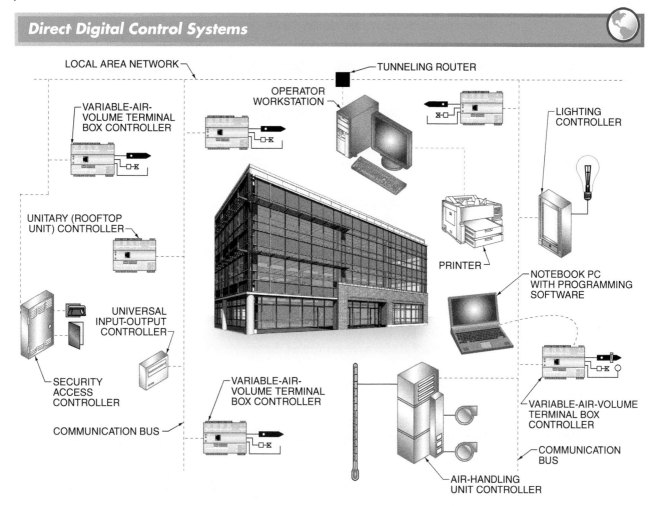

Figure 13-23. *In a direct digital control system, a building automation system controller is wired directly to controlled components.*

Direct digital control (DDC) systems contain controllers, sensors, and controlled components. The controller compares the actual values received to the programmed setpoint parameters. The controller also calculates the desired position of the controlled component and outputs a signal to the component. The sensors measure temperature, pressure, or humidity and provide the values measured as input signals to the controller. The controlled component(s) change position and supply heating, ventilation, air conditioning, lighting, or some other process variable. Most building automation systems are DDC systems.

Direct digital control strategies are not designed to be final safety devices. For example, low-temperature limit controls, boiler high-temperature limit controls, and flame safeguard controls are required with DDC systems. In addition, the installation of a DDC system does not reduce the need for regular preventive maintenance.

Direct digital control systems are often identified by the use of transducers in a building automation system. However, transducers only change signals from one form to another form. The signal change does not affect the building automation system, which still controls the pneumatic dampers or valve actuators. DDC systems include closed-loop control and open-loop control.

Closed-Loop Control

Closed-loop control is a type of control system where feedback occurs among the controller, sensor, and controlled component. *Feedback* is the measurement of the results of a controller action by a sensor or switch. **See Figure 13-24.**

For example, a hot water terminal device is being controlled by a thermostat. The controller (thermostat) positions the hot water valve to maintain the setpoint temperature in the building space. The thermostat in the building space contains the feedback to control the hot water valve and the temperature in the space.

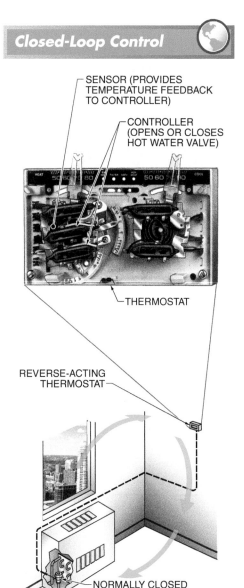

Closed-Loop Control

SENSOR (PROVIDES TEMPERATURE FEEDBACK TO CONTROLLER)

CONTROLLER (OPENS OR CLOSES HOT WATER VALVE)

THERMOSTAT

REVERSE-ACTING THERMOSTAT

NORMALLY CLOSED HOT WATER VALVE (CONTROLLED COMPONENT)

Figure 13-24. *In closed-loop control, feedback occurs among the controller, sensor, and controlled component(s).*

Open-Loop Control

Open-loop control is a type of control system where no feedback occurs among the controller, sensor, and controlled components. **See Figure 13-25.** For example, a controller cycles a chilled water pump ON when the outdoor-air temperature is above 65°F. In an open-loop control system, the controller has no feedback regarding the status of the pump, such as whether or not the pump started.

DIRECT DIGITAL CONTROL FEATURES

Direct digital control (DDC) features determine exactly how a particular energy-using component in a building is controlled. Direct digital control features include setpoint control, reset control, low-limit control, high-limit control, and high/low signal select.

Setpoint Control

The most common DDC feature is the ability to maintain a setpoint with a building automation system. A *setpoint* is the desired value to be maintained by a system. The setpoint can be one of many controlled variables, such as temperature, humidity, pressure, light level, dew point, and enthalpy.

The setpoint and the desired accuracy can be programmed into a BAS controller. For example, a building automation system in the summer is required to maintain a temperature of 72°F in a commercial building. The 72°F temperature is the setpoint of the DDC system. **See Figure 13-26.**

A *control point* is the actual value that a building automation system experiences at any given point in time. For example, a temperature sensor in a summer building space measures a temperature of 74°F. The control point is 74°F.

At any given time, a control point may differ from the setpoint. *Offset* is the difference between a control point and setpoint. Depending on the accuracy of the controller, the actual value (control point) in a building space may be different from the value the controller is programmed to maintain (setpoint). Due to the quality of sensors and digital offset control technology now available, an accuracy of ±0.2°F is a normal range for temperature control.

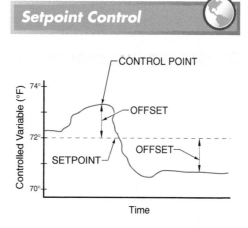

Figure 13-26. *Setpoint is the desired value to be maintained by a system with the desired accuracy to the setpoint being programmed into a BAS controller.*

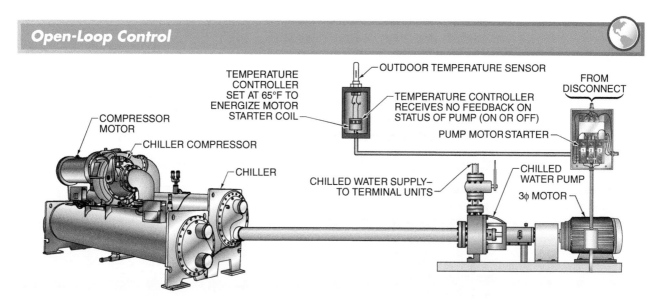

Figure 13-25. *In open-loop control, no feedback is provided from sensors to a building automation system controller.*

Setup and Setback Setpoints. Another common direct digital control feature is setup and setback setpoint control. Setup and setback setpoints are values that are active during the unoccupied mode of a building automation system. *Setup* is the unoccupied cooling setpoint. For example, a cooling setpoint is raised from 74°F during the day to 85°F at night. **See Figure 13-27.** The setup setpoint is 85°F. *Setback* is the unoccupied heating setpoint. For instance, if a heating setpoint is lowered from 70°F during the day to 55°F at night, the setback setpoint is 55°F.

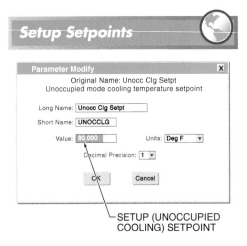

SETUP (UNOCCUPIED COOLING) SETPOINT

Figure 13-27. *Setup and setback setpoints are values that are active during the unoccupied mode of a building automation system.*

Setup and setback setpoint control saves energy by preventing heating and/or cooling systems from operating fully when a commercial building is unoccupied. Setup and setback setpoint control also protects commercial buildings from abnormal temperatures by enabling the heating and cooling systems to operate when the building space temperature becomes excessively hot or cold. Although the setup and setback setpoint control feature is commonly used with building space temperature setpoints, other setpoints, such as for hot water, chilled water, and static pressure, can also be controlled.

Reset Control

Reset control is a direct digital control (DDC) feature in which a primary setpoint is reset automatically as another value (reset variable) changes. The most common example of reset control is the reset of a hot water heating setpoint as the outdoor-air temperature changes. A *setpoint schedule* is a description of the amount of change that occurs when a reset variable resets the primary setpoint. Another example of reset control is the reset of a VAV system static pressure using the highest cooling demand from several zones.

Low-Limit Control

Low-limit control is a type of control that ensures that a controlled variable remains above a programmed low-limit value. The controlled component maintains a minimum value even when the controlled variable does not remain above the low-limit value. Low-limit control is commonly used with ventilation damper controls.

An excessively low outdoor-air temperature causes a low ventilation-air temperature. For example, when the ventilation-air temperature of a commercial building drops below 45°F, the controller wants to close the outside (ventilation)- air damper 100%. However, because of the low-limit control, the controller overrides the normal control logic and forces the outside (ventilation)- air damper not to close more than is required for a 45°F temperature. **See Figure 13-28.**

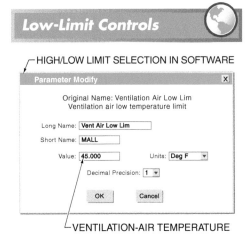

HIGH/LOW LIMIT SELECTION IN SOFTWARE

VENTILATION-AIR TEMPERATURE

Figure 13-28. *Low-limit controls stop outside-air dampers from closing 100% and still allow ventilation air into a building when ventilation-air temperatures are excessively low outside.*

High-Limit Control

High-limit control provides capabilities similar to low-limit control. *High-limit control* is a type of control that ensures that a controlled variable remains below a programmed high-limit value. High-limit control is commonly used in applications that require a temperature, pressure, or humidity not to exceed a programmed value. For example, a high-limit control can be used to prevent a hot water heating system water temperature from exceeding 210°F.

High/Low Signal Select

High/low signal select is a DDC feature in which a building automation system selects among the highest or lowest values from multiple input signals. The most common application of high/low signal select is the control of a building space temperature using multiple zone temperature sensors at various locations in a building. The highest signal represents the warmest building space and the lowest signal represents the coolest building space.

The signal selection results are communicated to the building automation system reset control to determine the setpoint. The highest signal may be used to reset a cooling function to satisfy the warmest building space. The lowest signal may be used to reset a heating function to satisfy the coolest building space. Many building automation systems use high/low signal select instead of outdoor-air temperature reset control.

DIRECT DIGITAL CONTROL ALGORITHMS

An *algorithm* is a mathematical equation used by a building automation system controller to determine a desired setpoint. DDC algorithms enable a building automation system to achieve a high level of accuracy. Common algorithms used in DDC systems include proportional, integration, derivative, and adaptive control algorithms.

Proportional Control Algorithms

A *proportional control algorithm* is a control algorithm that positions the controlled component in direct response to the amount of offset in a building automation system. Proportional control algorithms are the most common algorithms used in DDC systems. **See Figure 13-29.** Proportional control algorithms are also known as P control or P only control.

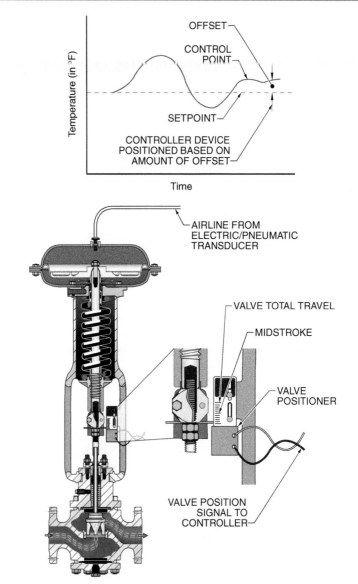

Proportional Control Algorithms

Figure 13-29. *Proportional control algorithms position a controlled component in response to the amount of offset experienced in a building automation system.*

The setpoint and the desired +/- accuracy (throttling range) are entered into the controller. The controller is programmed to provide a specific output signal when the input variable reaches the setpoint. This is often done by using the mid-stroke position of the controlled component. Proportional control algorithms do not provide precise control.

Integration Control Algorithms

An *integration control algorithm* is a control algorithm that eliminates any offset after a certain length of time. Integration algorithms are used in calculus to determine the area under a curve. Integration control algorithms are commonly used in conjunction with proportional control algorithms.

Proportional/integration (PI) control is the combination of proportional and integration control algorithms. When an offset remains after a specific length of time, the integration control algorithm repositions the controlled component to eliminate the offset. The integration control algorithm changes the position of the controlled component to meet the actual requirements of the load at a given time.

Derivative Control Algorithms

A *derivative control algorithm* is a control algorithm that determines the instantaneous rate of change of a variable. **See Figure 13-30.** Derivative control algorithms provide real-time data to a BAS controller. A derivative control algorithm acts against the integration control algorithm. The derivative control algorithm is usually used to increase the speed at which a controlled component eliminates an offset.

A *proportional/integration/derivative (PID) controller* is a direct digital control (DDC) system controller that uses proportional, integration, and derivative algorithms. Among heating, ventilation, and air conditioning (HVAC) systems, only extremely sensitive control applications require PID control. PI control is usually sufficient to achieve a setpoint.

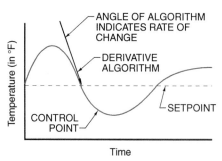

Figure 13-30. *Derivative control algorithms determine the instantaneous rate of change of a controlled variable.*

Adaptive Control Algorithms

The most sophisticated control algorithms used today are adaptive control algorithms. An *adaptive control algorithm* is a control algorithm that automatically adjusts its response time based on environmental conditions. Adaptive control algorithms are a self-tuning form of PID control but there is no substitute for manual loop tuning. **See Figure 13-31.**

Stationary engineers must accurately tune PI and PID controls by manually adjusting the response times in the BAS software to ensure accuracy and stability. Adaptive control algorithms can readjust to load changes or incorrectly programmed response times. Adaptive control algorithms reduce the amount of time a stationary engineer spends tuning a control loop during commissioning. Adaptive control algorithms are not found in all building automation systems.

IAQ STRATEGIES AND ENERGY CONSIDERATIONS

Many BAS and energy control strategies are generally compatible with indoor air quality (IAQ) strategies, provided that the strategies are instituted with certain IAQ protections. Energy retrofits and control strategies must include provisions to protect IAQ and provide additional outdoor air to meet the ventilation requirements of ANSI/ASHRAE Standard 62.1-2007, *Ventilation for Acceptable Indoor Air Quality.* Implementing IAQ strategies usually results in energy savings. **See Figure 13-32.**

Energy and operational strategies that are compatible with IAQ include:

- improving the energy efficiency of the building shell
- reducing internal loads (such as lighting and office equipment upgrades)
- upgrading fans, motors, and drive systems
- upgrading chillers and boilers
- using energy recovery ventilation (ERV) systems
- prudent equipment downsizing
- preventive maintenance (PM) of entire HVAC system
- using an air-side economizer
- night precooling
- reducing demand charges
- using supply-air temperature reset
- reducing light usage during unoccupied hours

Attempts in the past to save energy have needlessly compromised IAQ and resulted in occupant complaints and/or serious illness. Energy and operational strategies that are not compatible with IAQ include:

- reducing outdoor air ventilation below standards
- reducing HVAC operating hours
- relaxing temperature and/or humidity setpoints below standards

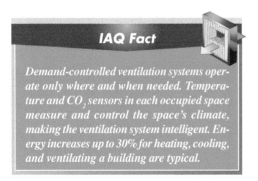

IAQ Fact

Demand-controlled ventilation systems operate only where and when needed. Temperature and CO_2 sensors in each occupied space measure and control the space's climate, making the ventilation system intelligent. Energy increases up to 30% for heating, cooling, and ventilating a building are typical.

Adaptive Control Algorithms

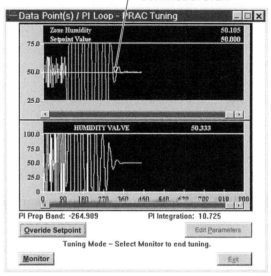

ADAPTIVE CONTROL ALGORITHMS CORRECT AN OVERCYCLING CONTROL SYSTEM

Figure 13-31. *Adaptive control algorithms are self-tuning algorithms that are often used to correct an overcycling control system.*

ANSI/ASHRAE Standard 62.1-2007

Figure 13-32. *Acceptable indoor air quality numbers for ventilation are detailed in ANSI/ASHRAE Standard 62.1-2007,* Ventilation for Acceptable Indoor Air Quality.

IAQ Strategies for Air-Handling Systems

Air-handling systems are classified as either constant or variable-air volume. The strategies used to control both types differ. In a constant volume (CV) ventilation system, variations in the temperature requirements of a space are satisfied by varying the temperature of the constant volume of air being delivered to the space.

A constant percentage of outdoor air (ventilation air) means that a constant volume of outdoor air is delivered to occupied spaces. The volume of outdoor air must be set to satisfy applicable ventilation standards. While CV systems are less energy efficient than variable-air-volume (VAV) systems, controlling the amount of outdoor air to a space is easier to do than varying the temperature of the air.

In a VAV ventilation system, temperature requirements of a space are satisfied by varying the volume of air that is delivered to the space at a constant temperature. VAV systems reduce HVAC energy costs by 10% to 20% in comparison to CV systems but complicate the delivery of outdoor air.

An inadequate outdoor air percentage, combined with an inadequate VAV box minimum setting, may result in inadequate outdoor airflow (ventilation) to occupant spaces. This occurs during partial-load conditions.

VAV systems also complicate pressure relationships in a building and make testing, adjusting, and balancing more difficult. Most of the year, the volume of outside air is reduced to about a third of the outdoor air volume at design load. This could result in indoor air quality problems.

Having separate controls to ensure adequate outdoor airflow (ventilation) to occupant spaces year-round does not increase energy costs. Some new VAV systems incorporate ventilation controls. Important air-handling strategies include carbon-dioxide (CO_2)-controlled ventilation, air-side economizer usage, and night precooling.

Carbon-Dioxide-Controlled Ventilation

Carbon-dioxide-controlled ventilation changes the outdoor-air supply in response to CO_2 levels in building spaces or zones, which is used as an indicator of occupancy. **See Figure 13-33.** CO_2 controls may be useful for reducing energy use for places where occupancy is highly variable and irregular such as general meeting rooms, studios, theaters, and educational facilities. A typical HVAC system increases ventilation when CO_2 levels rise to 600 parts per million (PPM) to 800 PPM so that levels do not exceed 1000 PPM. **See Figure 13-34.** The system must incorporate a minimum outdoor-air setting to dilute building-related contaminants during low-occupancy periods. Carbon dioxide sensors must be calibrated periodically and setpoints adjusted based on outdoor CO_2 levels around the building.

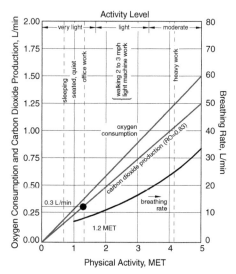

Figure 13-33. *CO_2 levels in building spaces or zones are used as an indicator of occupancy.*

IAQ Fact

The 20% rule for ventilation is often misunderstood. Setting outside-air dampers to a 20% stroke position will typically provide over 20% of outdoor air. The damper stroke position has almost no effect on ventilation airflow.

System Ventilation

Air-Side Economizer Usage

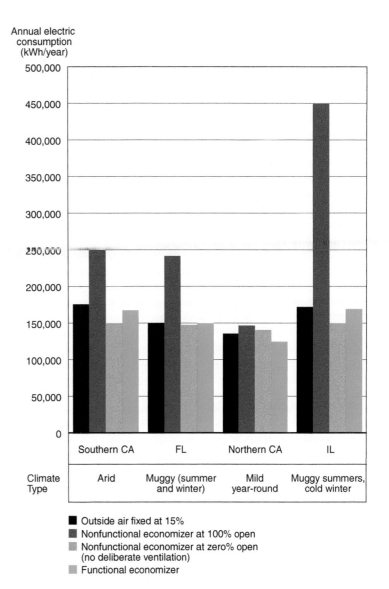

VENTILATION EFFECTS ON CO₂ LEVELS

MAXIMUM PERMISSIBLE VENTILATION LAG TIME

Figure 13-34. *A typical HVAC system increases ventilation when CO_2 levels rise to 600 parts per million (PPM) so that levels do not exceed 1000 PPM.*

Air-Side Economizers

Air-side economizers use outdoor air to provide free cooling. The operation of these economizers potentially improves IAQ by helping to ensure that the outdoor-air ventilation rate meets IAQ requirements. **See Figure 13-35.** With the exception of use in dry climates, moisture control must also be incorporated. Air-side economizers are not practical or advisable in hot-humid climates. An air-side economizer is disengaged when a problem occurs involving outdoor air pollution. These economizers usually reduce annual HVAC energy costs when used in cold or temperate climates.

Yearly Use of Air-Side Economizer			
Representative Cities	Available Hours of Full Economizer	Available Hours of Partial Economizer	No Economizer Availability
San Francisco	8,563	197	—
New York	6,634	500	1,626
Dallas	4,470	500	3,790
London	8,120	300	340

Calculation of available hours based on a 68°F DB/50°F dewpoint supply air

Figure 13-35. *Economizers use outdoor air to provide free cooling and are not practical or advisable in hot-humid climates.*

Night Precooling

Cool outdoor air at night is used to cool a building while simultaneously exhausting accumulated pollutants. This process is known as a building flush. However, the cool outdoor air may also have a high moisture content and could humidify the building at night, so caution must be used. **See Figure 13-36.** In addition to preventing microbiological growth, controls must stop precooling operations when the dew point of the outdoor air is high enough to cause condensation on building surfaces and equipment.

Night Precooling

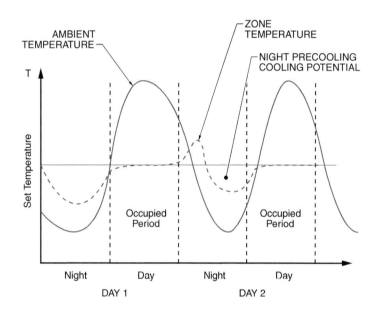

Figure 13-36. *In night precooling, cool outdoor night air is used to cool a building while simultaneously exhausting accumulated pollutants.*

Case Study – Office Building

IAQ Issues

A 40-year-old, 740,000 sq ft, 26-story commercial office building has a few retail shops on the first floor and individual offices and cubicles as office space. The basement has a two-level parking garage and space for the mechanical systems of the building. The existing mechanical systems and controls are no longer considered efficient and reliable.

Investigation

The perimeter of the building uses hot-water baseboard heating. The interior of the building uses heat supplied by air-handling heating coils in the ductwork of each floor. The hot water is generated by two electric boilers, each demanding 1250 kW of power. Electric baseboard heaters are used in the lobby and a few common areas of the building. The electric boilers cannot keep up with the heating demand of the building.

The cooling of the building is provided by a chilled water system. The system has two centrifugal chillers that operate at a 0.9 kW-per-ton efficiency. The existing air-supply boxes are constant volume units.

Building lighting mainly consists of T12 fluorescent tubes running on magnetic ballasts and some incandescent lamps.

Resolution

For heating, the old electric boilers were replaced by six high-efficiency, gas-fired, modular boilers. The new gas-fired boilers operate at a lower cost and, due to the modular design, some boilers will run steady at peak efficiency, while others cycle ON and OFF. This reduces the inefficiency of the ON/OFF cycling of the boilers. Also, the boilers now are of proper size to provide hot water to baseboard heaters, so all electric baseboard heaters were replaced with hot-water radiation-style heaters.

For cooling, the two centrifugal chillers were replaced with high-efficiency, 650-ton, water-cooled centrifugal chillers. The new chillers operate at 0.52 kW per ton efficiency. Variable-frequency drives were added to all air handlers, water pumps, and cooling tower fans. New actuators and controls were added to all dampers and air supply boxes.

All the old T12 fluorescent tubes were replaced with T8 tubes and electronic ballasts. A new central DDC automation system was installed not only to aid with security but also to provide the newest technology for an energy-efficient operation. The upgrades have saved the building 40% in energy costs every month ($500,000 annually). Building operating costs have also dropped by $120,000 a month.

Case Study

Definitions

Building Automation Systems for IAQ Definitions...

- An *adaptive control algorithm* is a control algorithm that automatically adjusts its response time based on environmental conditions.
- An *algorithm* is a mathematical equation used by a building automation system controller to determine a desired setpoint.
- An *air-handling unit (AHU) controller* is a controller that contains input terminals and output terminals required to operate large central-station air-handling units.
- An *analog humidity sensor* is a device that measures the amount of moisture in the air and sends a proportional signal to a controller.
- An *analog input (AI)* is a sensor that indicates a variable such as temperature, pressure, or humidity and causes a proportional electrical signal change at the building automation system controller.
- An *analog output component* is a piece of electrical equipment that receives a continuous signal that varies between two values.
- An *analog pressure sensor* is an analog input device that measures the pressure in a duct, pipe, or room and sends a signal to a controller.
- An *analog temperature sensor* is an analog input device that measures the temperature in a duct, pipe, or room and sends a signal to a controller.
- An *application-specific controller (ASC)* is a controller designed to control only one section or type of HVAC system.
- An *averaging temperature sensor* is used in large duct systems to obtain an accurate temperature reading.
- A *building automation system (BAS)* is a system that uses microprocessors (computer chips) to control the energy-using components in a building.
- A *building automation system input* is an object such as a sensor that indicates building conditions to a BAS controller.
- A *building automation system output component* is a piece of electrical equipment that changes state (ON or OFF) in response to a command from a building automation system controller.
- *Closed-loop control* is a type of control system where feedback occurs among the controller, sensor, and controlled component.
- A *control point* is the actual value that a building automation system experiences at any given point in time.
- A *control strategy* is a BAS software method used to control the energy-using equipment in a building.
- A *derivative control algorithm* is a control algorithm that determines the instantaneous rate of change of a variable.
- A *digital (binary) input* is a sensor that produces an ON or OFF signal.
- A *differential pressure switch* is a digital input device that opens or closes contacts because of the difference between two pressures.
- A *digital output component* is a piece of electrical equipment that receives an ON or OFF signal.
- A *direct digital control (DDC) system* is a building automation system in which controllers are wired directly to controlled components to turn them ON or OFF.

...Building Automation Systems for IAQ Definitions...

Definitions

- A *distributed direct digital control (DDC) system* is a building automation system that has multiple central processing units (CPUs) at the controller level.

- A *duct-mounted temperature sensor* is used to sense air temperatures inside ductwork.

- *Feedback* is the measurement of the results of a controller action by a sensor or switch.

- *Global data* is data needed by all controllers in a network.

- *High-limit control* is a type of control that ensures that a controlled variable remains below a programmed high-limit value.

- *High/low signal select* is a direct digital control (DDC) feature in which a building automation system selects among the highest or lowest values from multiple inputs.

- A *humidistat* is a digital input device that indicates that a humidity setpoint has been reached.

- A *hygroscopic element* is a sensor mechanism that changes its characteristics as the humidity level in the air changes.

- An *immersion (well-mounted) temperature sensor* is used to sense water temperature in piping systems.

- An *integration control algorithm* is a control algorithm that eliminates any offset after a certain length of time.

- *Low-limit control* is a type of control that ensures that a controlled variable remains above a programmed low-limit value.

- A *network communication module (NCM)* is a special controller that coordinates communication from controller to controller on a network and provides a location for operator interface.

- *Offset* is the differencunite between a control point and setpoint.

- *Open-loop control* is a type of control where no feedback occurs among the controller, sensor, and controlled components.

- An *operator interface* is a device that allows a technician to access and respond to building automation system information.

- A *proportional control algorithm* is a control algorithm that positions the controlled component in direct response to the amount of offset in a building automation system.

- A *proportional/integration/derivative (PID) controller* is a direct digital control (DDC) system controller that uses proportional, integration, and derivative algorithms.

- *Proportional/integration (PI) control* is the combination of proportional and integration control algorithms.

- *Reset control* is a direct digital control (DDC) feature in which a primary setpoint is reset automatically as another value (reset variable) changes.

- A *server computer* is a large-capacity, hard-drive computer attached to a network.

- *Setback* is the unoccupied heating setpoint.

- A *setpoint* is the desired value to be maintained by a system.

Definitions

...*Building Automation Systems for IAQ Definitions*

- A *setpoint schedule* is a description of the amount of change that occurs when a reset variable resets the primary setpoint.
- *Setup* is the unoccupied cooling setpoint.
- *Stand-alone control* is the ability of a controller to function on its own.
- A *thermostat* is a digital input device and controller.
- A *triac* is a solid-state switching device used to switch alternating current (AC).
- A *transducer* is an electrical component that changes one type of proportional control signal into another type of signal.
- A *unitary controller* is a type of controller designed for basic zone control using a standard wall-mount temperature sensor.
- A *variable-air-volume (VAV) air terminal unit controller* is a controller that modulates the damper inside a VAV air-handling unit or rpm of the blower motor(s) with signals to electric motor drives to maintain a specific building space temperature.
- A *wall-mounted temperature sensor* is the most common temperature sensor used in building automation systems and is used to sense air temperatures in building spaces.

For more information please refer to ATPeResources.com or http://www.epa.gov.

Review Questions

1. What is a distributed direct digital control system?

2. Explain the different methods an operator can use to interface with a building automation system.

3. Explain how the various types of analog input devices operate.

4. How do digital output components operate?

5. What is the difference between setup and setback?

HOMELAND SECURITY

14
chapter

Class Objective
Describe prevention strategies and plans for a chemical, biological, or radiological (CBR) agent release.

On-the-Job Objective
Develop a plan for a facility in the event of a CBR agent release.

Learning Activities:
1. Identify control strategies that may be used in the event of a CBR agent release.
2. Describe a typical walk-through of a building with a security professional.
3. List the requirements for a communications plan for a facility.
4. **On-the-Job:** Determine when the proper strategies described in this chapter are being implemented. Develop a plan for implementation.

Introduction

Stationary engineers are usually the first responders to any emergency within a building or facility. Therefore, stationary engineers must play a lead role in developing a homeland security plan. Reducing a building's vulnerability to an airborne chemical, biological, or radiological (CBR) attack requires a comprehensive approach. Decisions concerning which protective measures to implement must be based upon the threat profile and a security assessment of the building and its occupants. While physical security is the first layer of defense, other issues must also be addressed. The majority of the measures described in this chapter are recommended by The National Institute for Occupational Safety and Health (NIOSH) and can be accessed on-line at their website www.cdc.gov/niosh/topics/emres/.

HOMELAND SECURITY

The term "homeland security" became prominent in the United States following the terrorist attacks on September 11, 2001. *Homeland security* is a term that refers to a concerted national effort to prevent terrorist attacks within the United States, reduce the United States' vulnerability to terrorism, and minimize the damage and recovery time from attacks that do occur. Homeland security also refers to the effort by all levels of government, as well as civilian efforts, to protect the territory of the United States from internal or external threats that originate from either natural or man-made occurrences. The federal agency responsible for ensuring homeland security is officially called the United States Department of Homeland Security (DHS). The DHS is comprised of many different agencies and sub-departments, including the Federal Emergency Management Agency (FEMA). The DHS is responsible for preparedness and response to and recovery from domestic disasters, both natural and man-made. **See Figure 14-1.**

The scope of homeland security coordination includes the following:
• emergency preparedness and response including volunteer medical, police, emergency management, and fire personnel
• domestic intelligence activities
• critical infrastructure protection
• border security, including both land and maritime borders
• transportation security, including aviation and maritime transportation
• biodefense
• detection of nuclear and radiological materials
• research on next-generation security technologies

Building owners, managers, and stationary engineers can quickly implement some preventive measures to increase occupants' protection from a CBR attack. A *CBR attack* is an attack on a building or area that employs airborne chemical, biological, or radiological (CBR) agents. Many types of public, private, and government buildings such as offices, laboratories, hospitals, retail facilities, schools, and transportation terminals are possible CBR attack targets. **See Figure 14-2.** Also, public facilities such as sports arenas, malls, and coliseums are possible CBR attack targets. Facilities that may be considered high-risk targets for terrorist activity or that provide services essential to the welfare of citizens may require special security considerations.

Predicting the possibility of a specific building being targeted for terrorist activity is difficult. There is no specific formula that determines the risk level of a building. Building owners must make their own decisions about how to reduce their building's risk in the event of a CBR attack. Conducting a comprehensive building security assessment contributes to the decision-making process. Many government and private organizations have identified resources that will assist in conducting building security assessments.

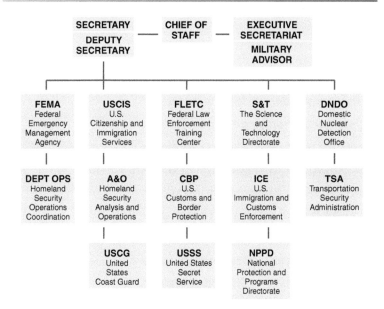

U.S. Department of Homeland Security Organizations

Figure 14-1. *Because the U.S. Department of Homeland Security (DHS) houses the Federal Emergency Management Agency (FEMA), the DHS is responsible for preparedness, response, and recovery to natural disasters as well as man-made events.*

Chemical, Biological, and Radiological (CBR) Targets

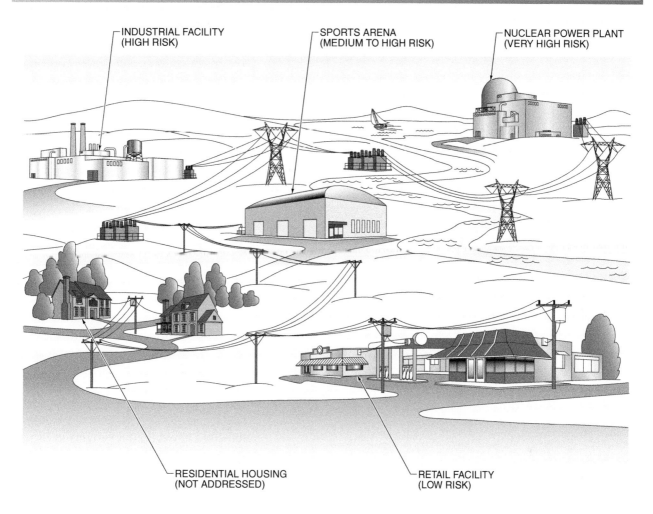

Figure 14-2. *A CBR attack employs chemical, biological, or radiological (CBR) agents to strike a building or area.*

A building's IAQ team can reduce the likelihood of an attack and develop other preventive measures to protect occupants in case of an attack. For instance, facility owners and managers can increase the difficulty of introducing a CBR agent into a building, increase the ability of security officials to detect terrorists before they carry out an intended release, and incorporate plans and procedures to reduce the effects of a CBR agent release. The decision of what protective measures to implement for a specific building is based on several factors, such as the perceived risk associated with the building,

engineering and architectural feasibility, and cost.

Background

The occurrence of recent terrorist attacks has prompted increased research into the vulnerability of workplaces, schools, and other occupied buildings to terrorist threats in the United States. The airflow patterns and dynamics in a building are of particular concern, specifically the heating, ventilating, and air conditioning (HVAC) systems of a building. **See Figure 14-3.** By their inherent design, HVAC systems are an entry point and a distribution system for hazardous contaminants, particularly CBR agents. Building owners need reliable information about how to modify buildings to decrease the likelihood or effects of a CBR event and to respond quickly and appropriately when a CBR incident occurs. Comprehensive guidance is needed in several areas, which include the following:

- modifying existing buildings for better air protection, security, and reviewing the preventive maintenance program
- designing new buildings to be more secure
- creating plans to execute in the event of an incident

CBR Agent-Building Flow Pattern Scenario

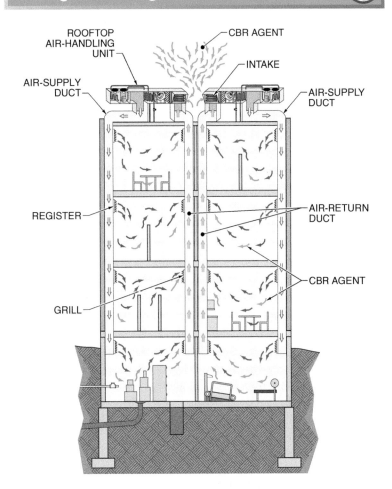

Figure 14-3. *The airflow patterns and dynamics in a building are of particular concern, specifically the heating, ventilating, and air conditioning (HVAC) systems of a building as an entry point and distribution system for hazardous contaminants, particularly CBR agents.*

Building Layout

Before initiating any plan to modify a building's system design or operation, the building's systems must first be understood. Information must be obtained on how the systems were designed to operate and how the systems currently operate. Understanding a building's operation can be accomplished by conducting a walk-through inspection of the building and its systems, including the HVAC, fire protection, and life-safety systems. During the inspection, it is important to compare the most up-to-date design drawings to the actual operation of the current systems. Without this baseline knowledge, it is difficult to accurately identify what impact any security modifications may have on building operation.

When the IAQ team finds unexplained discrepancies during the walk-through between the design drawings and the actual operations, corrections should be made and, if needed, an independent evaluation should be conducted by a qualified HVAC professional. **See Figure 14-4.** It is important for stationary engineers to understand how the mechanical systems of a building function; however, the systems need not operate according to design specifications before implementing security measures. A list of items to consider during a building walk-through includes the following:

- What is the mechanical condition of the equipment? What filtration systems are in place for the equipment? What are their efficiencies?
- Is all equipment appropriately connected and controlled? Are equipment access doors and panels in place and appropriately sealed?
- Are all dampers (outdoor air, return air, bypass, fire, and smoke) functioning? A physical check can help determine how well they seal when closed.
- How does the HVAC system respond to manual fire alarm, fire detection, or fire-suppression device activation?
- Are all supply and return ducts completely connected to their registers and grills?

- Are the variable-air-volume (VAV) boxes functioning?

- How is the HVAC system controlled? How quickly does it respond?

- How is the building zoned? Where are the air-handling units for each zone? Is the system designed for smoke control?

- How does air flow through the building?

- What are the pressure relationships between zones?

- Which building entryways are positively or negatively pressurized?

- Is the building connected to other buildings by tunnels or passageways?

- Are utility shafts and penetrations, elevator shafts, and fire stairs significant airflow pathways?

- Is there obvious air infiltration? If so, is it localized?

- Does the system provide adequate ventilation given the building's current occupancy and functions?

- Where are the outdoor air louvers? Are the louvers easily observable? Are they or other mechanical equipment accessible to the public?

SPECIFIC CBR ATTACK PREVENTION RECOMMENDATIONS

There are specific recommendations on actions to avoid when protecting a building against a CBR attack. In addition, recommendations for CBR attack prevention fall into one of the following categories: building physical security; securing outdoor-air intakes; ventilation and filtration; maintenance, administration, and training; and emergency plans, policies, and procedures. While reviewing the recommendations, it is important to consider their potential implications upon the necessary language for existing and future service contracts. Homeland security strategies are most effective when they are developed using a team approach. The team should include facility management, stationary engineers, security, cleaning personnel, tenant representatives, and other interested parties.

Building Walk-Through Considerations

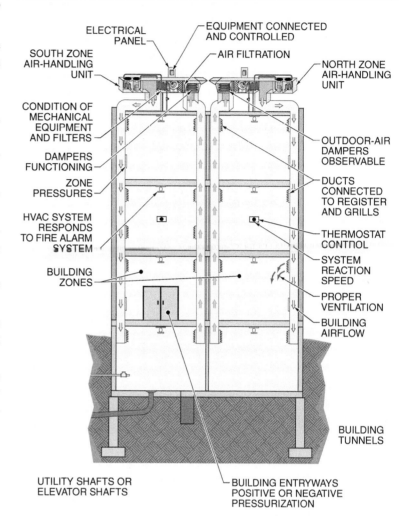

Figure 14-4. *When sufficient questions or surprises arise from a building walk-through, an independent evaluation by a qualified HVAC professional is used to establish a useful baseline.*

CBR Preventive Activities to Avoid

Building owners and managers, in consultation with their IAQ team, must ensure that any CBR preventive actions taken do not have a negative effect on the operation of building systems or occupants under normal building operation. Poorly executed efforts to protect a building from a CBR attack can have adverse effects on the building's indoor air quality. The IAQ team must understand how a building's systems operate and must assess the impact of security measures on IAQ systems. **See Figure 14-5.**

CBR Prevention Activities to Avoid

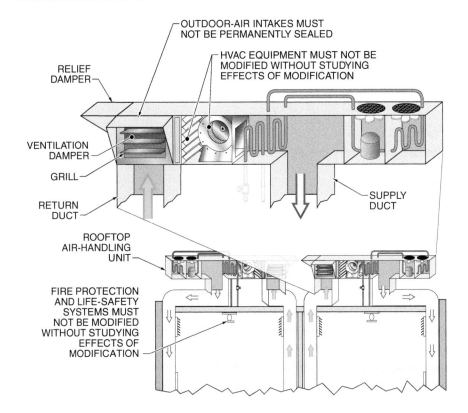

Figure 14-5. *Poorly executed efforts and security measures to protect a building from a CBR attack can have adverse effects on a building's indoor air quality.*

According to ANSI/ASHRAE Standard 62.1-2007, *Ventilation for Acceptable Indoor Air Quality,* outdoor-air intakes cannot be permanently sealed. Buildings require a steady supply of ventilation air that is appropriate to the occupancy levels and functions being performed in the building. The supply of ventilation air must be maintained at least at a minimal level during normal building operations. Closing off the outdoor-air supply inlet (grill) results in a decrease of IAQ. The decrease in ventilation air adversely affects the building's occupants, causing an increase in IAQ complaints and possible health problems.

An HVAC system must not be modified without first understanding the effects on the building systems and occupants. When there is uncertainty about the effects of a proposed modification, a quali-

fied professional must be consulted. Fire protection and life-safety systems must not be tampered with or interrupted. Fire protection and life-safety systems provide protection in the event of fire or other types of events.

Building Physical Security

Preventing terrorist access to a targeted building requires ensuring the physical security of all entry points, storage areas, roof, and mechanical areas, as well as securing access to the ventilation-air intakes of the building's HVAC system. The physical security needs of every building should be assessed, as the threat of a CBR attack varies considerably from building to building. For example, the threat to a large corporate headquarters may be considered greater than the threat to a small retail establishment.

Some physical security measures, such as locking the doors to mechanical rooms, are low cost and do not inconvenience the users of the building. **See Figure 14-6.** Locking doors is a type of preventive measure that can be implemented in most buildings. Other physical security measures, such as increased security personnel or CBR screening devices, are more costly and may inconvenience building occupants substantially. Nevertheless, costly measures, when merited, should be implemented.

The IAQ team must understand what building assets require protection and what characteristics about the building and/or occupants make it a potential target. By first assessing the vulnerabilities of facilities, building owners and managers can address physical security in an effective manner. The identification and resolution of building vulnerabilities are specific to each building. However, some physical security actions are applicable to many building types. One security action that should be performed is the prevention of access to outdoor-air intakes by relocating, extending, or establishing a security zone around the intakes.

Securing Outdoor-Air Intakes

One of the most important steps in protecting a building's indoor environment is the security of the outdoor-air intakes. Ventilation air enters the building through outdoor-intakes and is distributed throughout the building by the HVAC system. Introducing a chemical, biological, or radiological agent into the outdoor-air intakes allows a terrorist to use the HVAC system as a means of dispersing the agent throughout the building. Publicly accessible outdoor-air intakes located at or below ground level have the highest risk of being attacked. The high risk is due partly to accessibility. Also, CBR releases near a building and close to the ground allow vapors to stay as a cloud and not be dissipated as easily. Securing the outdoor-air intakes is a critical line of defense in limiting the effects of an external CBR attack on a building.

Physical Security Measures

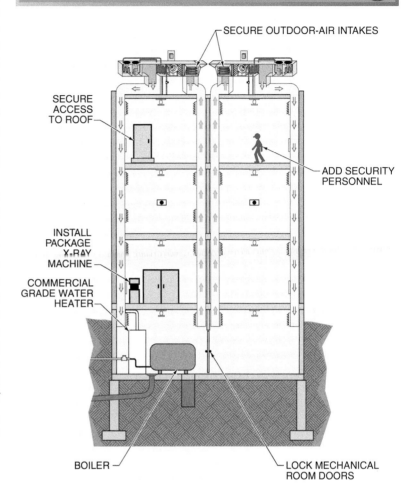

Figure 14-6. *Physical security measures such as locking the doors to mechanical rooms are low cost, but other physical security measures such as increased security personnel or packaged X-ray equipment are more costly.*

Relocating accessible air intakes to a publicly inaccessible location is preferable. Ideally, the intake should be located on a secure roof or high sidewall. **See Figure 14-7.** The lowest edge of the outdoor-air intakes should be placed at the highest feasible level above the ground or above any nearby accessible level, such as adjacent retaining walls, loading docks, or handrails. These measures are also beneficial in limiting the inadvertent introduction of other types of contaminants such as landscaping chemicals into the building air.

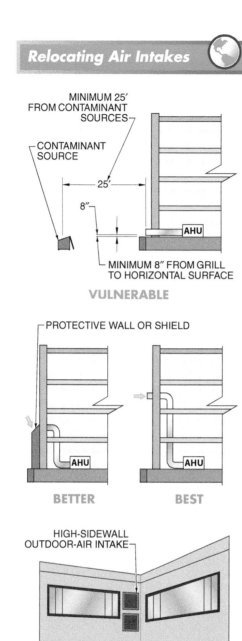

Relocating Air Intakes

MINIMUM 25′ FROM CONTAMINANT SOURCES

CONTAMINANT SOURCE

25′

8″

AHU

MINIMUM 8″ FROM GRILL TO HORIZONTAL SURFACE

VULNERABLE

PROTECTIVE WALL OR SHIELD

AHU

AHU

BETTER   **BEST**

HIGH-SIDEWALL OUTDOOR-AIR INTAKE

SIDEWALL

Figure 14-7. *Securing the outdoor-air intakes is a critical line of defense in limiting an external CBR attack on a building. Outside-air intakes must be located to allow for security.*

When relocation of outdoor-air intakes is not feasible, intake extensions can be constructed without creating adverse effects on HVAC performance. **See Figure 14-8.** Depending upon budget, time, or the perceived threat, the intake extensions may be temporary or permanent. The goal is to minimize public accessibility.

In general, higher intake extensions are better as long as other design constraints such as dynamic and static loads on structure and excessive pressure loss are appropriately considered. An extension height of 12′ or greater places the intake out of reach of individuals without some assistance. Also, the entrance to an outside-air intake is usually covered with a sloped metal hood to reduce the threat of objects being tossed into the intake. **See Figure 14-9.** A slope of 45° is generally adequate. The intake extension height must be increased where existing platforms or building features, such as loading docks or retaining walls, might provide access to the outdoor-air intake(s).

Fluke Corporation

The design of multiple sloped metal hoods for outside-air intakes has been used for many years.

Outdoor Air Intake Extensions

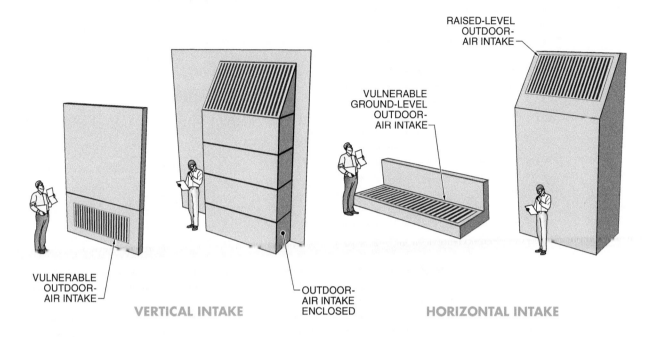

Figure 14-8. *When relocation of outdoor-air intakes is not feasible, intake extensions can be used to extend or enclose intakes in order to improve security without creating adverse effects on HVAC system performance.*

Outside Air Intake Hood Slope

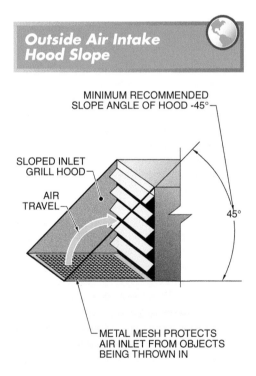

Figure 14-9. *Outside-air intakes have hoods sloped at 45° and a mesh screen that is used to prevent objects from being thrown into the intake.*

Physically inaccessible outdoor-air intakes are the preferred protection strategy. When outdoor-air intakes are publicly accessible and relocation or physical extensions are not a viable option, perimeter barriers that prevent public access to outdoor-air intake areas are an effective alternative. Iron fencing or similar see-through barriers that do not obscure visual detection of terrorist activities are preferred. **See Figure 14-10.**

IAQ Fact

The NOAA Weather Radio (NWR) is provided by the National Oceanic and Atmospheric Administration (NOAA). Working with the Federal Communications Commission's (FCC) Emergency Alert System (EAS), NWR is an all-hazards radio network. The NWR has 985 transmitters covering the United States, adjacent coastal waters, Puerto Rico, the U.S. Virgin Islands, and the U.S. Pacific Territories to keep the public informed on all weather conditions and hazards.

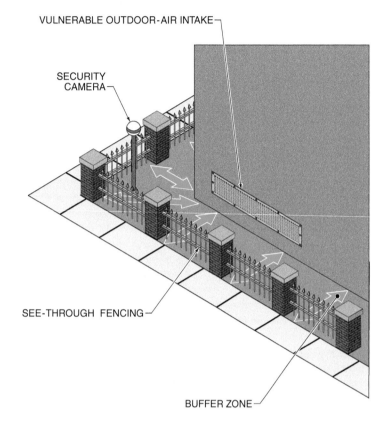

Figure 14-10. *Perimeter barriers such as iron fencing or other barriers that do not obscure visual detection of terrorist activities prevent public access to outdoor-air intake areas.*

The intake-restricted area must also include an open buffer zone between the public areas and the intake louvers. Thus, individuals attempting to enter these protected areas are more conspicuous to security personnel and the public. Monitoring the buffer zone by using physical security, closed-circuit television (CCTV), security lighting, or intrusion-detection sensors enhances this protective approach. **See Figure 14-11.**

Preventing Public Access to Mechanical Equipment Areas. Closely related to the relocation of outdoor-air intakes is the security of building mechanical areas. Mechanical areas may exist at one or more locations within a building. The mechanical areas provide access to the main mechanical systems (i.e., HVAC, elevator, water, and IAQ equipment). The IAQ equipment includes air filters, air-handling units, and exhaust systems. Such equipment is susceptible to tampering and could be used in a terrorist attack.

Buffer Zone Monitoring Methods

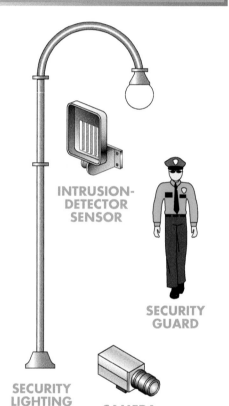

Figure 14-11. *All outdoor-air intake buffer zones must be monitored by physical security, closed-circuit television (CCTV), security lighting, or intrusion-detection sensors.*

Access to mechanical areas must be strictly controlled by security measures such as key-locked, keycarded, or key-coded doors. **See Figure 14-12.** Additional controls for access to keys, keycards, and key codes must be strictly maintained. Care must be taken when considering the possibility of providing access to subcontractor personnel.

Honeywell International, Inc.

Figure 14-12. *Access to secure areas of a building must be strictly controlled by key locks, keycards, or key-coded doors with the access to keys, keycards, and key codes strictly maintained.*

Security measures such as guards, alarms, and cameras are used to protect vulnerable areas. Difficult-to-reach outdoor-air intakes and mechanical rooms alone may not stop a terrorist or other intrusion activity. Security personnel, barriers that deter loitering, intrusion-detection sensors, and observation cameras further increase protection by quickly alerting security personnel to breaches of the outdoor-air intake areas or other vulnerable building locations.

It is important to ensure that areas such as lobbies, mailrooms, loading docks, and storage areas are isolated. Lobbies, mailrooms, loading docks, and other entry and storage areas should be physically isolated from the rest of the building. If bulk quantities of agents were released, these are the areas where the agents would likely enter occupied spaces of the building. Building doors, including vestibules and loading dock doors, should always remain closed when not in use.

To prevent widespread dispersion of a contaminant released within lobbies, mailrooms, and loading docks, the HVAC systems for these areas should also be isolated. An affected area of a building must be maintained at a negative pressure relative to the rest of the building, but at positive pressure in relation to the outdoors. **See Figure 14-13.** Physical

isolation of contaminated areas is critical to maintaining IAQ for the rest of the building. Usually, special attention is required to ensure airtight boundaries between contaminated areas and adjacent spaces. In some building designs, such as buildings with lobbies that have elevator access, establishing a negative pressure differential presents a challenge.

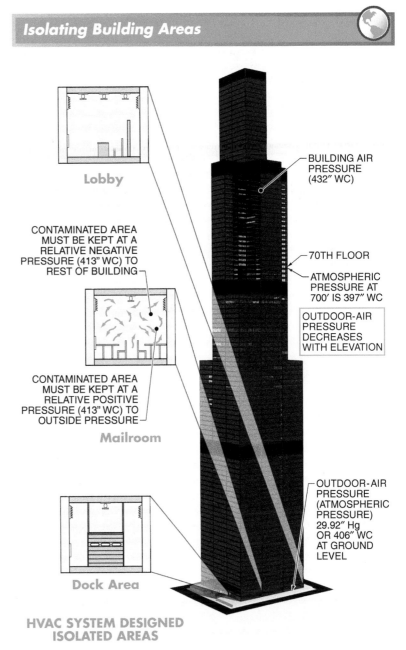

Lobby

CONTAMINATED AREA MUST BE KEPT AT A RELATIVE NEGATIVE PRESSURE (413" WC) TO REST OF BUILDING

CONTAMINATED AREA MUST BE KEPT AT A RELATIVE POSITIVE PRESSURE (413" WC) TO OUTSIDE PRESSURE

Mailroom

Dock Area

BUILDING AIR PRESSURE (432" WC)

70TH FLOOR

ATMOSPHERIC PRESSURE AT 700' IS 397" WC

OUTDOOR-AIR PRESSURE DECREASES WITH ELEVATION

OUTDOOR-AIR PRESSURE (ATMOSPHERIC PRESSURE) 29.92" Hg OR 406" WC AT GROUND LEVEL

HVAC SYSTEM DESIGNED ISOLATED AREAS

Figure 14-13. *To prevent the widespread dispersion of a contaminant released within a building's lobby, mailroom, or loading dock area, the HVAC system for these areas must be isolated.*

Lobbies, mailrooms, and loading docks should not share a return-air system or return pathway with other areas of the same building. Some contaminant prevention measures are more feasible for new construction or buildings undergoing major renovation. Any major modifications to these systems should be performed with input from a qualified HVAC professional.

The access to buildings from lobby areas must be controlled by performing security checks of individuals and packages prior to their entry into secure areas. Lobby isolation is particularly critical in buildings where the main lobbies are open to the public. Security checks of incoming mail should also occur before the mail is allowed to move into secure building areas. Side entry doors that circumvent established security checkpoints must also be strictly controlled.

All building return-air grills must be isolated. HVAC return-air grills that are publicly accessible and not easily observed by security are vulnerable to contaminant release. Public access facilities are the most vulnerable to this type of attack. A building security assessment can help determine, which, if any, protective measures to employ to isolate building return-air grills. Care must be taken to ensure that a selected measure does not adversely affect the performance of the HVAC system in a building.

Some return-air grill protective measures include relocating return-air grills to inaccessible yet observable locations, increasing security presence (through human or CCTV monitoring) near vulnerable return-air grills, directing public access away from return-air grill areas, and removing furniture and visual obstructions from areas near return air-grills. **See Figure 14-14.**

The access to building operational systems by outside personnel must be restricted. To deter tampering by outside personnel, a building staff member should escort outside personnel throughout their service visit and should visually inspect their work before final acceptance of the service. Alternatively, building owners and managers can ensure the reliability of outside personnel by using prescreened service personnel and trusted contractors.

Return-Air Grill Protecting Measures

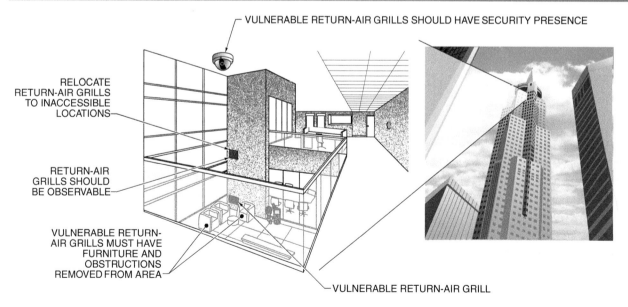

Figure 14-14. *Return-air grills must be inaccessible yet in an observable location, have increased security presence via more security personnel or security cameras, and have all furniture and visual obstructions removed from areas near return-air grills.*

Restricting access to building information is another important precaution. **See Figure 14-15.** Information on the following building systems must be controlled:

- mechanical
- electrical
- vertical transport
- fire and life-safety
- security system plans and schematics
- emergency operations procedures

Such information must only be released to authorized personnel, preferably by the use of a building access list that includes controlled copy numbering.

In addition to security measures for HVAC and other building operations, physical security upgrades can enhance the overall security of a building. A building or building complex might have security fencing and controlled access points. Some buildings such as museums are easily and necessarily accessible to the public. However, even in these buildings, areas such as mechanical rooms must remain off-limits to unauthorized individuals.

Information Security

RESTRICT WHO CAN OBSERVE
INFORMATION ON BUILDING SYSTEMS

Honeywell International, Inc.

BUILDING PERSONNEL CAN
SPECIALIZE IN ONE OF THE
FOLLOWING SYSTEMS: MECHANICAL,
ELECTRICAL, VERTICAL TRANSPORT,
FIRE AND LIFE-SAFETY, SECURITY,
AND EMERGENCY OPERATIONS

Figure 14-15. *Information on building mechanical, electrical, vertical transport, fire and life-safety, and emergency operation systems must be strictly controlled.*

Unless a building is regarded as open to the general public, owners and managers can institute a rule of not allowing visitors outside the lobby area without an escort. Layered levels of security access should be considered for some buildings. For example, access into the patient care areas of a hospital is usually less strict than access to hospital laboratories. Access is stricter for other areas, such as ventilation control rooms. Physical security is of prime concern in all building lobby areas.

Ventilation and Filtration

HVAC systems and their associated equipment must be evaluated in regards to their level of vulnerability in the event of a release of CBR agents. The evaluation should include examination of the HVAC system controls, the ability of the HVAC system to purge the building, the efficiency of installed filters, the capacity of the system relative to potential filter upgrades, and the significance of uncontrolled leakage into the building. Another consideration is the vulnerability of the HVAC system and components themselves, particularly when the facility is open to the public. For buildings under secure access, HVAC equipment may be considered less vulnerable, depending upon the perceived threat and the building management's confidence in the level of security.

Evaluate HVAC Control Options. Many central HVAC systems have building automation systems that are able to regulate airflow and pressures within a building in the event of an emergency. Some modern fire alarm systems also have capabilities that prove useful during CBR events. In some cases, the best response option (when a warning is available) might be to shut OFF the building's HVAC and exhaust systems. This can be accomplished with an emergency power off (EPO) switch. **See Figure 14-16.** Shutting down a building's HVAC system blowers and closing dampers can prevent the introduction of a CBR agent into a building from outside.

Interior space pressures and airflow control can also be used to prevent the spread of a CBR agent released in a building, and/or ensure the safety of areas that are used by occupants to leave the building. Any decision to install emergency HVAC control options should be made in consultation with a qualified HVAC controls professional. The HVAC controls professional will understand the ramifications of various HVAC operating modes on building operations and safety systems.

HVAC Control Options – Outdoor CBR Event

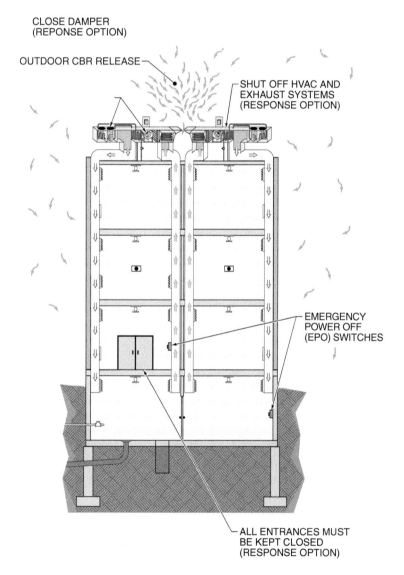

CLOSE DAMPER
(REPONSE OPTION)

OUTDOOR CBR RELEASE

SHUT OFF HVAC AND
EXHAUST SYSTEMS
(RESPONSE OPTION)

EMERGENCY
POWER OFF
(EPO) SWITCHES

ALL ENTRANCES MUST
BE KEPT CLOSED
(RESPONSE OPTION)

Figure 14-16. *Many building automation systems have the capability to regulate airflow and pressures or shut down an HVAC system within a building in the event of an outdoor CBR emergency.*

Depending upon the design and operation of the HVAC system and the nature of the CBR agent release, the HVAC controls in use may not be appropriate for all emergency situations. For example, lobbies, loading docks, and mailrooms might be provided with manually controlled exhaust systems activated by the appropriate personnel to remove contaminants in the event of a known release by exhausting air to an appropriate outside area. In other instances, manipulation of the HVAC system could minimize the spread of an agent.

When an HVAC control plan is pursued, building personnel must be trained to recognize a terrorist attack quickly and to recognize when to initiate certain control measures. For example, emergency egress stairwells must remain pressurized unless the stairwell is known to be the source of the CBR attack. **See Figure 14-17.** Other areas, such as laboratories, clean rooms, or pressure isolation rooms in hospitals, must remain ventilated. All procedures and training associated with the control of the HVAC system must be addressed in the building's emergency response plan.

Assess Filtration Efficiency. Increasing filter efficiency is one of the few measures that can be implemented in advance to reduce the consequences of both an interior and exterior release of an airborne CBR agent. However, the decision to increase filter efficiency should be made cautiously, with a careful understanding of the protective limitations resulting from a filter upgrade. In assessing the filtration requirements of a building, the objective is to implement the highest level of filtration efficiency that is compatible with the installed HVAC system and its required operating parameters.

In general, increased filter efficiency improves the indoor air quality of a building. However, increased protection from CBR aerosols occurs only when the increase in filtration efficiency applies to the particle size range and physical state of the contaminant. There are many possible types of CBR agents, each with unique characteristics.

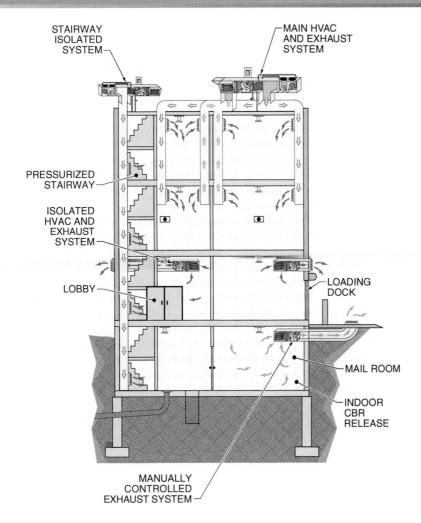

Figure 14-17. *Manipulation of a building's HVAC system could minimize the spread of a CBR agent. Building personnel must be trained to recognize a terrorist attack quickly and to know when to initiate certain control measures, such as maintaining pressure in stairwells, unless the stairwell is known to be the source of the CBR agent.*

Particulate air filters used for biological and radiological particles are not effective for gases and vapors typically used in chemical attacks. **See Figure 14-18.** Protection from gases and vapors requires adsorbent filters and often results in very high initial and ongoing maintenance costs. Effective adsorbent filters often contain activated carbon or another adsorbent-type medium.

Upgrading filtration is not as simple as just replacing a low-efficiency filter with a high-efficiency filter. High-efficiency filters have a higher pressure loss, which results in some airflow reduction through the HVAC system. The amount of this reduction is dependent on the design and capacity of the HVAC system and the ratings of the filter being installed. When the reduction of supply airflow is substantial, this reduction can result in inadequate ventilation, lower heating and cooling capacity, and possibly coil freeze-ups. In order to minimize pressure loss, filters having a larger nominal inlet area are sometimes feasible alternatives if space permits.

Filtering for CBR Agents

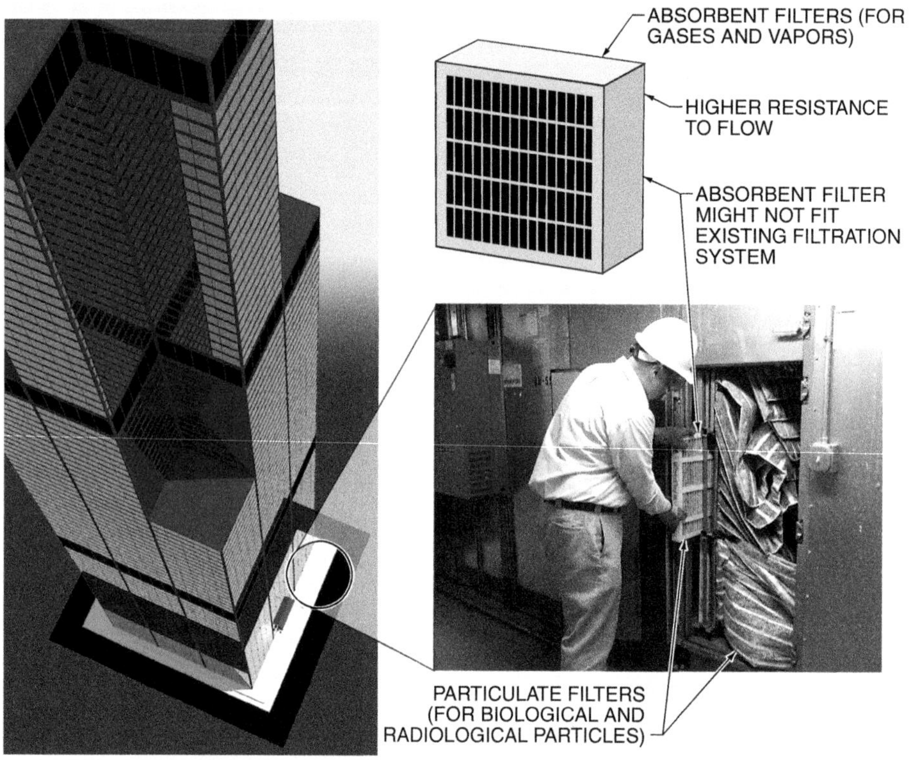

ABSORBENT FILTERS (FOR GASES AND VAPORS)

HIGHER RESISTANCE TO FLOW

ABSORBENT FILTER MIGHT NOT FIT EXISTING FILTRATION SYSTEM

PARTICULATE FILTERS (FOR BIOLOGICAL AND RADIOLOGICAL PARTICLES)

Figure 14-18. *Particulate air filters used for biological and radiological particles are not effective for gases and vapors typically used in chemical attacks. Gases and vapors require adsorbent filters that contain activated carbon or another adsorbent-type medium.*

In some cases, airflow and pressure losses can be avoided by either using prefilters or by more frequently changing filters. Pressure losses associated with adsorbent filters are usually even greater. The operation of a filter rack and frame system has a large impact upon the installed filtration efficiency. Usually, filter efficiency increases as the filter loads with contaminants. **See Figure 14-19.** Reducing the leakage of unfiltered air around filters, caused by a poor seal between the filter and the frame, can be as important as increasing filter efficiency. When a large amount of filter bypass is present, corrective actions must be taken.

Some filter systems have high-quality seals and frames that reduce the amount of bypass. During any filter efficiency upgrades, the filters and system should be evaluated by a qualified HVAC professional to verify proper performance. While higher filtration efficiency is encouraged and will likely provide IAQ benefits, the overall cost of filtration upgrades must also be considered.

IAQ Fact

According to the Centers for Disease Control and Prevention (CDC), some of the biological agents posing the greatest health threat in the case of a CBR agent release are anthrax, smallpox, plague, botulism, and viral hemorrhagic fever. Chemical agents such as VX, sarin, sulfur mustard, and cyanide are also possible materials terrorists may use. A dirty bomb containing radioactive material is yet another concern.

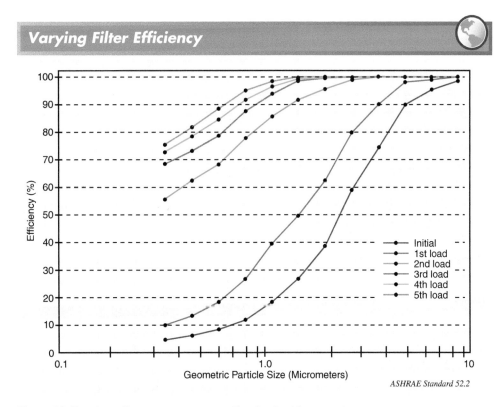

ASHRAE Standard 52.2

Figure 14-19. *Filter efficiency increases as a filter loads with contaminants.*

Filtration costs include the periodic cost of the filter medium, the labor cost to remove and replace filters, and the fan energy cost required to overcome the pressure loss of the filters being used. While higher-efficiency filters tend to have a higher life cycle cost than lower-efficiency filters, this is not always the case. With some high-efficiency filter systems, higher acquisition and energy costs can be offset by a longer filter life and a reduced labor cost for filter replacement. Also, improved filtration generally keeps heating and cooling coils cleaner, thus reducing energy costs through improvements in heat transfer efficiency. However, when high-efficiency particulate air (HEPA) filters and/or activated carbon filters are used, the overall costs of operating the system generally increase substantially.

Ducted and Nonducted Return-Air Systems. Ducted returns offer limited access points to introduce a CBR agent into the HVAC system. Return grills should be placed in conspicuous locations, reducing the risk of an agent being secretly introduced into the return system. *Nonducted return-air systems* are systems that commonly use hallways or spaces above dropped ceilings as return-air paths or plenums. **See Figure 14-20.** CBR agents introduced at any location above the dropped ceiling in a ceiling plenum return system will most likely migrate back to the HVAC unit and, without highly efficient filtration for the particular agent, be redistributed to occupied building spaces.

Buildings should be designed to minimize the mixing of air between building zones, which can be partially accomplished by limiting shared returns. Where ducted returns are not feasible or warranted, hold-down clips are used for accessible areas of dropped ceilings that serve as the return plenum. This issue is closely related to the isolation of lobbies and mailrooms, as shared returns are a common way for contaminants from these areas to be dispersed into the rest of the building. These types of modifications are more feasible for new building construction or buildings undergoing major renovations.

RETURN-AIR OPENING
ABOVE SUSPENDED CEILING

OPEN SPACE ABOVE
SUSENDED CEILING ACTS
AS PLENUM

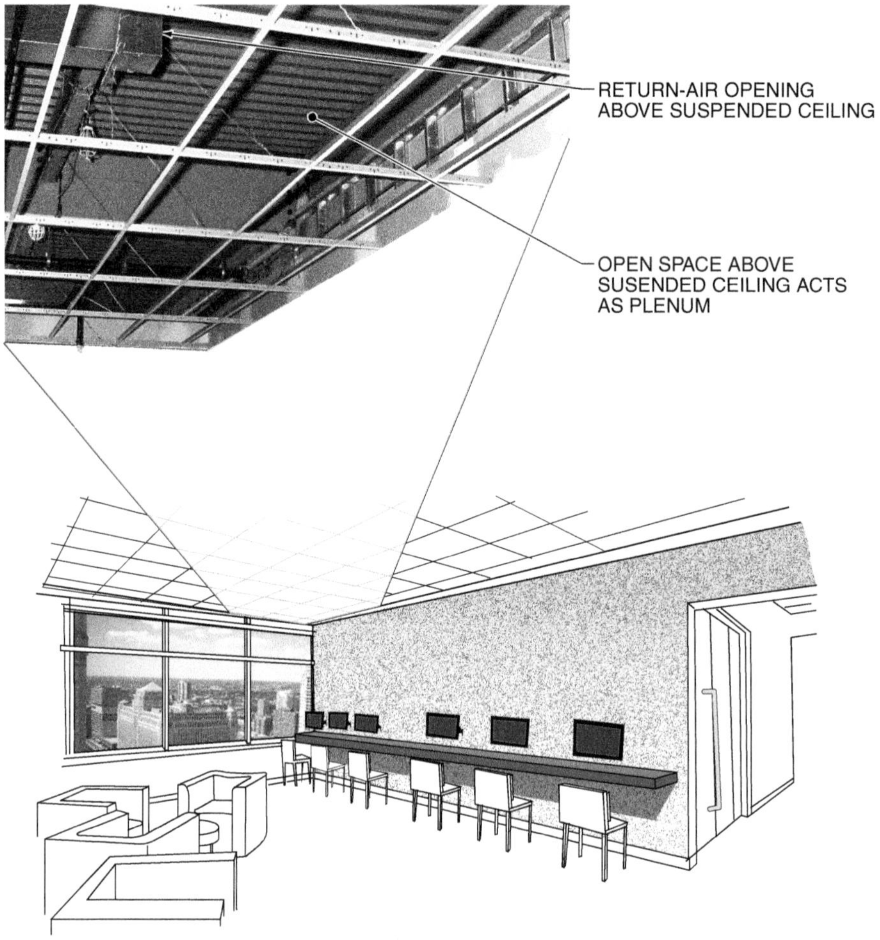

Figure 14-20. *Nonducted return-air systems commonly use hallways or spaces above dropped ceilings as return-air paths or plenums.*

Low-Leakage, Fast-Acting Dampers. Rapid responses, such as shutting down an HVAC system, may also involve closing various dampers, especially the dampers controlling the amount of ventilation airflow into a building (in the event of an exterior CBR agent release). When an HVAC system is turned OFF, the building pressure compared to the outdoor pressure can still be negative, allowing outdoor air into the building via many leakage pathways, including the HVAC system. Consideration should be given to installing low-leakage dampers to minimize this type of leakage pathway.

Damper and smoke damper leakage ratings are available from the manufacturer's specifications and from recommendations by Underwriters Laboratory (UL) and the Air Movement and Control Association International, Inc. (AMCA). **See Figure 14-21.** Leakage ratings for dampers range from UL Standard 555 Class I to Class III and AMCA Standard 511 of Class 1A to Class 3. When a warning is available prior to a direct CBR agent release, the speed with which dampers respond to a "close" instruction is important. From a protective standpoint, dampers that respond to instructions in 30 sec or less are preferred.

Tight Building Construction. Significant quantities of air can enter a building by means of infiltration through unintentional leakage paths in the envelope of a building. Such a leakage is more of a concern when a CBR agent is released at a distance from the building, such as a large-scale attack, than when a terrorist act is directed at the building. To minimize the risk of leakage, the amount of infiltration must be reduced. The reduction of air infiltration is a matter of tight building construction in combination with building space pressurization.

While building pressurization is a valuable CBR protection strategy for any building, it is much more likely to be effective in a tight building. However, to be effective, the type of filtration system for the building's supply air must be appropriate protection against the CBR agent that is released. Although increasing the air tightness of an existing building can be more challenging than planning for air tightness during new construction, air tightening a building should still be seriously considered. It is important to consult with an experienced HVAC industry professional to determine the possible risks involved with air tightening a location.

Maintenance, Administration, and Testing

The maintenance of ventilation systems and the training of staff are critical for controlling exposure to airborne contaminants, such as CBR agents. Stationary engineers should receive periodic training in system operation and maintenance. Training should include the procedures to be followed in the event of a suspected CBR agent release. Training must also cover health and safety aspects for maintenance personnel, as well as the potential health consequences to occupants of poorly performing HVAC systems. Development of current, accurate HVAC diagrams and HVAC system equipment labeling, along with procedural protocols, should be addressed. These procedural documents are of great value in the event of a CBR agent release.

CLASS	Differential Pressure†				
	4	**6**	**8**	**10**	**12**
I	8.0	9.5	11.0	12.5	14.0
II	20.0	24.0	28.0	31.5	35.0
III	80.0	96.0	112.0	125.0	140.0

SMOKE DAMPER LEAKAGE RATINGS*

* in CFM/Ft²
† in inches of water column

Source: Underwriter Laboratories Inc.

Figure 14-21. *Damper and smoke damper leakage ratings are available from manufacturers' specifications, Underwriters Laboratory (UL), and the Air Movement and Control Association International, Inc. (AMCA).*

Emergency Plans, Policies, and Procedures

All buildings must have current emergency plans that address common types of emergencies such as fire, weather-related crises, and on-the-job occupant injuries. Due to past experiences in the United States with anthrax and similar threats, homeland security plans should be updated to consider CBR attack scenarios. Homeland security plans must also address the associated procedures for communicating instructions to building occupants, identifying shelter-in-place areas, identifying appropriate use and selection of personal protective equipment, and directing emergency evacuations.

Fire-Lite Alarms

Many building CBR emergency plans are based on the technology of a building's fire alarm system.

A *shelter-in-place* is a small, interior room of a building with few or no windows where individuals can take refuge. **See Figure 14-22.** Individuals developing emergency plans and procedures must recognize that there are fundamental differences between chemical, biological, and radiological agents. In general, chemical agents cause a rapid onset of symptoms, while the symptoms caused by biological and radiological agents are usually delayed. Issues such as designated areas and procedures for chemical storage, HVAC control or shutdown, and communication with building occupants and emergency responders must all be addressed. Emergency plans should be as comprehensive as possible, but, as described earlier, protected by limited and controlled access. When appropriately designed, emergency plans, policies, and procedures will have a major impact on occupant survival in the event of a CBR agent release.

CBR Staff Training. Staff training, particularly for those with specific responsibilities during a CBR event, is essential and must encompass both inside and outside CBR attacks. Holding regularly scheduled practice drills allows for testing and rehearsal of the plan by occupants and key staff, which increases the likelihood for success in an actual CBR attack. For system protection in which HVAC control is done via an energy management control system, emergency procedures should be exercised periodically to guarantee that the various control options work as planned.

Shelter-in-Place

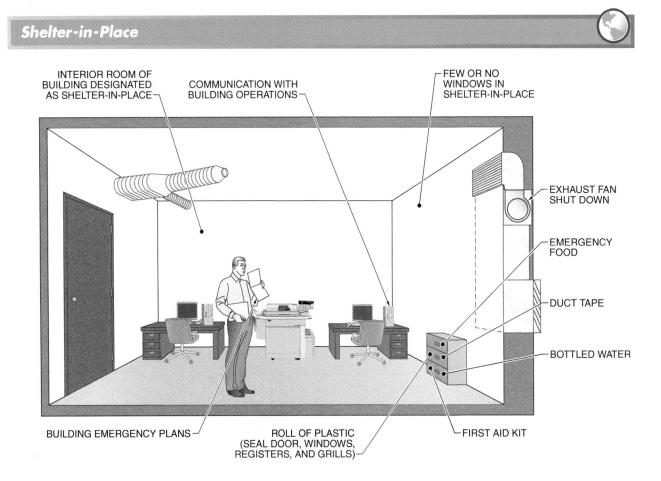

Figure 14-22. *A shelter-in-place is a small, interior room of a building with few or no windows where individuals can take refuge.*

Communication. In addition to staff training, effective communication procedures must be put in place. Internal communication should occur among the staff and occupants of the building. External communication involves regularly updating various government organizations.

As part of the planning process, contact names and phone numbers of relevant agencies such as local, state, and federal law enforcement should be readily available. This also applies to first responders such as police and medical care. A telephone script may be prepared beforehand to minimize confusion at the time of a CBR event.

Evacuation Procedures. An important aspect of planning for a possible CBR event is a building evacuation plan. A shelter-in-place plan may be needed as well. The appropriate procedure should be determined by the proper authorities. All staff and building occupants must be made aware of the specific logistics of a building evacuation plan.

Maintenance Procedures and Preventive Maintenance. Maintenance procedures and preventive maintenance schedules should already be in place for cleaning and maintaining ventilation system equipment. Replacement filters and system parts should be obtained from known manufacturers and examined prior to installation. It is important that ventilation systems be maintained and cleaned according to the manufacturer's specifications. Maintaining and cleaning a ventilation system requires information on HVAC system performance, flow rates, damper modulation and closure, sensor calibration, filter pressure loss, filter leakage, and filter change-out recommendations. These steps are critical to ensure that protection and mitigation systems, such as particulate filtration, operate as intended.

Wi-Fi

Wi-Fi is a wireless local area network technology that allows supported electronic devices to wirelessly connect to the Internet. Many stationary engineers can access building systems using cell phones or handheld PDA devices with Wi-Fi capabilities such as the BlackBerry®. Because many facilities have unencrypted wireless automation systems, terrorists could possibly connect to the automation system to prepare for and carry out an attack.

To combat this situation, fire, police, and National Guard personnel meet regularly with businesses to plan for such an attack. Every large commercial building and local union should have a representative seated at the meeting table. Many cities and towns have a crisis manager as part of homeland security measures. The local crisis manager helps a facility create a homeland security plan and coordinate a facility's needs with fire, police, and National Guard efforts.

Conclusions on Specific Recommendations

It is extremely important to prevent possible terrorist access to a facility's automation system, outdoor-air intakes, and mechanical rooms. Also, developing a CBR-contingent emergency plan should be addressed as soon as possible. Additional measures can provide further protection. A building security assessment should be done to determine if any additional measures are required.

Some precautions, such as improved maintenance and HVAC system controls, may also provide a payback in operating costs and/or improved building air quality. As new building designs or modifications are considered, designers should consider that practical CBR sensors may soon become available. Building system design features that are capable of incorporating this rapidly evolving technology will most likely offer a greater level of protection.

While it is not possible to completely eliminate the risk of a CBR terrorist attack, several precautions can be taken to reduce the likelihood and consequences of such an attack. The IAQ team should assess buildings by looking first for those items that are the most vulnerable and can be addressed easily. Additional measures should be implemented as practical. The goals are to make a facility an unattractive target for a CBR attack and to maximize occupant protection in the event that such an attack occurs.

Case Study

Case Study – It Is Possible for Terrorists to Enter the United States in Different Ways

Joe arrived in the United States with false documentation. Joe sought maintenance and/or cleaning jobs in major office buildings and facilities. Routine background checks did not detect that Joe was a terrorist. This is in part due to the fact that terrorists such as Joe attempt to behave as model citizens prior to carrying out attacks in order to avoid drawing attention to themselves.

Joe found work as an apprentice stationary engineer in a high-rise office building in a major metropolitan city. After months of on-the-job training and gathering building information regarding mechanical room floor plans, typical security operations, and times of maximum building occupancy, Joe decided to carry out his plans. He took the stairwell that led to the penthouse equipment room. He opened the access door to the outdoor-air intake and dispersed a thermos of 2-chlorobenzalmalononitrile (CS) tear gas, used for riot control, into the ventilation airstream. The tear gas was purchased legally off of the Internet.

The tear gas caused immediate breathing problems for building occupants, especially for individuals with asthma. Building operations reacted quickly by shutting down building systems to minimize the spread of the contaminant. However, two building occupants did require hospitalization, one of whom ended up on a ventilator. In the stairways, other occupants suffered from broken bones, scrapes, and cuts during the panic of trying to leave the building.

Definitions

Homeland Security Definitions

- A *CBR attack* is an attack on a building or area that employs airborne chemical, biological, or radiological (CBR) agents.
- *Homeland security* is a concerted national effort to prevent terrorist attacks within the United States, reduce America's vulnerability to terrorism, and minimize the damage and recovery time from attacks that do occur.
- *Nonducted return-air systems* are systems that commonly use hallways or spaces above dropped ceilings as return-air paths or plenums.
- A *shelter-in-place* is a small, interior room of a building with few or no windows where individuals can take refuge.

For more information please refer to ATPeResources.com or http://www.epa.gov.

1. What does CBR stand for?

2. Name three criteria to consider during a building walk-through to determine the building's vulnerability to a CBR attack.

3. List three actions to avoid in an effort to prevent a CBR attack.

4. Describe at least three physical security actions that can be implemented as preventive measures.

TROUBLESHOOTING AND MITIGATING IAQ PROBLEMS

15 chapter

Fluke Corporation

Class Objectives
Describe troubleshooting and mitigation strategies for IAQ problems. List the steps in troubleshooting an IAQ complaint.

On-the-Job Objective
Create a written step-by-step plan for troubleshooting IAQ complaints in your facility.

Learning Activities:
1. List the steps to take in troubleshooting IAQ complaints.
2. Why are recordkeeping and documentation so important in IAQ work?
3. How will you know when you have solved an IAQ problem?
4. **On-the-Job:** Work through the troubleshooting steps for an IAQ complaint in your facility.

Introduction

When properly implemented, a PM program, visual inspections, and scheduled air-sampling strategies will greatly reduce the potential for IAQ problems in a facility, ensure optimum building operations, and reduce occupant complaints. Despite a stationary engineer's best efforts, there are many building functions and activities that will be beyond their control. In addition, occupants may have complaints even under the best of IAQ conditions. It is essential, therefore, that a building have an established procedure to troubleshoot IAQ complaints when they occur.

TROUBLESHOOTING AND MITIGATING IAQ PROBLEMS

The usual method for solving IAQ problems is called a step-by-step troubleshooting process. A *step-by-step IAQ troubleshooting process* is a list of procedures that identifies steps to follow in chronological order until an IAQ problem is found. The troubleshooting process assumes that a facility has an established procedure for handling IAQ complaints. The procedure must include a communication tree that describes who receives what information and to whom each person reports. The EPA recommends and several proposed laws stipulate that each facility have an IAQ manager who coordinates the prevention and investigation of IAQ problems. An IAQ manager may be a building manager or a stationary engineer.

An IAQ complaint investigation procedure is written as if one person is complaining. The step-by-step troubleshooting process must work whether one person or many people have reported IAQ problems.

Step-by-Step Complaint Investigation Procedures

Stationary engineers usually learn of an IAQ problem from an occupant's complaint. Some occupants are loud and angry, others timid or embarrassed. The complaint may be vague or very precise. All complaints must be handled using the procedures established for a facility. **See Figure 15-1.** IAQ complaint investigation procedures should include the following steps:

1. Once an occupant complaint is issued to a stationary engineer, the complaint must be responded to immediately using proper channels. All complaints must be treated seriously. Failure to respond immediately to all complaints may allow an IAQ problem (contaminants) to spread and grow worse. By responding immediately, a stationary engineer stands a good chance of mitigating the problem at a manageable stage, which saves money and preserves good relations with building occupants. Liability is a major concern of management. Having standardized IAQ complaint investigation procedures can lessen the liability burden.

2. Talk to the person who issued the complaint. Be polite and concerned. Treat the person with dignity and respect. Ask the person to provide as much detail as they can on the nature of the complaint. Stationary engineers must be attentive and take notes on what contributing factors must be checked. It is especially important to determine what time of the day symptoms occur. Having the time of day when occupants become symptomatic is helpful when evaluating time and space conditions.

3. Document everything and record every piece of information received. Check the building automation system to document the history of the equipment and to determine if there are any current problems. Establish a holding file to contain the investigation records. As the investigation proceeds, record all ideas and actions on IAQ paperwork such as a work order ticket.

4. Conduct an initial visual inspection of the space where the person works. Checking space conditions (obstructed grills, grill location, occupancy level, activity, and equipment in the space, odors, lighting, and noise) provides a thumbnail sketch of the space and points out obvious problems. As building personnel proceed through the complaint investigation step-by-step, personnel will investigate much of the required complaint information at a later time in further detail.

5. Check the temperature and relative humidity at the thermostat and humidity sensor and where the person issuing the complaint feels the symptoms are the strongest. Perform any necessary HVAC calibrations and repairs. High temperatures and high relative humidity can cause a high emission of some volatile organic compounds (VOCs), like formaldehyde. Relative humidity that is too low (generally below 30%) can cause irritation to eyes, nose, and throat.

Step-by-Step Complaint Investigation Procedures

RESPOND TO COMPLAINTS IMMEDIATELY ❶

SPEAK TO PERSON WHO GENERATED COMPLAINT ❷

CHECK AIRFLOW FROM COMPLAINT SPACE DIFFUSERS ❻

❼ INVESTIGATE CONDITION OF COMPLAINT SPACE FOR WATER, MOLD, AND OTHER PROBLEMS

❽ INVESTIGATE COMPLAINT SPACE ACTIVITIES AND TIMING PATTERNS

REVIEW ANY PRINTS AND POSSIBLE CONTAMINANT PATHWAYS ❿

Person Generating Complaint

Stationary Engineer

INVESTIGATE CHANGES SUCH AS NEW PARTITIONS, ADDITION OF LASER PRINTERS, AND PEOPLE IN COMPLAINT AREA ❾

❸ DOCUMENT ALL WORK AND RECORD INFORMATION

❹ CONDUCT VISUAL INSPECTION OF COMPLAINT SPACE

❺ CHECK TEMPERATURE AND RELATIVE HUMIDITY OF COMPLAINT SPACE

⓫ REVIEW ALL REPAIRS AND WORK ORDERS FOR COMPLAINT SPACE AND ADJOINING AREAS

⓬ CLEAN, BALANCE, AND CALIBRATE LOCAL HVAC SYSTEM

⓭ FOLLOW FACILITY PROCEDURES FOR REPORTING FINDINGS TO BUILDING OCCUPANTS

⓮ FILL OUT AND FILE ALL FORMS DEALING WITH RESOLUTION OF OCCUPANT'S COMPLAINT

> **NOTE:** SOLVING AN IAQ PROBLEM MAY REQUIRE CLEANING, BALANCING, AND CALIBRATION OF ENTIRE HVAC SYSTEM.

Figure 15-1. All indoor air quality complaints must be handled using the procedures established by a building or company.

6. Check the airflow from nearby diffusers and at the complaining person's location. When airflow is not at the proper level, check the airflow at every diffuser on the HVAC loop and consider rebalancing the entire loop.

7. Investigate space conditions such as dust on surfaces, water damage, mold or mildew, odors, open chemical products, and damaged walls, which may allow gas or vapor penetration from other sources. **See Figure 15-2.**

8. Investigate space activities and timing patterns. List the activities that occur before and during the time when symptoms are noticed. Identify any operational products used during these time periods. Include custodial activities and cleaning products used, office activities and products (e.g., glues, correction fluid, and lab work), and personal products (e.g., strong perfumes and tabletop air fresheners). Identify possible contaminants generated by building activities. From the chemical inventory of the building, gather the MSDSs for products used in the complaining person's space. All of this investigation information must be saved for future reference.

Mold and Mildew Damage

Figure 15-2. Water entering an occupied space will cause damage to materials and cause mold and mildew to grow at a rate that is dependent on the humidity levels in the space.

Pollutant Pathway Records

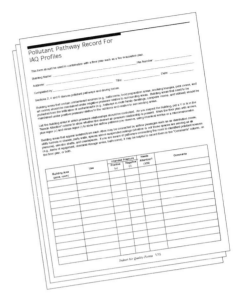

Figure 15-3. Airflow pathways can change with changes in furniture location and modifications to space layout. Pollutant pathway records indicate the normal airflow paths through spaces of a building for future reference.

9. Investigate space changes such as new office partitions being used and any source of VOC emissions. Investigate whether multiple laser printers have recently been installed (source of ozone) or whether there are more people in the space than the space was originally designed for, leading to inadequate ventilation.

10. If no obvious contaminants are found in the complaint space, follow pathways leading out of the area. Review the Pollutant Pathway Record and accompanying prints. **See Figure 15-3.** Check spaces connected to the complaint area for contaminants.

11. Evaluate the HVAC system. Review all PM work orders for the HVAC system. Review any repair or PM work orders for the immediate complaint area or adjoining areas. Check the HVAC system using an HVAC Checklist. Correct any obvious problems by issuing repair work orders. Correct any problems with airflow and pressure relationships. Remove any contaminant sources and clean the pathways between the source and problem area. If no contaminants or equipment problems are found within the HVAC system, stationary engineers must check the outdoor air being brought into the system for contaminant sources. Fill out an Engineer's Watch/Shift/Incident Log on a regular basis to document activities being performed.

12. Solving an IAQ problem may include cleaning, balancing, and calibrating the entire HVAC system. Airflows may be increased or decreased to bring them back in line with design specifications. The HVAC system start times may need rearranging. Certain products or activities may be modified, banned, or rescheduled. Surface areas may require cleaning and/or sealing.

13. A facility's IAQ investigation procedures must be followed when reporting investigation information to the person issuing the complaint and other building occupants. If problems reoccur, ask the person issuing the complaint to notify authorities through the proper channels.

14. When an IAQ investigation solves an IAQ problem, all files and forms used and a brief account of the actions taken must be entered in the Engineer's Watch/ Shift/Incident Log for future reference.

IAQ Complaint Solutions

To determine when an IAQ problem has been solved, four points must be considered. **See Figure 15-4.** The factors to consider in determining if an IAQ problem has been solved include the following:

- The HVAC system is operating within design parameters.
- The investigation yields no problem sources or environmental conditions, or it yields sources of contaminants or conditions that could be resolved within the scope of the facility.
- The investigation yields no unusual sources, or it yields environmental conditions or sources with solutions beyond the scope of the facility.
- The complaints do not continue.

Reducing or eliminating complaints alone may not mean that an IAQ problem is solved. People often stop complaining when work is seen being done to solve the IAQ problem. Also, an IAQ problem could have been solved, but complaints may still continue because of unrelated factors such as stress, distrust, or various working conditions.

IAQ Complaint Solution Conditions

ALL OCCUPANTS APPEAR TO BE SATISFIED WITH BUILDING CONDITIONS

TEMPERATURE, HUMIDITY, AND VENTILATION ALL MEASURED AND WITHIN TOLERANCES

NO FURNITURE, CARPETING, WALLPAPER, OR CONSTRUCTION PRODUCTS WERE FOUND TO BE GIVING OFF CONTAMINANT VAPORS

NO SPILLS, OPEN CONTAINERS, OR CONTRABAND PRODUCTS WERE FOUND

Figure 15-4. To accept that an IAQ problem has been solved, all requirements should be met: the HVAC system operates as designed, no contaminant sources that were found remain, and the occupant complaints about a space have stopped.

Just measuring for contaminants in the complaint area air is not a reliable way to decide when a problem is solved. Indoor contaminant levels vary greatly over time, and the contaminant being measured may not be the contaminant causing the problem. The EPA asserts that measurements of airflow, ventilation rates, and air distribution patterns are better ways to assess the results of IAQ control efforts. When air measurements are taken, the stationary engineer must make sure various conditions are consistent for all the measurements. Conditions that should be similar from measurement to measurement include the time of day, occupancy levels, heating load, outdoor conditions, and system operation.

OVERVIEW OF NON-IAQ PROBLEMS

Occupant complaints are not always clearly attributable to the factors that caused a person to complain. Complaints can arise from other factors besides IAQ problems, including other environmental factors and physical, emotional, and social factors. Sometimes, a lack of mental and social well-being can contribute to employee complaints. Building occupants who are unhappy with their jobs may think there is "something in the air" that makes them feel bad. In some cases, occupant revenge or sabotage is possible since IAQ investigations can be costly to a company.

There is nothing that stationary engineers can do about some of the factors that generate employee complaints. A stationary engineer may never know if a non-IAQ factor is the cause of an occupant's complaint. At best, most complaints unrelated to IAQ will stop when a stationary engineer investigates the complaint. At worst, the stationary engineer may have a continuing complaint for which no source or resolution can be found. When every reasonable step has been taken and all actions are documented, the unresolved complaint is then referred to management.

Work-Related Complaints

There are work-related complaints, however, that stationary engineers can improve or that can be brought to management's attention for corrective action. During PM and IAQ surveys, stationary engineers must stay on the lookout for any work-related conditions that might lead to employee complaints, such as computer monitor eyestrain, improper lighting, and/or high levels of noise.

Computer Monitor Eyestrain. Working with computer monitors can cause eyestrain and eye irritation due to improper lighting, glare from the screen, poor screen position, and difficult-to-read material. Light should not shine directly into the eyes of a person looking at a monitor. Auxiliary lighting should be installed to allow building occupants to read material comfortably. Bright light washes out monitor displays, making them difficult to see and read. Solutions for reducing glare include changing computer background colors, using antiglare screens, practicing eye exercises, and increasing the frequency of breaks. Breaks away from the computer monitor will also help reduce fatigue caused by holding the same posture for a long time. Normal office illumination is usually in the 50 foot-candles (fc) to 100 fc range. **See Figure 15-5.** Illumination in office areas can be reduced to the 30 fc to 50 fc range. Finally, occupants should be able to adjust the position of the monitor to ensure maximum comfort.

Improper Lighting. Lighting should always conform to the original design specifications unless changes in design or space use warrant otherwise. When lighting does conform to specifications, occupancy levels should be checked to ensure that they are in line with the original specifications. When discrepancies are found with space use or lighting, stationary engineers should report the findings to management. Lighting should be tailored to the specific tasks to be performed in the space. It is important to note any changes in the use of the space, which might indicate that lighting needs have changed.

Some types of light also cause psychological effects. When deprived of natural light, some people develop symptoms such as depression and fatigue. The condition caused by a lack of natural light is known as seasonal affective disorder. When lighting appears to meet specifications and associated complaints continue, management must be advised of the situation.

IAQ Fact

Window glare causes a variety of complaints in high-rise buildings, disrupting the ability of building occupants to use their monitor displays. Poor IAQ is sometimes incorrectly believed to be the cause of headaches that are actually caused by window glare.

Reducing Occupant Eyestrain

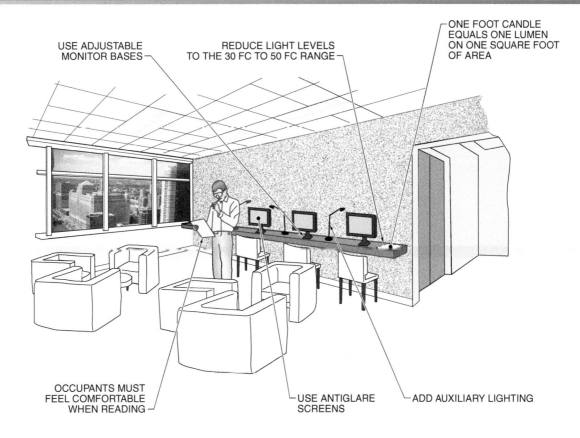

USE ADJUSTABLE MONITOR BASES

REDUCE LIGHT LEVELS TO THE 30 FC TO 50 FC RANGE

ONE FOOT CANDLE EQUALS ONE LUMEN ON ONE SQUARE FOOT OF AREA

OCCUPANTS MUST FEEL COMFORTABLE WHEN READING

USE ANTIGLARE SCREENS

ADD AUXILIARY LIGHTING

Figure 15-5. To reduce eyestrain, glare can be reduced by changing the computer monitor's background colors, using antiglare screens, increasing the frequency of breaks, and reducing the normal lighting levels in computer spaces to the 30 fc to 50 fc range.

Noise. Noise can cause discomfort or distraction in an office environment. Most noise standards are set to prevent hearing loss. Still, lower levels of noise may cause discomfort when the noise interferes with concentration, conversation, or privacy. Frequency and tone may cause discomfort or stress. For example, some sounds and frequencies can cause headaches and fatigue. Common noise sources are certain types of office equipment, mechanical rooms, HVAC ducting, and street sounds. It is important to pay attention to noise-producing situations that may cause discomfort to building occupants. When necessary, time should be spent in affected areas to observe noise conditions that might have been missed during noise level testing.

Office copy machines, printers, fax machines, paper-shredding machines, and some computer hard drives generate noises that individually or together create discomfort for building occupants.

INDOOR AIR QUALITY CONTROL STRATEGIES

Indoor air quality problems result from interactions among various factors, including contaminant sources, the building site, the building structure, activities within the building, mechanical equipment, the climate, and building occupants. Efforts to control indoor air contaminants change the relationships between these factors. There are many ways that people can intervene to prevent or control indoor air contaminant problems. Control strategies can be categorized as the following:
- contaminant source control
- ventilation
- modification of HVAC systems
- air cleaning
- exposure control
- mitigation strategies

Contaminant Source Control

All efforts to prevent or correct IAQ problems must include an effort to identify and control pollutant sources. Source control is generally the most cost-effective approach to mitigating IAQ problems when point sources of contaminants can be identified. A *point source* is a type of contamination that results from a single or small source. A *nonpoint source* is a type of contamination that results from multiple sources or a large-area source. In the case of a strong source, source control may be the only solution that will work. The two main areas of action for source control are to seal or cover the source, or to remove or reduce the source. **See Figure 15-6.**

In order to seal or cover the source, the following steps can be taken:
- Improve the storage of materials that produce contaminants.
- Seal surfaces of building materials that emit VOCs such as formaldehyde.
- Modify the environment.

In order to remove or reduce the source, the following steps can be taken:
- Prohibit smoking indoors or limit smoking to areas from which air is exhausted, not recirculated. (NIOSH regards smoking areas as an interim solution.)
- Relocate contaminant-producing equipment to an unoccupied or exhaust-only ventilated space, or to a better-ventilated space.
- Select products that produce fewer or less potent contaminants while maintaining adequate safety and effectiveness.
- Modify other occupant activities.

For example, after cleaning and disinfecting an area that is contaminated by fungal or bacterial growth, controlling humidity can make conditions inhospitable for fungus or bacterial growth. Source removal or reduction can sometimes be accomplished by a one-time effort such as a thorough cleaning of a spill. In other cases, source control requires an ongoing process, such as establishing and enforcing a nonsmoking policy.

Sealing or covering the source of a contaminant can be a solution in some cases. The application of a barrier (slip cover) over formaldehyde-emitting building material is a prime example. Sealing may also involve educating staff or building occupants about the contaminant-producing features of materials and supplies, and the inspection of storage areas to ensure that containers are properly covered. Sealing can be a difficult technique to implement because of hidden airflow pathways above drop ceilings, under raised flooring, and against brick or block walls. **See Figure 15-7.** However, sealing can have other benefits such as energy savings and effective pest control by eliminating pathways used by vermin.

In some cases, modification of the environment is necessary for effective mitigation. If an indoor air problem arises from microbiological contaminants, just disinfecting the affected area may not eliminate the problem. The regrowth of microbiologicals can occur unless humidity control or other steps are taken, such as adding insulation to prevent surface condensation in order to make the environment inhospitable to microbiologicals.

Contaminant Source Control

SEAL ALL SURFACES

IMPROVE STORAGE OF CONTAMINANT-PRODUCING PRODUCTS

SELECT PRODUCTS WITH FEWER CONTAMINANTS

SEAL OR COVER CONTAMINANT SOURCE

REMOVE OR REDUCE CONTAMINANT SOURCE

MODIFY ENVIRONMENT

PROHIBIT INDOOR SMOKING

RELOCATE CONTAMINANT-PRODUCING EQUIPMENT

MODIFY OCCUPANT ACTIVITIES

Figure 15-6. The available options when dealing with a strong contaminant source are to seal or cover the source, or remove or reduce the source.

Ventilation

Ventilation modifications, whether temporary or permanent, are often used to correct or prevent indoor air quality problems. Ventilation as a contaminant control system can be effective either where buildings are underventilated or where a specific contaminant source cannot be identified. The following ventilation methods can be used to control indoor-air contaminants:

- diluting contaminants with outdoor air
- increasing the total quantity of supply air (including outdoor air)
- increasing the proportion of outdoor air to total air
- improving air distribution

Hidden Airflow Pathways

ABOVE DROP CEILINGS

FROM WALL OPENINGS

AGAINST BRICK WALLS

UNDER DOORS

Figure 15-7. Sealing a space can be a difficult technique to implement because of hidden airflow pathways above drop ceilings and against brick or block walls, but space sealing has other benefits such as energy savings and pest control.

- isolating or removing contaminants by controlling air-pressure relationships
- installing effective local exhaust at the location of the source
- avoiding recirculation of air that contains contaminants
- locating occupants near supply diffusers and locating contaminant sources near exhaust registers
- using air-tightening techniques to maintain pressure differentials and eliminate pollutant pathways
- making sure that doors are closed where necessary to separate zones

Diluting contaminants by increasing the flow of outdoor air can be accomplished by increasing the total supply airflow in the complaint area. Total airflow can be increased by opening supply diffusers, adjusting dampers, adjusting the air-handling unit, and/or by replacing dirty filters. An alternative is to increase the proportion of outdoor air by adjusting the outdoor air intake damper and installing minimum settings on the variable-air-volume (VAV) boxes so that they satisfy the outdoor air requirements of ASHRAE Standard 62.1-2007, *Ventilation for Acceptable Indoor Air Quality.* Studies have shown that increasing ventilation rates to meet this ASHRAE standard does not significantly increase the total annual energy consumed. **See Figure 15-8.** The energy increase to meet ventilation flow rates (5 cfm/person, 7.5 cfm/person, 10 cfm/person, or 20 cfm/person) appears to be less than 5% in typical commercial buildings. The cost of ventilation is generally overshadowed by lighting costs. Further, improved maintenance can produce energy savings to balance ventilation costs.

Diluting Contaminants

MEETING ASHRAE STANDARD 62.1-2007 DOES NOT CAUSE A SIGNIFICANT ENERGY CONSUMPTION PROBLEM

INCREASING VENTILATION FLOW RATES INCREASES ENERGY CONSUMPTION BY 5% FOR MOST NONINDUSTRIAL BUILDINGS

ASHRAE STANDARD 62.1-2007	
OCCUPANCY CATEGORY	**OUTDOOR AIR RATE***
EDUCATIONAL FACILITIES	10
HOTELS, MOTELS, RESORTS, DORMITORIES	5
OFFICE BUILDINGS	5
PUBLIC ASSEMBLY	5
RETAIL	7.5
SPECIAL CONDITION	
HOTEL LOBBIES	7.5
MUSEUMS	7.5
BEAUTY SALONS	20

* typical CFM/person

Figure 15-8. *Diluting contaminants by increasing the flow of outdoor air to meet ASHRAE Standard 62.1-2007,* Ventilation for Acceptable Indoor Air Quality *causes an energy increase of less than 5% in most commercial buildings.*

IAQ Fact

Per the requirements of the Energy Policy Act, ANSI/ASHRAE/IESNA Standard 90.1-2004, **Energy Standard for Buildings Except Low-Rise Residential Buildings** *must be adopted by all states, or states must adopt a commercial energy code that exceeds 90.1. If a conflict arises involving health and safety between standards 62.1-2004 and 90.1-2004, the provisions of Standard 62.1-2004 will govern.*

Modification of HVAC Systems

The cost of modifying an existing HVAC system to condition additional outdoor air can vary widely depending upon the specific situation. In some buildings, HVAC equipment may not have sufficient capacity to allow successful mitigation using this approach. Original equipment is often oversized so that it can be adjusted to handle the increased load, but in some cases additional capacity is required. Most ventilation deficiencies appear to be linked to inadequate quantities of outdoor air. However, inadequate distribution of ventilation air can also produce IAQ problems. Diffusers must be properly selected, located, installed, and maintained so that supply air is evenly distributed and blends thoroughly with room air in the breathing zone. Short circuiting occurs when clean supply air is drawn into the return-air plenum before it has mixed with the dirtier room air and therefore fails to dilute contaminants. Mixing problems can be aggravated by temperature stratification. **See Figure 15-9.** Stratification can occur, for example, in a space with high ceilings in which ceiling-mounted supply diffusers distribute heated air.

Increased ventilation side effects include the following situations:

- Mitigation by increasing the circulation of outdoor air requires good outdoor air quality.
- Increased supply air at the problem location might mean less supply air in other areas.
- Increased total air in the system and increased outdoor air will tend to increase energy consumption and may require increased equipment capacity.

Any approach that affects airflow in a building can change pressure differences between rooms (or zones), indoors and outdoors, and might lead to increased infiltration of unconditioned outdoor air. Increasing air in a VAV system may overcool an area to the extent that terminal reheat units are needed.

Ventilation equipment can be used to isolate or contain contaminants by controlling air-pressure relationships. When the contaminant source has been identified, this strategy can be more effective than dilution. Techniques for controlling air-pressure relationships range from adjustment of dampers to installation of local exhaust. In special applications, using local exhaust confines the spread of contaminants by capturing them near the source and exhausting them to the outdoors. It also dilutes the contaminant by drawing cleaner air from surrounding areas into the exhaust airstream.

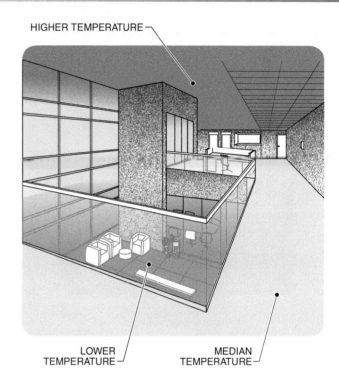

Temperature Stratification

HIGHER TEMPERATURE

LOWER TEMPERATURE

MEDIAN TEMPERATURE

Figure 15-9. Temperature stratification typically occurs in building spaces with high ceilings in which ceiling-mounted supply diffusers distribute the heated air.

When there are return grills in a room equipped with local exhaust, the local exhaust should exert enough suction to prevent recirculation of contaminants. Properly designed and installed local exhaust results in far lower contaminant levels in a building than could be accomplished by a general increase in dilution ventilation, with the added benefit of costing less. **See Figure 15-10.** Importantly, replacement air must be able to flow freely into the area from which the exhaust air is being drawn. It may be necessary to add door or wall louvers in order to provide a path for the makeup air. It is important to ensure that the addition of louvers does not violate any fire codes.

Isolating Contaminants Using Local Exhaust

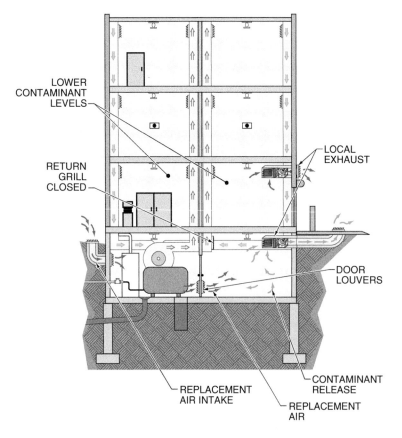

LOWER
CONTAMINANT
LEVELS

RETURN
GRILL
CLOSED

LOCAL
EXHAUST

DOOR
LOUVERS

REPLACEMENT
AIR INTAKE

CONTAMINANT
RELEASE

REPLACEMENT
AIR

Figure 15-10. Properly designed and installed local exhaust results in far lower contaminant levels in a building than could be accomplished by a general increase in dilution ventilation.

Correct identification of the pollutant source and installation of the local exhaust is critically important. For example, an improperly designed local exhaust can draw other contaminants through the occupied space and make the problem worse. The physical layout of grills and diffusers relative to room occupants and pollutant sources can be important. If all supply diffusers are at one end of a room and all returns are at the other end, the people located near the supplies may be provided with relatively clean air, while those located near the returns may breathe air that has already picked up contaminants from all the sources in the room that are not served by local exhaust.

Air Cleaning

Another common IAQ mitigation strategy is to clean the air. Air cleaning is usually most effective when used in conjunction with either source control or ventilation; however, it may be the only approach to use when the source of pollution is outside of the building. Most air cleaning in large buildings is aimed primarily at preventing contaminant buildup in HVAC equipment and enhancing equipment efficiency.

Air cleaning equipment intended to provide better indoor air quality for occupants must be properly selected. That is, it must be designed to remove the particular pollutants of interest. (For example, gaseous contaminants can be removed only by gas sorption.) Once installed, the equipment requires regular maintenance in order to ensure good performance; otherwise, the equipment may become a major pollutant source in itself. If an air cleaning system involving a large number of units is under consideration for a large building, this maintenance requirement should be planned for in advance. When room units are used, the installation must be designed for proper air recirculation. There are four technologies in general use that remove contaminants from the air:
• particulate filtration
• electrostatic precipitation
• negative ion generators
• gas sorption

Particulate filtration, electrostatic precipitation, and negative ion generators are designed to remove particulates, while gas sorption is designed to remove gases. **See Figure 15-11.** Particulate filtration removes suspended liquid or solid materials whose size, shape, and mass allow them to remain airborne for the air velocity conditions present. Filters are available in a range of efficiencies, with a higher efficiency indicating removal of a greater proportion of particles and of smaller particles. Moving to medium efficiency pleated filters is advisable to improve IAQ and increase protection for equipment. However, the higher the efficiency of the filter, the more it will increase the pressure drop within the air distribution system and reduce total airflow (unless other adjustments are made to compensate). It is important to select an appropriate filter for the specific application and to make sure that the HVAC system will continue to perform as designed.

Electrostatic Precipitation. Electrostatic precipitation is another type of particulate control that uses the attraction of charged particles to oppositely charged surfaces in order to collect airborne particulates. In this process, the particles are charged by ionizing the air with an electric field. The charged particles are then collected by a strong electric field generated between oppositely charged electrodes. This provides relatively high-efficiency filtration of small respirable particles at low air pressure losses. *Respirable particles* are airborne particles that can be inhaled into the lungs when breathing.

Electrostatic precipitators may be installed in air distribution equipment or in specific usage areas. As with other filters, they must be serviced regularly. However, all electrostatic precipitators produce some amount of ozone. Because ozone is harmful at elevated levels, the EPA, NIOSH, and OSHA have set standards for ozone concentrations in outdoor air. Standards and guidelines have been established for ozone in indoor air as well. The amount of ozone emitted from electrostatic precipitators varies from model to model. **See Figure 15-12.**

Negative Ion Generators. Negative ion generators use static charges to remove particles from the indoor air. When the particles become charged, they are attracted to surfaces such as walls, floors, tabletops, draperies, and occupants. Some designs include collectors to attract charged particles back to the unit. Negative ion generators are not available for installation in ductwork, but are sold as portable or ceiling-mounted units. As with electrostatic precipitators, negative ion generators may produce ozone, either intentionally or as a by-product of use.

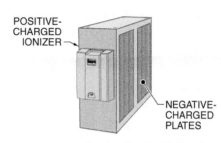
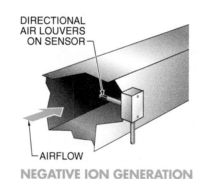

Figure 15-11. Air cleaning removes contaminants from the air. Particulate filtration, electrostatic precipitation, and negative ion generators are designed to remove particulates, while gas sorption is designed to remove gases.

Electrostatic Precipitator-Created Ozone

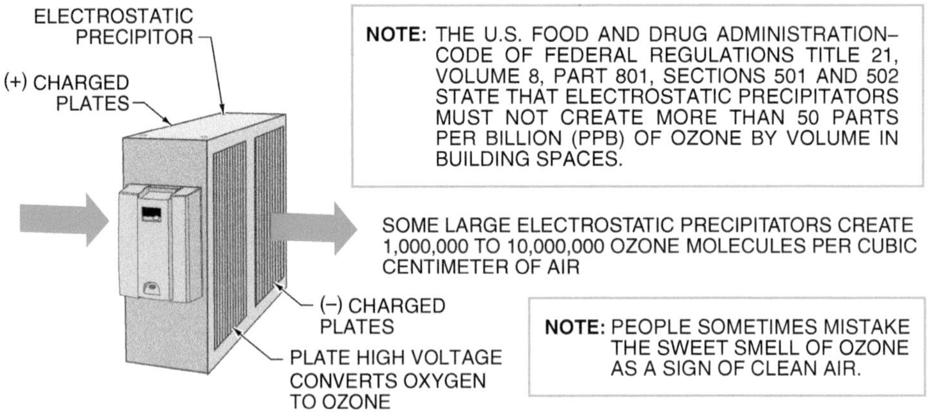

ELECTROSTATIC PRECIPITOR

(+) CHARGED PLATES

NOTE: THE U.S. FOOD AND DRUG ADMINISTRATION–CODE OF FEDERAL REGULATIONS TITLE 21, VOLUME 8, PART 801, SECTIONS 501 AND 502 STATE THAT ELECTROSTATIC PRECIPITATORS MUST NOT CREATE MORE THAN 50 PARTS PER BILLION (PPB) OF OZONE BY VOLUME IN BUILDING SPACES.

SOME LARGE ELECTROSTATIC PRECIPITATORS CREATE 1,000,000 TO 10,000,000 OZONE MOLECULES PER CUBIC CENTIMETER OF AIR

(–) CHARGED PLATES

PLATE HIGH VOLTAGE CONVERTS OXYGEN TO OZONE

NOTE: PEOPLE SOMETIMES MISTAKE THE SWEET SMELL OF OZONE AS A SIGN OF CLEAN AIR.

Figure 15-12. Because ozone is harmful at elevated levels, standards and guidelines have been established for ozone in indoor air. The amount of ozone emitted from electrostatic precipitators varies from model to model.

Gas Sorption. Gas sorption is used to control compounds that behave as gases rather than as particles. These gaseous contaminants include formaldehyde, sulfur dioxide, ozone, and oxides of nitrogen. Gas sorption has activated carbon or chemically treated active clays in the sorption material. A chemical reaction between the pollutant and the sorbent causes a binding of the pollutant and the sorbent, or diffusion of the contaminant from areas of higher concentration to areas of lower concentration.

Gas sorption units are installed as part of the air distribution system. Each type of sorption material performs differently with different gases. Gas sorption is not effective for removing carbon monoxide. There are no standards for rating the performance of gaseous air cleaners, making the design and evaluation of such systems problematic. The operating expenses of these units can be quite high, and the units may not be effective if there is a strong source nearby.

Exposure Control

Controlling exposure by relocating susceptible individuals may be the only practical approach in a limited number of exposure control cases. Relocating susceptible individuals is the least desirable option and should only be used when all other strategies are ineffective in resolving a complaint.

Mitigation Strategies

Over the years, many types of mitigation (corrective) strategies have been implemented to solve IAQ problems. Stationary engineers must have an understanding of basic approaches to mitigation as well as the various solutions that can be effective in treating commonly encountered IAQ problems. It is impossible to provide detailed instructions for using each type of mitigation approach. However, guidance can be given in selecting a mitigation strategy and in judging proposals from in-house staff or outside consultants. Mitigation of IAQ problems will likely require the involvement of building management and staff representatives from various areas, including the following:
• facility operation and maintenance
• housekeeping
• shipping and receiving
• purchasing
• policy makers
• staff trainers

Successful mitigation of IAQ problems also requires the cooperation of other building occupants, including the employees of building tenants. Occupants must be educated about the cause or causes of IAQ problems and about actions that must be taken to avoid or prevent a recurrence of problems.

Mitigation strategies for exposure control is an administrative approach that uses behavioral methods, such as the following:

1. Carefully schedule contaminant-producing activities during times when complaints will be minimized.

   • Schedule contaminant-producing activities for unoccupied periods whenever possible. This may be the best way to limit complaints about activities such as roofing or demolition, which unavoidably produce odors and dust.

   • Notify susceptible individuals about upcoming activities such as roof repairs and pesticide application so that the individuals can avoid contact with the contaminants.

   • Relocate space occupants.

2. Move susceptible individuals away from the area where they experience symptoms.

Remedies for Complaints Not Attributed to Poor Air Quality

Specific lighting deficiencies or localized sources of noise or vibration can sometimes be readily identified. Remedial action may be fairly straightforward such as turning lights ON or OFF; making adjustments for glare; or relocating, replacing, or acoustically insulating a noise or vibration source. Similarly, blatant ergonomic or psychosocial stress may be apparent even to an untrained observer. In other cases, however, problems may be more subtle or solutions more complex. Since specialized knowledge, skills, and instrumentation are usually needed to evaluate lighting, noise, vibration, ergonomic stress, or psychosocial stress, such evaluations should gener-

ally be performed by a qualified individual from the particular field. **See Figure 15-13.**

JUDGING PROPOSED MITIGATION DESIGNS AND THEIR SUCCESS

Mitigation efforts should be evaluated at the planning stage by considering the following criteria:

• permanence
• operating principles
• degree to which a strategy fits the job
• ability to institutionalize the solution
• durability
• installation and operating costs
• conformity with codes

Permanence

Permanence mitigation is a type of mitigation effort that creates a permanent solution to an indoor air problem compared to efforts that provide temporary solutions (unless the problems are also temporary). Permanence mitigation is clearly the superior method of mitigation. Opening windows or running air handlers on full outdoor air may be suitable mitigation strategies for a temporary problem such as outgassing of volatile compounds from new furnishings, but these would not be good ways to deal with emissions from a print shop. A permanent solution to microbiological contamination involves not only cleaning and disinfecting, but also modification of the environment to prevent regrowth.

IAQ Fact

It is better to carry out a building investigation and learn the specific facts of a case than to adopt a mitigation approach that might not be appropriate. Attempting to correct an IAQ problem without understanding the cause of the problem is ineffective and expensive.

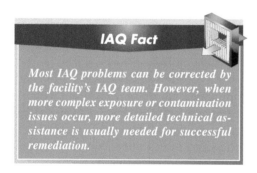

Specialized Remedial Action Meters

LIGHT LEVEL METER

SOUND LEVEL METER

VIBRATION METER

Figure 15-13. Because specialized knowledge, skills, and instrumentation are needed to evaluate lighting, noise, vibration, ergonomic stress, and psychosocial stress, such evaluations should be performed by a qualified individual in that field.

IAQ Fact

Most IAQ problems can be corrected by the facility's IAQ team. However, when more complex exposure or contamination issues occur, more detailed technical assistance is usually needed for successful remediation.

Operating Principles

The most economical and successful solutions to IAQ problems are those in which the operating principle of the correction strategy is suited to the problem. For example, when a specific point source of contaminants has been identified, treatment at the source such as by removal, sealing, or local exhaust is almost always a more appropriate correction strategy than dilution of the contaminant by increased general ventilation. Likewise, when an IAQ problem is caused by the introduction

of outdoor air that contains contaminants, increased general ventilation will only make the situation worse (unless the outdoor air is sufficiently cleaned).

Degree to Which a Strategy Fits the Job

It is important to make sure that stationary engineers thoroughly investigate and understand an IAQ problem to properly select the correction strategy that fits the job. When odors from a special-use area such as a kitchen are causing complaints in a nearby office, increasing the ventilation rate in the office may not be a successful approach. **See Figure 15-14.** The mitigation strategy should address the entire affected area. When mechanical equipment is needed to correct the IAQ problem, it must be powerful enough to accomplish the task. For example, a local exhaust system should be strong enough and close enough to the contaminant source so that none of the contaminant is drawn into nearby return grills and recirculated back into the HVAC system.

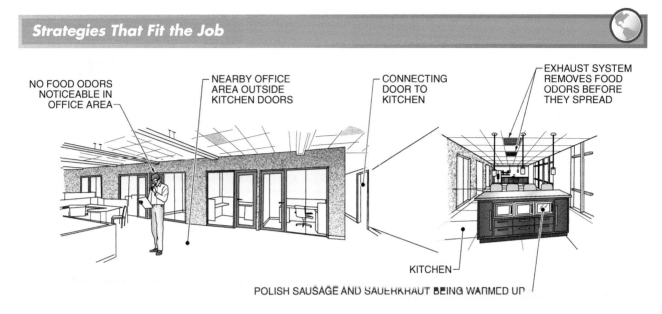

Figure 15-14. When a local exhaust system is being used, the system must be strong enough and close enough to the contaminant source so that none of the contaminant is drawn into nearby return grills and recirculated back into the HVAC system.

Ability to Institutionalize the Solution

A mitigation strategy will be most successful when it is institutionalized as part of normal building operations. Solutions that require relatively familiar equipment are more likely to be successful in the long run than approaches that involve unfamiliar concepts or delicately maintained systems. When maintenance, housekeeping procedures, or supplies must change as part of the mitigation process, it may be necessary to plan for additional staff training, new inspection checklists, or modified purchasing practices. Operating schedules for HVAC equipment may also require modification.

Durability

IAQ mitigation strategies that have proven to be sound and cost-effective are easier to sell to building owners than approaches that are untried, time-consuming, and/or require special maintenance skills. Stationary engineers will keep any equipment running no matter the condition. However, some equipment designs are more efficient than others for some applications.

Installation and Operating Costs

The approach with the lowest initial cost may not be the least expensive over the long run. Other economic considerations include energy costs for equipment operation, increased staff time for maintenance, the differential cost of alternative materials and supplies, and higher hourly rates when odor-producing activities such as cleaning must be scheduled for unoccupied periods. Although these costs will almost certainly be less than the cost of letting the problem continue, they are more readily identifiable, so a thorough presentation to management may be required.

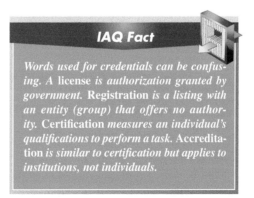

IAQ Fact

Words used for credentials can be confusing. A **license** *is authorization granted by government.* **Registration** *is a listing with an entity (group) that offers no authority.* **Certification** *measures an individual's qualifications to perform a task.* **Accreditation** *is similar to certification but applies to institutions, not individuals.*

Conformity to Codes

Conformity to codes and any modification to building components or mechanical systems must be designed and installed in keeping with applicable fire, electrical, and other building codes.

Judging the Success of a Mitigation Effort

Many day-to-day IAQ problems can be corrected by using common sense. Other more severe contamination problems may require special technical expertise for successful remediation. Two criteria are usually used to judge the success of an investigation and mitigation effort for correction of an indoor air problem. One criterion is reduced complaints and the other is measurement of indoor air properties (often only of limited usefulness). **See Figure 15-15.**

Persistent Problems. Solving an IAQ problem requires repeated data collection and hypothesis testing. Sometimes, a deeper and more detailed investigation is required that will suggest new hypotheses after unsuccessful or partially successful control attempts are made. Even the best-planned investigations and mitigation actions may not produce a resolution to an IAQ problem. Stationary engineers may have performed a careful investigation, found one or more apparent causes for the problem, and implemented a control system. Nonetheless, the correction strategy chosen may not have caused a noticeable reduction in the concentration of the contaminant or improvement in ventilation rates or efficiency. Worse, the complaints may persist even if there is documented improvement in ventilation and in control of all of the contaminants that were identified.

When stationary engineers have pursued source control options and have increased ventilation rates and efficiency to the limits of their expertise, they must decide how important it is to pursue the problem further. When a stationary engineer has made several unsuccessful efforts to control a problem, then it may be advisable to seek outside assistance. The problem is probably complex and may only occur intermittently, or cross the borders that divide traditional fields

JUDGING THE SUCCESS OF A MITIGATION EFFORT	
Mitigation Task	**Judging Points**
Identify the extent of contamination	Locate the source of a contaminant release. What is thought to be the source might only be the tip of the iceberg. Contaminants travel (migrate) through a building and collect in the unwanted areas (contaminant sinks). Specific sampling or testing of surfaces may be required to identify secondary sources of contamination.
Develop a specific plan on how a remediation will be performed	Depending on the type of contamination, specific technical expertise such as chemistry, microbiology, building science, and/or health and safety may be required.
Monitor remediation efforts to guarantee work is preceding correctly	Air sampling and testing is required along with regular inspection during remediation of contaminants from building spaces.
Perform "all clear" air sampling and testing	After an area has been decontaminated, air samples and testing must be performed to guarantee that all contaminant indicators have returned to background levels.

Figure 15-15. Reduced occupant complaints and good air measurement readings are used to judge the success of an investigation and mitigation effort to correct an indoor air problem.

of knowledge. It is even possible that poor IAQ is not the actual cause of the complaints. Bringing in a new perspective at this point can be very effective.

An interdisciplinary team, such as people with engineering, medical, or health backgrounds, may be needed to solve particularly difficult IAQ problems. When a stationary engineer has made several unsuccessful efforts to control a problem, then it may be advisable to seek outside assistance.

SAMPLE CASE STUDY PROBLEMS AND SOLUTIONS

There are a variety of IAQ problems found in buildings. IAQ problems fall into various categories. When a sample IAQ problem is provided, the example should be followed by a solution that has been used for that category of problem. Most of the problems used to represent real-world IAQ problems do not have serious, life-threatening consequences. However, some IAQ problems are very serious and can have severe health impacts. The basic correction principles that apply to sample case study problems are usually similar to those used in less critical situations.

Case study problems help with the thinking process to determine what is the best way to solve specific IAQ problems. It is important to remember that case study problems are brief sketches, and apparent parallels to any specific building could be misleading. It is better to carry out a building investigation and learn the specific facts in a case rather than adopt a mitigation approach that might not be appropriate. Attempting to correct IAQ problems without understanding the cause of the problems can be both ineffective and expensive. Stationary engineers will note that some solutions are simple and low-cost, while others are complex and expensive. The reader should not assume that each solution listed would be an effective treatment for all of the problems in its category.

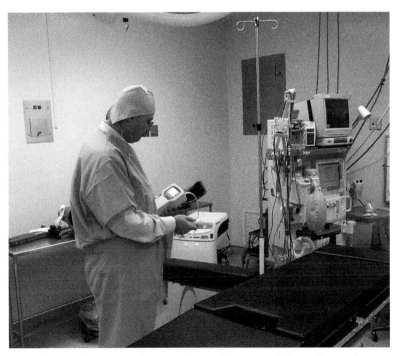

Depending on the situation, the contaminants of concern and the quality requirements for the indoor air can vary greatly.

Case Study Problem #1: Outdoor-Air Ventilation Rate Too Low

Problem Reported:

Several occupants in a commercial office building report that the air seems to be "stuffy."

Documentation:

- The IAQ complaint is logged.
- The IAQ complaint form is filled out.
- If needed, the IAQ complaint interview diary is used.
- At the end of the investigation, the IAQ complaint summary is completed.
- All records are then stored properly.

Troubleshooting Steps:

After documentation is complete, a survey of the building spaces and the HVAC system is taken, and air quality measurements are taken.

Possible Pollutants and Problems:

- Routine odors from normal occupant activities are present.
- Measured outdoor-air ventilation rates indicate that ventilation rates do not meet guidelines for the outdoor-air supply (e.g., design specifications, applicable codes, and ASHRAE Standard 62-1989).
- Measured CO_2 concentrations that are above 1000 ppm indicate inadequate ventilation.

Possible Solutions:

1. Open, adjust, or repair the air distribution system, including:
   - outdoor-air intakes
   - mixing and relief dampers
   - supply diffusers
   - fan casings
2. Increase outdoor-air ventilation within the design capacity of the following:
   - air handler
   - heating and air conditioning equipment
   - distribution system
3. Modify the heating and cooling coil components of the HVAC system as needed to allow increased outdoor air capacity.
4. Design and install an updated ventilation system.
5. Reduce the pollutant and/or thermal load on the HVAC system in the following ways:
   - Reduce the occupant density by relocating some occupants to other spaces.
   - Redistribute the load on the ventilation system.
   - Relocate or reduce the use of heat-generating equipment.

Resolution:

In Case Study Problem #1, *Outdoor-Air Ventilation Rate Too Low*, the problem was resolved by the combination of modifying heating and cooling components and retrofitting the ventilation system.

Case Study Problem #2: Contaminant Enters Building from Outdoors

Problem Reported:

Recently, strange odors are being reported in different spots in the building. There had been no previous reports of this nature.

Documentation:

- The IAQ complaint is logged.
- The IAQ complaint form is filled out.
- If needed, the IAQ complaint interview diary is used.
- At the end of the investigation, the IAQ complaint summary is completed.
- All records are then stored properly.

Troubleshooting Steps:

After documentation is complete, a walk-through survey of the building spaces and the HVAC system is taken. During the walk-through survey, it is noticed that odors (possible contaminants) are present at the outside-air intakes.

Possible Pollutants and Problems:

- soil gases (radon, gasoline from leaking tanks, and methane from landfills)
- contaminants from nearby activities (roofing, dumpster, and construction)
- outdoor-air intakes near contaminant sources (parking, loading dock, and building exhaust)
- outdoor-air contains pollutants or excess moisture (cooling tower mist entrained in intake air)

Possible Solutions:

1. Remove the source if it can be moved easily. For example:
   - Remove debris around outdoor-air intake.
   - Relocate the dumpster.
   - Reduce the source. (For example, shift time of activity to avoid occupied periods.)
   - Remove painting, roofing, and demolition work.
   - Change the hours and location of housekeeping and pest control.
2. Relocate elements of the ventilation system that contribute to entry of outdoor air contaminants:
   - Separate outdoor-air intakes from sources of contaminant odors.
   - Separate exhaust-fan outlets from operable windows, doors, and air intakes.
   - Make rooftop exhaust outlets taller than intakes.
3. Change air-pressure relationships to control pollutant pathways:
   - Install sub-slab depressurization to prevent entry of soil gas contaminants (radon, gases from landfills, and underground tanks).
   - Pressurize the building interior relative to the outdoors (this will not prevent contaminant entry at outdoor-air intakes).
   - Close pollutant pathways (seal cracks and holes).
4. Add special equipment to the HVAC system:
   - Use filtration equipment to remove pollutants (select to fit the specific application).

Resolution:

In Case Study Problem #2, *Contaminant Enters Building from Outdoors*, it was found that a dumpster had been inadvertently relocated to an area too close to an outside-air intake. The dumpster was moved and the odor problem disappeared.

Case Study Problem #3: Occupant Activities Contribute to Airborne Contaminants or to Space Comfort Problems

Problem Reported:

Several occupants in an office building report various odors and cigarette smoke in various parts of the building.

Documentation:

- The IAQ complaint is logged.
- The IAQ complaint form is filled out.
- If needed, the IAQ complaint interview diary may be used.
- At the end of the investigation, the IAQ complaint summary is completed.
- All records are then stored properly.

Troubleshooting Steps:

After documentation is complete, a survey of the building spaces and the HVAC system indicates that illegal activities are occurring in the building. Personal HVAC devices, such as desktop humidifiers, have been brought in by building occupants. In addition, building occupants were observed smoking in lightly occupied areas of the building, in violation of building policy.

Possible Pollutants and Problems:

- smoking
- fumes due to special activities in areas such as print shops, laboratories, and kitchens
- interference with HVAC system operation
- blockage of supply diffusers to eliminate drafts
- turning off exhaust fans to eliminate noise
- use of space heaters and desktop humidifiers to remedy local discomfort

(*Note:* While interference in HVAC system operation can cause IAQ problems, it often occurs in response to previously unresolved ventilation or temperature control problems.)

Possible Solutions:

1. Remove the contaminant source by eliminating the activity:
   - Restrict the use of desktop humidifiers and other types of personal HVAC equipment.
   - Eliminate unsupervised manipulation of HVAC system.
   - Strictly enforce the building's nonsmoking policy.

   (*Note:* This step may require a combination of policy-setting and educational outreach.)

2. Reduce the source in the following ways:
   - Select materials and processes that minimize the release of contaminants while maintaining adequate safety and efficacy (e.g., solvents and art materials).
   - Install a new or improved local exhaust system to accommodate the activity, adjust the HVAC system to ensure adequate makeup air, and verify the effectiveness of these measures.
   - Eliminate smoking lounges or storage areas that contain contaminant sources.
   - Check all laboratory hoods and kitchen range hoods to ensure the hoods are venting to the outdoors and not just recirculating air.

Resolution:

In Case Study Problem #3, *Occupant Activities Contribute to Airborne Contaminants or to Space Comfort Problems*, the occupants received education regarding their role in improving the building's IAQ. The smoking policy was enforced throughout the building and all humidifiers were removed.

Case Study Problem #4: HVAC System as a Source of Biological Contaminants

Problem Reported:

Over the past few months, it is noticed that there has been excessive absenteeism due to illness, which has spread to all parts of the building.

Documentation:

- The IAQ complaint is logged.
- The IAQ complaint form is filled out.
- If needed, the IAQ complaint interview diary may be used.
- At the end of the investigation, the IAQ complaint summary is completed.
- All records are then stored properly.

Troubleshooting Steps:

After documentation is complete, a survey of the building spaces and the HVAC system is taken, and it is discovered that the HVAC system is acting as a source of contaminants and is distributing the contaminants. Personnel inspect the surfaces of the air handler, drain pans, and interior of the ductwork.

Possible Pollutants and Problems:

- HVAC system ductwork is providing an environment for microorganism growth.
- The air handler and drain pan are contaminated by mold and bacteria.
- The air filters are full of debris.

Possible Solutions:

1. Remove the source by improving maintenance procedures in the following ways:
   - Inspect equipment for signs of corrosion and high humidity.
   - Replace corroded parts.
   - Clean drip pans, outdoor-air intakes, and other affected locations.
   - Use biocides, disinfectants, and sanitizers with extreme caution to ensure that occupant exposure is minimized.
   - Provide the maintenance personnel access to all the items that must be cleaned, drained, or replaced periodically.

Resolution:

In Case Study Problem #4, *HVAC System as a Source of Biological Contaminants*, after the discovery of the contaminated HVAC system parts, a thorough cleaning of the parts and the replacement of needed parts eliminated most of the problem.

Case Study Problem #5: Surface Contamination Due to Poor Sanitation or Accidents

Problem Reported:
After a water line break, a flood occurred and an outbreak of humidifier fever occurred.

Documentation:
- The IAQ complaint is logged.
- The IAQ complaint form is filled out.
- If needed, the IAQ complaint interview diary may be used.
- At the end of the investigation, the IAQ complaint summary is completed.
- All records are then stored properly.

Troubleshooting Steps:
After documentation is complete, a survey of the building spaces and the HVAC system is taken, and it is found that a large percentage of the floor's carpeting was flooded.

Possible Pollutants and Problems:
- biological contaminants, resulting in allergies or other diseases
- fungal, viral, or bacterial pollutants (whole organisms or spores)
- bird, insect, or rodent parts, or droppings, hair, or dander (in HVAC, crawlspace, building shell, or near outdoor-air intakes)
- building operation accidents
- spills of water, beverages, cleansers, paints, varnishes, mastics, or specialized products (e.g., printing and chemical art supplies)
- fire damage resulting in soot, odors, or chemicals

Possible Solutions:
1. Clean the HVAC system components as follows:
   - Clean some materials and furnishings, though others may have to be discarded. (*Note:* Use biocides, disinfectants, and sanitizers with caution and ensure that occupant exposure is minimized.)
   - Remove the sources of microbiological contamination.
2. Inspect water-damaged carpet, furnishings, and building materials as follows:
   - Modify the environment to prevent the recurrence of microbiological growth.
   - Improve HVAC system maintenance.
   - Control humidity or surface temperatures to prevent condensation.
   - Provide access to all items that require periodic maintenance.
   - Use local exhaust where corrosive materials are stored.
   - Adjust the HVAC system to provide adequate makeup air, and test to verify performance.

Resolution:
In Case Study Problem #5, *Surface Contamination Due to Poor Sanitation or Accidents*, after the wet, moldy carpet was found, the carpet was removed, the area was thoroughly disinfected and cleaned, and a new carpet was installed.

Case Study Problem #6: Mold and Mildew Growth Due to Moisture from Condensation

Problem Reported:
Several occupants report a musty, moldy smell on the third floor of the building.

Documentation:
- The IAQ complaint is logged.
- The IAQ complaint form is filled out.
- If needed, the IAQ complaint interview diary may be used.
- At the end of the investigation, the IAQ complaint summary is completed.
- All records are then stored properly.

Troubleshooting Steps:
After documentation is complete, a survey of the building spaces and the HVAC system is taken. Mold is discovered growing on the interior walls of an exterior room.

Possible Pollutants and Problems:
- carpeting on floors that become ice cold
- building locations where high surface humidity promotes condensation

Possible Solutions:
1. Clean and disinfect the walls to remove mold and mildew. (*Note:* Follow up by taking actions to prevent recurrence of microbiological contamination. Use biocides, disinfectants, and sanitizers with caution and ensure that occupant exposure is minimized.)
2. In the following ways, increase surface temperatures to treat locations that are subject to condensation:
   - Insulate thermal bridges.
   - Improve air distribution.
3. Reduce moisture levels in locations that are subject to condensation in the following ways:
   - Repair leaks.
   - Increase ventilation (in cases where outdoor air is cold and dry).
   - Dehumidify (in cases where outdoor air is warm and humid).
4. Dry carpeting or other textiles promptly after steam cleaning. (*Note:* Increase ventilation to accelerate drying.)
5. Discard contaminated materials.

Resolution:
In Case Study Problem #6, *Mold and Mildew Growth Due to Moisture from Condensation*, the wall was thoroughly cleaned and disinfected and insulation was added to the wall to decrease heat loss.

Case Study Problem #7: Building Materials and Furnishings Produce Contaminants

Problem Reported:

Dizziness and headaches are reported after a number of offices are freshly painted and renovated with new carpeting and furnishings.

Documentation:

- The IAQ complaint is logged.
- The IAQ complaint form is filled out.
- If needed, the IAQ complaint interview diary may be used.
- At the end of the investigation, the IAQ complaint summary is completed.
- All records are then stored properly.

Troubleshooting Steps:

After documentation is complete, a survey of the building spaces and the HVAC system is taken. During the walk-through survey, heavy odors from the newly installed carpets, furniture, and wall coverings are noticed. Also, the drapes were dry-cleaned, causing odors to be released.

Possible Pollutants and Problems:

- formaldehyde from furniture and carpeting
- UV protective chemical used on drapes to maintain color
- new paint

Possible Solutions:

1. Remove contaminant source with the appropriate cleaning methods, such as the following:
   - Steam clean the carpeting and upholstery. Dry the carpeting and upholstery quickly, and ventilate the area to accelerate the drying process.
   - Accept only fully dried and odorless dry-cleaned products.
2. Encapsulate the source by sealing surfaces of building materials that emit formaldehyde.
3. Reduce the source by scheduling the installation of carpet, furniture, and wall coverings to occur during periods when the building is unoccupied.
4. Increase outdoor air ventilation in the following ways:
   - total air supplied
   - proportion of fresh air
5. Remove the materials that are producing the emissions and replace them with lower-emission alternatives.

   (*Note:* Only limited information on emissions from materials is available at this time. Purchasers can request that suppliers provide emissions test data but should use caution in interpreting the test results.)

Resolution:

In Case Study Problem #7, *Building Materials and Furnishings Produce Contaminants*, additional outdoor air was introduced over a long weekend until the new materials had given off most of their outgassing vapors. The drapes were replaced with older ones that gave off no gases.

Case Study Problem #8: Housekeeping or Maintenance Activities Contribute to Problems

Problem Reported:

Strange odors are reported in the vicinity of housekeeping storage areas.

Documentation:

- The IAQ complaint is logged.
- The IAQ complaint form is filled out.
- If needed, the IAQ complaint interview diary may be used.
- At the end of the investigation, the IAQ complaint summary is completed.
- All records are then stored properly.

Troubleshooting Steps:

After documentation is complete, a survey of the building spaces and the HVAC system is taken, and a janitor's closet is found to be close to where the odor is reported. The inspection finds that a spill of cleaning materials has taken place and not been removed. Also, the cleaning material containers are not capped properly, allowing fumes to escape.

Possible Pollutants and Problems:

- Cleaning products emit chemical odors.
- Particulates become airborne during cleaning (sweeping and vacuuming) activities.
- Contaminants are released from painting, caulking, and lubricating activities.
- The frequency of maintenance is insufficient to eliminate the contaminants.

Possible Solutions:

1. Remove the contaminant source by modifying standard procedures or frequency of maintenance in the following ways:
   - Improve storage practices.
   - Shift the times of painting, cleaning, pest control, and other contaminant-producing activities to avoid occupied building periods.
   - Make maintenance easier by improving access to filters, coils, and other components.
     (*Note:* Changing procedures may require a combination of policy-setting and the training of staff regarding the connection between IAQ and staff activities.)
2. Reduce the contaminant source by using the following methods:
   - Select materials to minimize emissions of contaminants while maintaining adequate safety and efficacy.
   - Use portable HEPA (high-efficiency particulate arrestance) vacuums instead of low-efficiency paper bag collector vacuums.
3. Use local exhaust in the following ways:
   - on a temporary basis to remove contaminants from work areas
   - as a permanent installation in spaces where contaminants are stored

Resolution:

In Case Study Problem #8, *Housekeeping or Maintenance Activities Contribute to Problems*, the cleaning agent chemical spill is absorbed and disposed of properly. All other chemical containers are sealed tightly. A small portable exhaust fan and filtering system is used to temporarily reduce the vapors escaping the room. Also, an in-house training session is used to show cleaning staff the correct method to clean up spills and store cleaning chemicals.

Case Study Problem #9: Remodeling or Repair Activities that Produce Problems

Problem Reported:

Occupants who work in an area adjacent to another area that has recently been renovated report experiencing headaches and nausea.

Documentation:

- The IAQ complaint is logged.
- The IAQ complaint form is filled out.
- If needed, the IAQ complaint interview diary may be used.
- At the end of the investigation, the IAQ complaint summary is completed.
- All records are then stored properly.

Troubleshooting Steps:

After documentation is complete, a survey of the building spaces and the HVAC system is taken, and it is found that heavy demolition work took place. Also, during the renovation, new particleboard and partitions were installed, along with new carpet and furnishings. All walls were painted.

Possible Pollutants and Problems:

- demolition dust and odors
- new construction material off gasing
- new construction obstructing area airflow

Possible Solutions:

1. Modify ventilation in the following ways to prevent recirculation of contaminants:
   - Install temporary local exhaust in work area, adjust HVAC system to provide makeup air, and test to verify that the HVAC system returns are sealed off in work areas.
   - Close the outside-air damper during reroofing.
2. Reduce the contaminant source by scheduling work for unoccupied periods and keeping the ventilation system in operation to remove odors and contaminants.
3. Reduce the contaminant source by careful material selection and installation:
   - Select materials to minimize emissions of contaminants while maintaining adequate safety and efficacy.
   - Have the supplier store new furnishings in a clean, dry, well-ventilated area until VOC off gasing has diminished.
   - Request installation procedures (adhesives) that limit the emissions of contaminants.
4. Ventilation modifications can be used to isolate the work area and prevent pollutant buildup in occupied spaces; proper storage practices can minimize the release of contaminants.
5. Modify HVAC or wall partition layout if necessary as follows:
   - Partitions should not interrupt airflow.
   - Relocate supply and return diffusers.
   - Adjust supply and return air quantities.
   - Adjust the total air and/or outdoor air supply to serve the new occupancy level.

Resolution:

In Case Study Problem #9, *Remodeling or Repair Activities that Produce Problems*, temporary plastic sheeting was used to isolate the area being renovated. Also, temporary portable exhaust fans were brought in. After these changes, the reports of headaches and nausea diminished dramatically.

Case Study Problem #10: Combustion Gases

Problem Reported:
Complaints of headaches, dizziness, and nausea are reported after heating equipment has been started in the fall.

Documentation:
- The IAQ complaint is logged.
- The IAQ complaint form is filled out.
- If needed, the IAQ complaint interview diary may be used.
- At the end of the investigation, the IAQ complaint summary is completed.
- All records are then stored properly.

Troubleshooting Steps:
After documentation is complete, a survey of the building spaces and the HVAC system is taken and a carbon monoxide meter is used in the mechanical room to take measurements. High concentrations of carbon monoxide are indicated. Combustion odors can indicate the existence of a serious problem. One combustion product, carbon monoxide, is an odorless gas. Carbon monoxide poisoning can be life threatening.

Possible Pollutants and Problems:
1. Vehicle exhaust:
   - offices above (or connected to) an underground parking garage
   - rooms near (or connected by pathways to) a loading dock or service garage
2. Combustion gases from the heating equipment (spillage or leakage from inadequately vented appliances, cracked heat exchanger, or re-entrainment because local chimney is too low):
   - Combustion gases can be found in areas near a mechanical room.
   - Gas is distributed throughout the zone or the entire building.

Possible Solutions:
1. Seal to remove the pollutant pathway:
   - Close openings between the contaminant source and the occupied space.
   - Install well-sealed doors with automatic closers between the contaminant source and the occupied space.
2. Remove the contaminant source in the following ways:
   - Improve the maintenance of combustion equipment.
   - Modify the venting or HVAC system to prevent back drafting.
   - Relocate the holding area for vehicles at the loading dock and parking area.
   - Turn off the engines of vehicles that are waiting to be unloaded.
3. Modify the ventilation system:
   - Install the local exhaust in the underground parking garage (adjust the HVAC system to provide makeup air and test it to verify performance).
   - Relocate the fresh air intake (move it away from source of contaminants).
4. Modify the pressure relationships:
   - Pressurize spaces around the area containing the source of combustion gases. Air intakes are frequently located near the loading dock for aesthetic reasons. Unfortunately, this air intake placement can draw car and truck exhaust into the building, causing a variety of indoor air quality complaints.

Resolution:
In Case Study Problem #10, *Combustion Gases*, the heating equipment is checked with a CO meter. One of the boilers is spilling CO due to incomplete combustion. The boiler is shut down, and repairs are completed. Upon startup, the boiler is checked and no carbon monoxide is present. The health complaints then cease.

Case Study Problem #11: Serious Building-Related Illness

Problem Reported:

A number of occupants throughout the building report having difficulty breathing, as well as flu-like symptoms.

Documentation:

- The IAQ complaint is logged.
- The IAQ complaint form is filled out.
- If needed, the IAQ complaint interview diary may be used.
- At the end of the investigation, the IAQ complaint summary is completed.
- All records are then stored properly.

Troubleshooting Steps:

After documentation is complete, a survey of the building spaces and the HVAC system is taken, and a large amount of a green, slimy substance is found on the surface of the water in the cooling tower sump.

(Note: Some building-related illnesses can be life threatening. Even a single confirmed diagnosis, which involves results from specific medical tests, should provoke an immediate and vigorous response.)

Possible Building-Related Illnesses:

1. Legionnaire's disease

    (Note: If you suspect Legionnaire's disease, call the local public health department, check for obvious problem sites, and take corrective action. There is no way to be certain that a single case of this disease is associated with building occupancy; therefore, public health agencies usually do not investigate single cases. Public health agencies will watch for more cases.)

2. Hypersensitivity pneumonitis

    (Note: Affected occupant(s) should be removed and may not be able to return unless the causative agent is removed from the affected person's environment.)

Possible Solutions:

1. Work with public health authorities.
    - Evacuation may be recommended or required.
2. Remove the source:
    - Drain, clean, and decontaminate drip pans, cooling towers, room unit air conditioners, humidifiers, dehumidifiers, and other habitats of *Legionella,* fungi, and other organisms using appropriate protective equipment.
    - Install drip pans that drain properly.
    - Provide access to all the items that must be cleaned, drained, or replaced periodically.
    - Modify the schedule and procedures for improved maintenance.
3. Discontinue processes that deposit potentially contaminated moisture in the air distribution system, such as:
    - air washing
    - humidification
    - nighttime shutdown of air handlers

Resolution:

In Case Study Problem #11, *Serious Building-Related Illness*, public health authorities are contacted. The cooling tower water is tested and found to be contaminated. The cooling tower is drained and thoroughly cleaned. The water treatment program is reviewed and changed to prevent future outbreaks.

HIRING PROFESSIONALS TO SOLVE AN IAQ PROBLEM

Many IAQ problems are simple to resolve when facility staff has been educated about the complaint investigation process. In other cases, however, a time comes when outside assistance is needed. Professional help might be necessary or desirable in the following situations:

- Mistakes or delays could generate serious consequences such as health hazards, liability exposure, or regulatory sanctions.

- Building management feels that an independent investigation would be better received or more effectively documented than an in-house investigation.

- Investigation and mitigation efforts by facility staff have not relieved the IAQ problem.

- Preliminary findings by staff suggest the need for measurements that require specialized equipment and training beyond in-house capabilities.

Stationary engineers may be able to find additional help from professional engineers, environmental service organizations, laboratories, or industrial hygienists. Local, state health, or air pollution agencies may have lists of firms offering IAQ services in the area needed. It may also be useful to seek out referrals from other building management firms or chief engineers. Local, state, or federal government agencies may be able to provide expert assistance or direction in solving IAQ problems. It is particularly important to contact the local or state health department when building personnel suspect that there is a building-related illness potentially linked to biological contamination in the building. In some states, government agencies have personnel with the appropriate skills to assist in solving IAQ problems.

Note: Even certified professionals from disciplines closely related to IAQ issues such as industrial hygienists, ventilation engineers, and toxicologists may not have the specific expertise needed to investigate and resolve indoor air problems. Individuals or groups that offer services in this evolving field should be questioned closely about their related experience and their proposed approach to an IAQ problem.

As with any hiring process, the better the hiring managers know their own needs, the easier it will be to select a firm or individual to service those needs. Firms and individuals working in IAQ may come from a variety of disciplines. Typically, the skills of HVAC engineers and industrial hygienists are useful for this type of investigation, although input from other disciplines such as chemistry, chemical engineering, architecture, microbiology, or medicine may also be important. When problems other than indoor air quality are involved, experts in lighting, acoustic design, interior design, psychology, or other fields may be helpful in resolving occupant complaints about the indoor environment.

Hiring Professionals with an Effective Approach

Indoor air quality is still a developing area of knowledge. Most consultants working in the IAQ field received their primary training in other areas. A variety of IAQ investigative methods can be employed, many of which are ineffective for resolving any but the most obvious IAQ problems.

Inappropriately designed studies may lead to conclusions that are either false negative (falsely concluding that there is no problem associated with the building) or false positive (incorrectly attributing the cause to building conditions).

Diagnostic outcomes to avoid include:

- an evaluation that overemphasizes measuring concentrations of pollutants and comparing those concentrations to numerical standards

- a report that lists a series of major and minor building deficiencies and links all the deficiencies to the problem without considering their actual association with occupant complaints

- hastily hiring an IAQ consultant who may subscribe to strategies that are unwanted in your facility; a qualified IAQ investigator must have appropriate IAQ experience, demonstrate a broad understanding of IAQ problems and the conditions that create them, and use a phased diagnostic approach

Selection Criteria

Most of the criteria used in selecting a professional to provide indoor air quality services are similar to those used for the selection of other professionals and include the following:

- experience in solving similar indoor air quality problems, including appropriate training and experience for the individuals who would be responsible for performing the work
- quality of the interview and proposal
- company reputation
- knowledge of local codes and regional climate conditions
- costs

Experience. An EPA survey of firms that provide indoor air quality services found that almost half of all the firms had been providing IAQ diagnostic or mitigation services in nonindustrial settings for ten or fewer years. Questions to ask an indoor air quality firm before hiring are:

- How much IAQ work and what type of IAQ work has your firm done?
- Has the firm identified the personnel who would be responsible for the work in my building? What are their specific experiences and related qualifications?

Similarly, it is important to contract only for the services of those individuals needed, and require "sign-off approval" for any substitute personnel the firm may want to use.

The quality of an interview and proposal guidelines are of assistance when hiring IAQ professionals. Interview and proposal guidelines include the following:

- Competent professionals will ask questions to see whether they feel they can offer services that will assist you. The causes and potential remedies for IAQ problems vary greatly. An experienced IAQ consultant or firm will need at least a preliminary understanding of the facts about what is going on in your building to evaluate if it has access to the professional IAQ skills necessary to address your concerns and to make effective use of its personnel from the outset. Often, an IAQ problem requires a multidisciplinary team of professionals.

- The proposal for the investigation should emphasize observations rather than measurements. There are four types of investigative information that must be gathered in order to resolve an IAQ problem: occupant complaints, specifics of building HVAC system, pollutant pathways, and pollutant sources. Nonroutine measurements such as relatively expensive air sampling for VOCs should not be provided without site-specific justification.

- The staff responsible for a building's IAQ investigation should have a good understanding of the relationship between IAQ and the building structure, mechanical systems, and occupant activities. For example, the lack of adequate ventilation is at least a contributing factor in many IAQ problem situations.

Evaluating the performance of the ventilation system depends on understanding the interaction between the mechanical system and the human activity within the building. In some cases, building investigators may have accumulated a breadth of knowledge. For example, a mechanical

IAQ Fact

IAQ-related certifications that are Council of Engineering and Scientific specialty boards (CESB)-accredited include the American Board of Industrial Hygiene (ABIH) Certified Industrial Hygienist (CIH), the Board of Certified Safety Professionals (CSP), the AmIAQ Council-Certified Indoor Environmental Consultant (CIEC), the Council-Certified Mold Remediation Supervisor (CMRS), and the Council-Certified Microbial Consultant (CMC).

engineer and an industrial hygienist see buildings differently. However, a mechanical engineer with several years of experience in IAQ problem investigations may have seen enough health-related problems to cross the gap.

Likewise, an industrial hygienist with years of experience studying problems in an office setting may have considerable expertise in HVAC and other building mechanical systems. Either in the proposal or in discussion, the consultant or firm should do the following:

- Describe the goal(s), methodology, and sequence of the IAQ investigation, the information that will be obtained, and the process of hypothesis development and testing, including criteria for decision-making about further data gathering.
- Submit a proposal that includes an explanation for the need of any air quality measurements.

The goal when hiring an IAQ professional or firm is to reach a successful resolution to any IAQ complaints, not to simply generate data. Additional information that must be spelled out in a proposal from an IAQ professional or firm wishing to perform work in a building includes:

- identifying any elements of the work that will require a time commitment from building staff, including information to be collected by building staff
- identifying additional tasks (and costs) that are part of solving the IAQ problem but are outside the scope of the contract; for example, medical examination of building occupants, laboratory fees, and contractor's fees for mitigation work
- a document that describes the work schedule, costs, and product(s) that will be used to perform contract work, specifications of work, and plans for mitigation work, such as who will supervise the mitigation work and what training will be provided for building staff
- how communication will occur between the IAQ professionals and building management; how often the contractor will

discuss the progress of the work with building personnel

- how building occupants will be notified of test results and other data; whether communications will be in writing, by telephone, or face-to-face; whether IAQ consultants will meet with building occupants to collect information; whether the consultants will meet with occupants to discuss findings if requested to do so

Reputation. There are no federal regulations covering professional services in the field of indoor air quality, although some disciplines such as engineering and industrial hygiene, whose practitioners work with IAQ problems, have licensing and certification requirements. Building owners and managers who suspect that a problem exists with a specific pollutant such as radon, asbestos, or lead may be able to obtain assistance from local and state health departments. Government agencies and affected industries have developed training programs for contractors who diagnose or mitigate problems with these particular contaminants.

Firms should be asked to provide references from clients who have received comparable services. In exploring references, it is useful to ask about long-term follow-up. After the contract was completed, did the IAQ professional or firm remain in contact with the client to ensure that problems did not reoccur? The goal of an IAQ investigation is to reach a successful resolution to all complaints.

Knowledge of Local Codes and Regional Climate Conditions. The IAQ professional or firm chosen must have a familiarity with state and local codes and regulations to avoid problems during mitigation. For example, when making changes to an HVAC system, it is important to follow all local building codes. Heating, cooling, and humidity control requirements are different in various geographic regions and can affect the selection of the appropriate mitigation to use. Receiving assurances that all firms under consideration have this knowledge is particularly important if it becomes necessary to seek expertise from outside the local area.

Cost. It is impossible for any proposal to provide the exact costs of professional IAQ services that will be incurred. When projected costs jump suddenly during the IAQ investigation process, the IAQ professional or firm must be able to justify the added cost. An IAQ investigation and mitigation budget will be influenced by a number of factors, including:
- complexity of the IAQ problem
- size and complexity of the building and its HVAC system(s)
- quality and extent of recordkeeping by building staff and management
- type of investigation report or other product required
- number of meetings required between IAQ professionals and building management (formal presentations can be quite expensive)
- the type of air-sampling instruments and laboratory analysis used

---

Definitions

Troubleshooting and Mitigating Definitions

- A *nonpoint source* is a type of contamination that results from multiple sources or a large-area source.
- *Permanence mitigation* is a type of mitigation effort that creates a permanent solution to an indoor air problem compared to efforts that provide temporary solutions (unless the problems are also temporary).
- A *point source* is a type of contamination that results from a single or small source.
- *Respirable particles* are airborne particles that can be inhaled into the lungs when breathing.
- A *step-by-step IAQ troubleshooting process* is a list of procedures that identifies steps to follow in chronological order until an IAQ problem is found.

For more information please refer to ATPeResources.com or http://www.epa.gov.

Review Questions

1. Summarize the steps involved in the step-by-step method of troubleshooting IAQ complaints.

2. Name two mitigation strategies for exposure control.

3. Name two standards used to judge mitigation design success.

4. List three categories of IAQ control strategies.

5. What are the four factors that may indicate an IAQ problem is solved?

APPENDIX

OUTLET AIR BALANCE REPORT

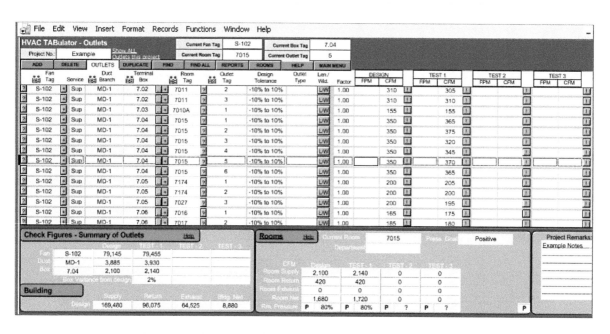

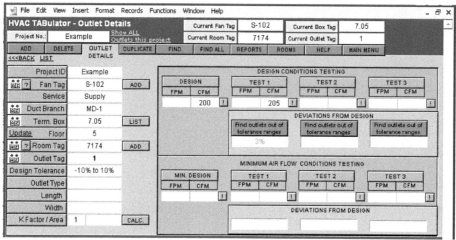

BUILDING COMMISSIONING

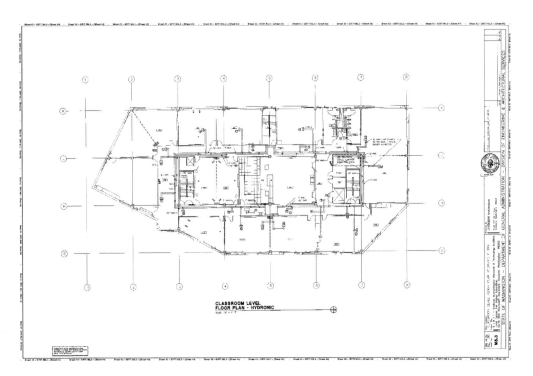

ONGOING COMMISSIONING

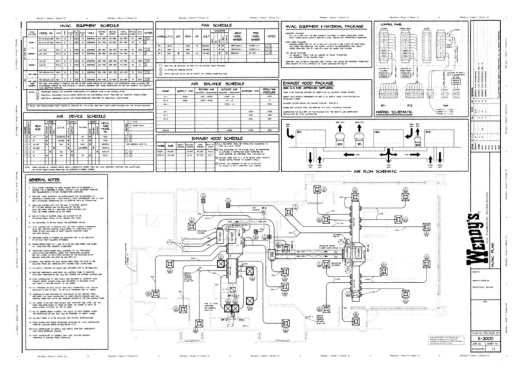

RETRO COMMISSIONING

EPA CONTACT INFORMATION

EPA Regional Offices

US EPA/Region 1
(CT, ME, MA, NH, RI, VT)
1 Congress Street, Suite 1100
Boston, MA 02114-2023
(617) 918-1639
(617) 918-1505 fax

US EPA/Region 2
(NJ, NY, PR, VI)
290 Broadway
New York, NY 10007-1866
(212) 637-3660
(212) 637-4942 fax

US EPA/Region 3
(DC, DE, MD, PA, VA, WV)
1650 Arch Street **(3PM52)**
Philadelphia, PA 19103-2029
(215) 814-5000
(215) 814-2101 fax

US EPA/Region 4
(AL, FL, GA, KY, MS, NC, SC, TN)
61 Forsyth Street, SW
Atlanta, GA 30303-8960
(404) 562-9900
(404) 562-8174 fax

US EPA/Region 5
(IL, IN, MI, MN, OH, WI)
77 West Jackson Boulevard
Chicago, IL 60604-3590
(312) 353-2000
(312) 886-0617 fax

US EPA/Region 6
(AR, LA, NM, OK, TX)
1455 Ross Avenue, Suite 1200
Dallas, TX 75202-2733
(214) 665-6444
(214) 665-6762 fax

US EPA/Region 7
(IA, KS, MO, NE)
901 North 5th Street
Kansas City, KS 66101
(913) 551-7003
(913) 551-7065 fax

US EPA/Region 8
(CO, MT, ND, SD, UT, WY)
80C-EISC
1595 Wyakoop Street
Denver, CO 80202-6312
(303) 312-6312
(303) 312-6044 fax

US EPA/Region 9
(AZ, CA, HI, NV, Pacific Islands)
75 Hawthorn Street
San Francisco, CA 94105
(415) 947-4192
(415) 947-3583 fax

US EPA/Region 10
(AK, ID, OR, WA)
1200 6th Avenue, Suite 900
Seattle, WA 98101-9797
(206) 553-1200
(206) 553-0110 fax

Additional Resources

U.S. Environmental Protection Agency

www.epa.gov/iaq/schools
Download the *IAQ Tools for Schools* Action
Kit from the EPA's web site.

www.epa.gov/asthma
Find more asthma resources on the EPA's
web site.

www.epa.gov/cleanschoolbus
Learn about EPA's clean school bus
initiative on the EPA's web site.

**U.S. EPA's National Service
Center for Environmental
Publications (NSCEP)**

(800) 490-9198
(301) 604-3408 Fax
Call to request the *IAQ Tools for
Schools* Action Kit.

INDOOR AIR QUALITY COMPLAINT FORM

Use this form to report problems related to the indoor environment in this building. Indoor air quality problems include concerns with temperature, humidity, ventilation, odors, or air pollutants that may be causing health or discomfort symptoms. The information that you provide will remain confidential.

Occupant Name_____ Date_____

Building Name_____ Address_____

Room Number/Location_____ Phone_____

Completed by_____ Title _____ Phone_____

What is the nature of this problem?

Where is the problem experienced (in one or more locations)?

When was the problem first experienced?

When does it occur or when is it the worst (time of day, day of week, related to certain activities/events)?

Other comments:

For Office Use Only

MATERIAL SAFETY DATA SHEET (MSDS)...

MIDSUN GROUP ™

P.O. Box 864 • 135 Redstone Street
Southington, CT 06489 U.S.A

Toll Free: (800)-4-midsun (U.S.A. only)
(860) 378-0100 • (860) 378-0103 (Fax)
www.midsungroup.com

Acetone Material Safety Data Sheet (MSDS)

MANUFACTURER'S CONTACT INFORMATION:

Sunoco, Inc. (R&M)   EMERGENCY
1735 Market Street LL   Sunoco: (800) 964-8861
Philadelphia, Pennsylvania 19103-7583   Chemtrec: (800) 424-9300
Product Safety: (610) 859-1120

I. Product Identification

Trade Name	Acetone
Product Use	Chemical Intermediate

II. Hazardous Ingredients of Material

Components	Amount (Vol. %)	CAS No.	ACGIH TLV
Acetone	100	67-64-1	–

Exposure Limits (See Section VI for additional Exposure Limits)

Governing Body	CAS No.	Exposure Limits
ACGIH	67-64-1	STEL 750 ppm
ACGIH	67-64-1	TWA 500 ppm
OSHA	67-64-1	TWA 1,000 ppm

Emergency Overview:

Danger! Extremely flammable liquid and vapor. Vapors may cause flash fire or explosion. Harmful if inhaled. Vapor concentrations may cause drowsiness. Causes skin and eye irritation. Harmful if swallowed. May cause target organ or system damage to the following: Eye, skin, respiratory system, central nervous system.

HAZARD RATINGS

Key: 0 = Least   1 = Slight   2 = Moderate   3 = High   4 = Extreme

	Health	Fire	Reactivity	PPI
NFPA	1	3	0	
HMIS	1	3	0	X

III. Physical/Chemical Data

Appearance & Odor	Colorless liquid
Boiling Point	133° F
Melting Point	-137.2° F
Specific Gravity	0.79
Molecular Weight g/mole	58.08
pH	7
Odor	Sweet, pungent
Odor Threshold	62 ppm
Vapor Pressure (mm Hg @20° C)	181
Solubility in Water	Complete
Volatile (wt %)	100%

Acetone MSDS   Page 1 of 1   Rev. 01/25/06

. . . *MATERIAL SAFETY DATA SHEET (MSDS)* . . .

Acetone Material Safety Data Sheet (MSDS)

IV. Fire and Explosion Data

Flash Point	1.4
Flammable Limits in Air (% By Volume)	
Lower	2.5%
Upper	12.8%
Auto Ignition Temperature	869° F
Unusual Fire & Explosion Hazards	Use water spray. Use water spray to cool fire exposed tanks and containers. Acetone/water solutions that contain more than 2.5% acetone have flash points. When the acetone concentration is greater than 8% (by weight) in a closed container, it would be within flammable range and cause fire or explosion if a source of ignition were introduced.
Fire Extinguishing Media	Water spray, alcohol resistant foam, dry chemical or carbon dioxide.

V. Reactivity Data

Stability	Stable
Conditions to Avoid	Avoid heat, sparks and open flame.
Incompatibility	Acetone may form explosive mixtures with chromic anhydride, chromyl alcohol, hexacholromelamine, hydrogen peroxide, permonosulfuric acid, potassium terbutoxide and thioglycol. Strong oxidizers.
Hazardous Decomposition	May produce carbon dioxide, carbon monoxide and other asphyxiants.
Hazardous Polymerization	Will not occur.

VI. Health Hazard and Toxicological Data

Pre-existing Medical Conditions: The following diseases or disorders may be aggravated by exposure to this product. Skin, eye, lung (asthma-like conditions).

Chronic Exposure	Effects of Exposure
Eyes	Contact with the eye may cause moderate to severe irritation.
Skin	Moderately irritating to the skin. Prolonged or repeated contact can result in defatting and drying of the skin which may result in skin irritation and dermatitis (rash). LD50 mg/kg Rabbit, 20,000 Draize Skin Score: no data Out of 8.0
Inhalation	High concentrations may lead to central nervous system effects (drowsiness, dizziness, nausea, headache, paralysis and loss of consciousness and even death).High vapor concentrations are irritating to the eyes, nose, throat and lungs. LC50 (mg/l) no data LC50 (mg/m^3) Rat 8 hrs. 50,000 LC50 (ppm) no data
Ingestion	Product may be harmful or fatal if swallowed. Pulmonary aspiration hazard. After ingestion, may enter lungs and produce damage. May produce central nervous system effects, which may include dizziness, loss of balance and coordination, unconsciousness, coma and even death. LD50 (g/kg) Rat 5.8

Acetone MSDS   Page 2 of 2   Rev. 01/25/06

MidSun Group, Inc.

. . . MATERIAL SAFETY DATA SHEET (MSDS) . . .

Acetone Material Safety Data Sheet (MSDS)

VII. First Aid Procedures	
Inhalation	Remove to fresh air. If not breathing, give artificial respiration. If breathing is difficult, give oxygen and continue to monitor. Get immediate medical attention.
Eye Contact	Flush eye(s) with water for 15 minutes. Get medical attention.
Skin Contact	Immediately flush skin with plenty of water. Remove clothing. Get medical attention immediately. Wash clothes separately before reuse.
Ingestion	If swallowed, DO NOT INDUCE VOMITING. Give victim a glass of water or milk. Call a physician or poison control center immediately. Never give anything by mouth to an unconscious person. Get medical attention immediately. See Section X for additional first aid information.

VIII. Preventive Measures Consult with a Health and Safety Professional for Specific Selections	
A. PERSONAL PROTECTIVE EQUIPMENT	
Respiratory Protection	Concentrations in air determines the level of respiratory protection needed. Use only NIOSH certified respiratory equipment. Half-mask air purifying respirator with organic vapor cartridges is acceptable for exposure to ten (10) times the exposure limit. Full-face air purifying respirator with organic vapor cartridges is acceptable for exposures to fifty (50) times the exposure limit. Exposure should not exceed the cartridge limit of 1000 ppm. Protection by air purifying respirators is limited. Use a positive pressure-demand full-face supplied air respirator or SCBA for exposures greater than fifty (50) times the exposure limit. If exposure is above the IDLH (Immediately Dangerous to Life and Health) or there is a possibility of an uncontrolled release, or exposure levels are unknown, then use a positive pressure-demand full-face air respirator with escape bottle or SCBA. Wear a NIOSH-approved (or equivalent) full-face piece airline respirator in the positive pressure mode with emergency escape provisions.
Eye/Face Protection	Splash proof chemical goggles or full-face shield recommended to protect against splash of product.
Clothing/Gloves	The glove(s) list below may provide protection against permeation. Gloves or other chemically resistant materials may not provide adequate protection. Protective gloves are recommended to protect against contact with product. Neoprene, Natural rubber.
Engineering Controls	Use with adequate ventilation. Ventilation is normally required when handling or using this product to keep exposure to airborne contaminants below the exposure limit. Use explosion-proof ventilation equipment.
Other	The following materials are acceptable for use as protective clothing; Neoprene, Natural rubber. Facilities storing or utilizing this material should be equipped with an eyewash facility and a safety shower. Remove contaminated clothing and wash before reuse.
B. STORAGE AND HANDLING	
Storage Conditions	Keep away from heat, sparks and flame. Store in a cool, dry place. Keep container closed when not in use.
Handling Procedure	Use only in a well-ventilated area. Ground and bond containers when transferring material. Avoid breathing (dust, vapor, mist, gas). Avoid contact with this material. Wash thoroughly after handling. Do not use air pressure to unload containers.

Continued on Next Page

MidSun Group, Inc.

. . . MATERIAL SAFETY DATA SHEET (MSDS) . . .

Acetone Material Safety Data Sheet (MSDS)

VIII. Preventive Measures (Continued)

C. ENVIRONMENTAL PROTECTION

Spill and Leak Procedure	Prevent ignition, stop leak and ventilate the area. Contain spilled liquid with sand or earth. DO NOT use combustible materials such as sawdust. Use appropriate personal protective equipment as stated in Section VIII of this MSDS. Advise the Environmental Protection Agency (EPA) and appropriate state agencies, if required. U.S. regulations require reporting spills of this material could that reach any surface waters. The toll-free number for the U.S. Coast Guard National Response Center is (800) 424-8802. After removal, flush contaminated area thoroughly with water.
Waste Disposal	Follow federal, state and local regulations. In Canada, follow federal, provincial and local regulations. This material is a RCRA hazardous waste. DO NOT flush material to drain or storm sewer. Contract to authorized disposal service.
Ecological Information	This product is not expected to persist in the environment.

D. TRANSPORTATION INFORMATION

Governing Body	U.S. DOT
Proper Shipping Name	Acetone
Mode	Ground
Hazard Class	3 (Flammable Liquid)
UN/NA Number	UN1090

IX. Regulatory Information/Classifications

Regulatory List	Component	CAS Number
ACGIH – Occupational Exposure Limits – Carcinogens	Acetone	67-64-1
ACGIH – Occupational Exposure Limits – TWAs	Acetone	67-64-1
ACGIH – Short Term Exposure Limits	Acetone	67-64-1
CAA (Clean Air Act) – HON Rule – SOCMI Chemicals	Acetone	67-64-1
Canada – WHMIS – Ingredient Disclosure	Acetone	67-64-1
CERCLA/SARA – Hazardous Substances and their RQs	Acetone	67-64-1
CERCLA/SARA – Hazardous Substances and their RQs	Acetone	67-64-1
CERCLA/SARA – Hazardous Substances and their RQs	Acetone	67-64-1
Inventory – Australia – (AICS)	Acetone	67-64-1
Inventory – Canada – Domestic Substances List	Acetone	67-64-1
Inventory – China	Acetone	67-64-1
Inventory – European – EINECS Inventory	Acetone	67-64-1
Inventory – Japan – (ENCS)	Acetone	67-64-1
Inventory – Korea – Existing and Evaluated	Acetone	67-64-1
Inventory – Philippines – (PICCS)	Acetone	67-64-1
Inventory – TSCA – Section 8(b) Inventory	Acetone	67-64-1
Massachusetts – Right to Know List	Acetone	67-64-1
New Jersey – Department of Health RTK List	Acetone	67-64-1
New Jersey – Special Hazardous Substances	Acetone	67-64-1
OSHA – Final PELs – Time Weighed Averages	Acetone	67-64-1
Pennsylvania – Right to Know List	Acetone	67-64-1
TSCA – Section 12(b) – Export Notification	Acetone	67-64-1
TSCA – Section 4 – Chemical Test Rules	Acetone	67-64-1

Continued on Next Page

MidSun Group, Inc.

Acetone Material Safety Data Sheet (MSDS)

IX. Regulatory Information/Classifications - Continued

Regulatory Information/Classifications Title III, Sections 311, 312

Acute	Chronic	Fire	Reactivity	Sudden Release of Pressure
YES	NO	YES	NO	NO

X. Other Information

If swallowed, acetone should be removed by emesis and/or gastric lavage. Mechanical assisted ventilation may be necessary. In severe cases, an initial period of hypoglycemia may require correction by intravenous solutions of dextrose. In some cases, an initial period of hyperglycemia has occurred during the recovery phase and has lasted for a few days. Treatment with insulin may be beneficial but should be used cautiously. Empty containers retain product residue (liquid and/or vapor) and can be dangerous. DO NOT pressurize, cut, weld, braze, solder, drill, grind or expose such containers to heat, flame, sparks, static electricity or other sources of ignition. They may explode and cause injury or death. Empty drums should be completely drained, properly bunged and promptly returned to a drum reconditioner or properly disposed of. This product is subject to the Chemical Division and Trafficking Act of 1988 and subject to specific record keeping requirements. WHMIS Classification: Class B, Division 2 – Flammable Liquids.

REPORT OF CALIBRATION

FLUKE.

Report of Calibration

Manufacturer:	BW	Temperature:	24.1 C
Model:	975 AirMeter	Humidity:	7.6 RH
Serial Number:	93490029	Barometric Pressure:	26.6 inHg
Cal Date:	2/1/2007	Test Results:	PASS
Cal Tech:	Eddy W		

The Fluke Corporation, ISO Certification No. U0018, certifies that the instrument identified above was calibrated in accordance with applicable Fluke Corporation procedures. Its calibration processes are ISO-9001 controlled and are designed to certify that the instrument was within its published specifications at the time of calibration.

The measurement standards and instruments used during the calibration of this instrument are traceable to the United States National Institute of Standards and Technology (NIST), other reputable National Institutes that are signatories to the Mutual Recognition Arrangement (MRA), natural physical constants, consensus standards, or by ratio type measurements.

Standards

Manufacturer	Model	Description	Cal Date	Due Date
Thunder Scientific	2500	Temperature/humidity chamber	3/28/06	3/28/07
AirGas		100 ppm CO	2/27/06	
AirGas		500 ppm CO	5/17/06	
AirGas		500 ppm CO2	4/17/06	
AirGas		2500 ppm CO2	4/13/06	

Test Data

Parameter	Result	- Limit	+ Limit	Pass/Fail
CARBON MONOXIDE				
99.7 ppm	100	95	105	PASS
500.9 ppm	503	476	526	PASS
CARBON DIOXIDE				
502.0 ppm	503	413	591	PASS
2501.0 ppm	2583	2357	2645	PASS
TEMPERATURE				
15 deg C	-0.09	-0.50	0.50	PASS
35 deg C	0.16	-0.50	0.50	PASS
HUMIDITY				
30 %RH	-1.28	-3.00	3.00	PASS
75 %RH	-1.45	-3.00	3.00	PASS

Fluke Corporation
PO Box 9090 Everett WA 98206.9090 USA

Telephone
425.347.6100

Facsimile
425.446.5116

Internet
www.fluke.com

Building Air Quality: A Guide for Building Owners and Facility Managers

The Building Air Quality, developed by the EPA and the National Institute for Occupational Safety and Health, provides practical suggestions on preventing, identifying, and resolving indoor air quality (IAQ) problems in public and commercial buildings. This guidance provides information on factors affecting indoor air quality; describes how to develop an IAQ profile of building conditions and create an IAQ management plan; describes investigative strategies to identify causes of IAQ problems; and provides criteria for assessing alternative mitigation strategies, determining whether a problem has been resolved, and deciding whether to consult outside technical specialists. Other topics included in the guide are key problem causing factors; air quality sampling; heating, ventilation, and air conditioning systems; moisture problems; and additional sources of information.

- You can **select just the form or section that you need** to download and view/print (see the table of contents below).
- You can <u>download the entire PDF version of the guide</u> (PDF, 228 pp, 2.69MB, <u>About PDF</u>).
- **EPA 402-F-91-102, December 1991**

- For more information on IAQ in Large Buildings, please see our Indoor Air Quality Education and Assessment Model (I-BEAM) which updates this guidance at <u>www.epa.gov/iaq/largebldgs/i-beam/index.html</u>

Building Air Quality Action Plan

The Building Air Quality Action Plan meets the needs of building owners and managers who want an easy-to-understand path for taking their building from current conditions and practices to the successful institutionalization of good IAQ management practices. It emphasizes changing <u>how</u> you operate and maintain your building, <u>not increasing the amount of work or cost</u> of maintaining your building. The BAQ Action Plan follows 8 logical steps and includes a 100-item Checklist that is designed to help verify implementation of the Action Plan. This guidance is available here only as a PDF file.

- <u>PDF Version</u> (PDF, 28 pp, 845KB, <u>About PDF</u>)
- **EPA 402-K-98-001, June 1998**

. . . IAQ IN LARGE BUILDINGS . . .

1. **Table of Contents** (PDF, 3 pp, 12KB)
2. **Forward** (PDF, 1 page, 10KB)
3. **Note to Building Owners and Facility Managers** (PDF, 4 pp, 16KB)
4. **Acknowledgements** (PDF, 3 pp, 13KB)

TAB I: Basics

Section 1: About this Document (PDF, 4 pp, 16KB)

Section 2: Factors Affecting Indoor Air Quality (PDF, 8 pp, 65KB)
Sources of Indoor Air Contaminants; HVAC System Design and Operation; Pollutant Pathways and Driving Forces; Building Occupants

Section 3: Effective Communication (PDF, 5 pp, 44KB)
Communicating to Prevent IAQ Problems; Communicating to Resolve IAQ Problems

TAB II: Preventing IAQ Problems

Section 4: Developing an IAQ Profile (PDF, 11 pp, 177KB)
Skills Required to Create an IAQ Profile; Steps in an IAQ Profile

Section 5: Managing Buildings for Good IAQ (PDF, 13 pp, 154KB)
Developing an IAQ Management Plan

TAB III: Resolving IAQ Problems

Section 6: Diagnosing IAQ Problems (PDF, 35 pp, 434KB)
Overview: Conducting an IAQ Investigation; Initial Walkthrough; Collecting Additional Information; Collecting Information About Occupant Complaints; Using the Occupant Data; Collecting Information about the HVAC System; Using the HVAC System Data; Collecting Information about Pollutant Pathways and Driving Forces; Using Pollutant Pathway Data; Collecting Information on Pollutant Sources; Using Pollutant Source

TAB IV: Appendices

Appendix A: Common IAQ Measurements - A General Guide (PDF, 11 pp, 235KB)
Overview of Sampling Devices; Simple Ventilation/Comfort Indications; Air Contaminant Concentrations

Appendix B: HVAC Systems and Indoor Air Quality (PDF, 19 pp, 514KB)
Background; Types of HVAC Systems; Basic Components of an HVAC System; ASHRAE Standards and Guidelines

Appendix C: Moisture, Mold and Mildew (PDF, 6 pp, 198KB)
Background on Relative Humidity, Vapor Pressure, and Condensation; Taking Steps to Reduce Moisture; Identifying and Correcting Common Problems from Mold and Mildew

Appendix D: Asbestos (PDF, 4 pp, 38KB)
EPA and NIOSH Positions on Asbestos; Programs for Managing Asbestos In-Place; Where to Go for Additional Information

Appendix E: Radon (PDF, 2 pp, 12KB)
Building Measurement, Diagnosis and Remediation; Where to Go for Additional Information

Appendix F: Glossary and Acronyms (PDF, 4 pp, 15KB)

Appendix G: Resources (PDF, 11 pp, 29KB)
Federal Agencies with Major IAQ Responsibilities; Other Federal Agencies with Indoor Air Responsibilities; State and Local Agencies; Private Sector Contacts; Publications; Training

TAB V: Indoor Air Quality FORMS

* **IAQ Management Checklist** (PDF, 4 pp, 14KB)
* **Pollutant Pathway Record for IAQ Profiles** (PDF, 1 page, 8KB)
* **Zone/Room Record** (PDF, 1 page, 10KB)

Data; Sampling Air for Contaminants and Indicators; Complaints Due to Conditions Other Than Poor Air Quality; Forming and Testing Hypotheses

Section 7: Mitigating IAQ Problems (PDF, 24 pp, 590KB)
Background: Controlling Indoor Air Problems; Sample Problems and Solutions; Judging Proposed Mitigation Designs and Their Successes

Section 8: Hiring Professional Assistance to Solve an IAQ Problem (PDF, 4 pp, 18KB)
Make Sure That Their Approach Fits Your Needs; Selection Criteria

- **Ventilation Worksheet** (PDF, 1 page, 9KB)
- **Indoor Air Quality Complaint Form** (PDF, 1 page, 11KB)
- **Incident Log** (PDF, 1 page, 13KB)
- **Occupant Interview** (PDF, 2 pp, 9KB)
- **Occupant Diary** (PDF, 1 page, 10KB)
- **Log of Activities and System Operation** (PDF, 1 page, 9KB)
- **HVAC Checklist - Short Form** (PDF, 4 pp, 22KB)
- **HVAC Checklist - Long Form** (PDF, 15 pp, 42KB)
- **Pollutant Pathway Form for Investigations** (PDF, 1 page, 8KB)
- **Pollutant and Source Inventory** (PDF, 7 pp, 18KB)
- **Chemical Inventory** (PDF, 1 page, 15KB)
- **Hypothesis Form** (PDF, 2 pp, 18KB)

All Forms in one zipped file (91.7KB file)

INDEX (PDF, 5 pp, 16KB)

ASTHMA TRIGGERS

TRIGGER	COMMON SOURCES	SOLUTION
Mold	Plumbing leaks; excessive humidity	Fix leaks; control humidity levels; and remediate mold
Secondhand smoke	Smoking lounges; smoking near entrances	Develop smoking ban and/or a designated smoking location away from entrance and air intakes
Dust mites	Carpeting and furniture	Vacuum and clean hard surfaces often
Outdoor air pollution	Outdoor-air intakes	Divert outdoor pollution sources; check air filter system
Pests	Storing food improperly	Remove trash often; keep food well sealed

RECOMMENDATIONS FOR INDOOR AIR QUALITY OF EDUCATIONAL FACILITIES

SCHOOL SPACE	WINTER TEMP*	SUMMER TEMP*	VENTILATION RATE†	1ST STAGE AIR FILTER	2ND STAGE AIR FILTER
Classrooms, auditoriums, libraries, and administrative areas	72	78	15	20%-30%	60%-85%
Corridors	68	80	5	20%-30%	–
Laboratories	72	78	–	20%-30%	85%
Locker rooms and shower areas	75	–	5	20%-30%	–
Mechanical rooms	60	–	5	20%-30%	–
Industrial technologies	72	78	5	20%-30%	60%-85%
Storage	65	–	5	20%-30%	–
Restrooms	72	–	20	20%-30%	–

* in degrees Fahrenheit
† flow rate in cubic feet per minute per person

NOTE: Corridors, mechanical rooms, industrial technology shops, and storage areas are not usually air conditioned. Shops, laboratories, and restrooms usually have additional exhaust system requirements.

CONTAMINANT CATEGORIES

CATEGORY	HEALTH EFFECTS	EXAMPLES
Irritants	Inflammation, redness, and discoloration	Fiberglass, pesticides, cleaning agents, and perfume
Asphyxiates	Drowsiness, headaches, and slowed breathing	Carbon monoxide and refrigerants
Allergens	Extreme fever and asthma	Pollen, molds, and animal dander
Neurotoxins	Cardiac problems, convulsions, and loss of memory	Lead and chemicals in paint and glue
Pathogens	Fever, chills, and chest congestion	Some molds and bacteria
Carcinogens	Cancer	Radon gas, asbestos fibers, and environmental tobacco smoke

EXPOSURE TO RADON

Radon Level*	If 1000 people who smoked were exposed to this level over a lifetime†...	The risk of cancer from radon exposure compares to‡...
20	About 260 people could get lung cancer	250 times the risk of drowning
10	About 150 people could get lung cancer	200 times the risk of dying in a home fire
8	About 120 people could get lung cancer	30 times the risk of dying in a fall
4	About 62 people could get lung cancer	5 times the risk of dying in a car crash
2	About 32 people could get lung cancer	6 times the risk of dying from poison
1.3	About 20 people could get lung cancer	(Average indoor radon level)
0.4	About 3 people could get lung cancer	(Average outdoor radon level)

* in pCi/L
† Lifetime risk of lung cancer deaths from EPA Assessment of Risks from Radon in Homes (EPA 402-R-03-003)
‡ Comparison data calculated using the Centers for Disease Control and Prevention's 1999-2001
  National Center for Injury Prevention and Control Reports

Source: Environmental Protection Agency (EPA)

NITROGEN DIOXIDE EXPOSURE LIMITS

CLASSIFICATION	10 MINUTES*	30 MINUTES*	1 HOUR*	4 HOURS*	8 HOURS*
Nondisabling	0.50	0.50	0.50	0.50	0.50
Disabling	20	15	12	8.2	6.7

EXPOSURE	SYMPTOMS	PREVENTION	FIRST AID TREATMENT
Inhalation	Burning sensation, cough, headache, dizziness, sore throat, nausea, and vomiting	Ventilation–area exhaust Respiratory protection–supplied air	Half-upright position with rest and fresh air; artificial respiration may be required.
Skin	Burns, redness, and pain	Protective clothing and neoprene gloves	Rinse with water for 15 min; remove clothes (discard); rinse again for 15 min
Eyes	Severe burns, redness, and pain	Nonvented eye protection or face shield	Rinse eyes with water for 15 min
Ingestion	—	Do not drink or eat in nitrogen dioxide environment	Rinse mouth for 15 min

* in ppm
Sources: Occupational Safety & Health Administration (OSHA), U.S. Department of Health and Human Services–Food and Drug Administration (FDA), and Praxair, Inc.

CARBON MONOXIDE EXPOSURE LIMITS

CLASSIFICATION	10 MINUTES*	30 MINUTES*	1 HOUR*	4 HOURS*	8 HOURS*
Nondisabling	2500	1500	600	150	35–50
Disabling	6000	3000	800	400	80

EXPOSURE	SYMPTOMS	PREVENTION	FIRST AID TREATMENT
Inhalation	Asphyxiation	Ventilation–area exhaust Respiratory protection–supplied air	Artificial respiration and cardiopulmonary resuscitation
Skin	Possible frostbite	Appropriate gloves–not rubber or neoprene	Rinse with warm water
Eyes	Mechanical injury	Nonvented eye protection or face shield	Force eyes open and rinse with lukewarm water
Ingestion	—	—	—

* in ppm

Sources: Occupational Safety & Health Administration (OSHA), Spectra Gases, Inc., Washington State Department of Labor and Industries, and International Program on Chemical Safety (IPCS)

CARBON DIOXIDE EXPOSURE LIMITS

CLASSIFICATION	10 MINUTES*	30 MINUTES*	1 HOUR*	4 HOURS*	8 HOURS*
Nondisabling	15,000	10,000	7500	6000	5000
Disabling	30,000	20,000	15,000	10,000	10,000

EXPOSURE	SYMPTOMS	PREVENTION	FIRST AID TREATMENT
Inhalation	Asphyxiation–headaches and drowsiness	Ventilation–area exhaust Respiratory protection–supplied air	Artificial respiration using oxygen
Skin	Possible frostbite	Cuffless trousers–no rubber or neoprene gloves	Do not remove clothing–rinse with warm water
Eyes	Stinging	Face shield	Force eyes open and rinse with lukewarm water
Ingestion	—	—	—

* in ppm

Sources: Occupational Safety & Health Administration (OSHA), Praxair, Inc., and Valero Marketing & Supply Company

OZONE EXPOSURE LIMITS

CLASSIFICATION	10 MINUTES*	30 MINUTES*	1 HOUR*	4 HOURS*	8 HOURS*
Nondisabling	0.3	0.2	0.12	0.11	0.1
Disabling	25	10	2	1	0.5

EXPOSURE	SYMPTOMS	PREVENTION	FIRST AID TREATMENT
Inhalation	Nose and throat irritation, pain in chest, nausea, drowsiness, and death	Full-face supplied air respirator or self-contained breathing apparatus	Artificial respiration using oxygen–do not breathe air from victim
Skin	Irritation and possible chemical burns	Plastic clothes and gloves	Remove clothing and shoes–rinse with warm water
Eyes	Stinging	Face shield	Force eyes open and rinse with lukewarm water
Ingestion	—	—	—

* in ppm

Sources: Environmental Protection Agency (EPA), Occupational Safety & Health Administration (OSHA), Illinois Department of Public Health (IDPH), and Praxair, Inc.

ASBESTOS EXPOSURE LIMITS

CLASSIFICATION	10 MINUTES	30 MINUTES	1 HOUR	4 HOURS	8 HOURS
Chrysotile	0.9*	0.7	0.5	0.3	0.2
Non-chrysotile	0.6	0.5	0.4	0.2	0.1

EXPOSURE	SYMPTOMS	PREVENTION	FIRST AID TREATMENT
Inhalation	May cause asbestosis, pulmonary fibrosis, and/or other lung disorders	Exhaust ventilation and full-face self-contained breathing apparatus	Rest in well ventilated area–oxygen possibly needed
Skin	Irritation	Special full body suit, hood, gloves, and boots	Wash with plenty of water and disinfectant soap
Eyes	Scratching of eyes	Splash goggles–do not wear contacts	Remove contacts; do not use ointments; flush eyes with water
Ingestion	May cause gastrointestinal irritation or cancer	Handling of materials, food, or drink prior to washing is prohibited	Do not induce vomiting–rinse mouth with water

* in fiber/cc

Sources: Occupational Safety & Health Administration (OSHA), National Institute of Occupational Safety and Health (NIOSH), National Institute of Standards and Technology (NIST), and Structure Probe, Inc. (SPI)

LEAD EXPOSURE LIMITS

CLASSIFICATION	10 MINUTES	30 MINUTES	1 HOUR	4 HOURS	8 HOURS*
Nondisabling	—	—	—	—	0.03
Disabling	—	—	—	—	0.05

EXPOSURE	SYMPTOMS	PREVENTION	FIRST AID TREATMENT
Inhalation	Irritation of upper respiratory tract and lungs; possible chest and abdominal pain	Half-face high-efficiency particulate respirator or full-face positive-pressure air supplied respirator	Move to fresh air and provide oxygen; if not breathing, give artificial respiration
Skin	Short exposure may cause local irritation, redness, and pain; prolonged exposure absorbed through skin; lead posoning	Use impervious protective clothing that includes boot, gloves, lab coat, apron, or coveralls	Flush skin with soapy water for 15 min; remove contaminated clothing and shoes
Eyes	Local irritation, blurred vision, and possibly severe tissue burns	Ventless safety goggles or full-face shield; eye wash station	Flush eyes with water for at least 15 min; move lower and upper eyelids while flushing
Ingestion	Poisoning that includes abdominal pain, spasms, nausea, vomiting, and headache; acute poisoning causes muscle weakness, insomnia, dizziness, loss of appetite, coma, and/or death	—	Induce vomiting immediately

* in mg/m³

Sources: Occupational Safety & Health Administration (OSHA), National Institute of Occupational Safety and Health (NIOSH), American Conference of Governmental Industrial Hygienists (ACGIH), and Mallinckrodt Baker, Inc.

MOLD DENSITY LABORATORY TERMINOLOGY

DENSITY OF GROWING MOLD SPORES	
Level <1	Very few mold spores and mold parts observed. Spores cover less than 10% of surface. Up to 20,000 spores/cm²
Level 1	Few spores and mold parts observed. Spores cover 10% to 25% of surface. Up to 200,000 spores/cm²
Level 2	A number of spores and mold parts observed. Spores cover 25% to 50% of surface. Up to 2 million spores/cm²
Level 3	Many spores and mold parts observed. Spores cover 50% to 75% of surface. Up to 4 million spores/cm²
Level 4	Many spores and mold parts observed. Spores cover more than 75% of surface. Up to and more than 4 million spores/cm²

DENSITY OF NON-GROWING MOLD SPORES	
Very Few	Very few mold spores detected. Densities range up to 10 spores/cm²
Few	Few mold spores detected. Densities range up to 25 spores/cm²
Moderate	A number of mold spores detected. Densities range up to 50 spores/cm²
Many	Many mold spores detected. Densities range up to or more than 50 spores/cm²

FORMALDEHYDE EXPOSURE LIMITS

CLASSIFICATION	10 MINUTES*	30 MINUTES*	1 HOUR*	4 HOURS*	8 HOURS*
Nondisabling	2	1.5	1.25	1	0.5
Disabling	10	5	4	2	1

EXPOSURE	SYMPTOMS	PREVENTION	FIRST AID TREATMENT
Inhalation	Respiratory tract irritant; short exposure causes headache, vomiting, dizziness, and drowsiness; prolonged exposure causes sensory impairment	Wear full-face self-contained breathing apparatus	Remove to fresh air; if not breathing, provide artificial respiration; personnel may give oxygen
Skin	Short exposure may cause burns; prolonged exposure allows harmful amounts to be absorbed into body	Wear plastic or rubber gloves; wear coated protective clothing and footwear	Remove clothing, gloves, and shoes; flush skin with large amounts of warm water
Eyes	Severe irritation; vapors may cause tearing; liquid causes burns and permanent eye injury	For short exposure, wear ventless safety glasses and a full-face shield	Flush eyes with water for at least 15 min; move eyelids from eyeballs to ensure all surfaces are flushed
Ingestion	Causes severe irritation and inflammation of mouth, throat, and stomach; may cause nausea, blindness, and death	—	Do not induce vomiting; if conscious, provide water or milk

* in ppm

Sources: Occupational Safety & Health Administration (OSHA), Office of Environmental Health & Safety (EH&S), and Praxair, Inc.

INDOOR AIR QUALITY UNIT CONVERSIONS

MULTIPLY	BY	TO OBTAIN
Volume		
Cubic feet	1728	Cubic inches
Cubic feet	7.4805	Gallons
Pressure		
Inches of water	0.0361	Pounds per square inch
Inches of water	0.0735	Inches of mercury
Inches of mercury	0.4912	Pounds per square inch
Pounds per square inch	2.036	Inches of mercury
Pounds per square inch	2.307	Feet of water
Pounds per square inch	14.5	
Pounds per square inch	27.67	Inches of water
Pounds per square inch	0.06804	Atmospheres
kpa	100	
Miscellaneous		
Pounds	453.5924	Grams
Pounds	0.4536	Kilograms
HP (steam)	42,418	Btu
Therm	100,000	Btu
Btu	0.252	Calories
Cubic feet per minute	7.481	Gallons per minute

Atmosphere Equivalent		
29.921 in. Hg	=	1 Atmosphere
406.782 in. WC	=	1 Atmosphere
14.696 psi	=	1 Atmosphere

RADON EXPOSURE LIMITS

CLASSIFICATION	10 MINUTES	30 MINUTES	1 HOUR*	4 HOURS	8 HOURS*
Nondisabling	—	—	4	—	0.2
Disabling	—	—	8	—	0.4

* in pCi/L

Sources: Occupational Safety & Health Administration (OSHA), Environmental Protection Agency (EPA), and Agency for Toxic Substances & Disease Registry (ATSDR)

PARTICULATE EXPOSURE LIMITS

CLASSIFICATION	10 MINUTES	30 MINUTES	1 HOUR	4 HOURS	8 HOURS*
Nondisabling	—	—	—	—	0.03
Disabling	—	—	—	—	0.05

EXPOSURE	SYMPTOMS	PREVENTION	FIRST AID TREATMENT
Inhalation			
Skin	Symptoms, prevention measures, and first aid vary depending on the type of matter the particulate originated from		
Eyes			
Ingestion			

* in mg/m^3

Sources: Occupational Safety & Health Administration (OSHA) and the National Institute of Occupational Safety and Health (NIOSH)

VENTILATION AIR REQUIREMENTS*

	OUTDOOR AIR RATE†	AREA OUTDOOR AIR RATE‡	DEFAULT VALUES	
			OCCUPANT DENSITY§	COMBINED OUTDOOR AIR RATE†
Office Building				
Office space	5	0.06	5	17
Reception area	5	0.06	30	7
Telephone/data entry	5	0.06	60	6
Main entry lobbies	5	0.06	10	11
Miscellaneous Spaces				
Computer room (not printing)	5	0.06	4	20
Electrical equipment rooms	—	0.06	—	
Telephone closets	—	0.00	—	
Public Assembly Spaces				
Lobbies	5	0.06	150	5
Libraries	5	0.12	10	17
Retail				
Mall common areas	7.5	0.12	40	9

* Partial ASHRAE Table 6.1
† cfm/person
‡ cfm/sq ft
§ #/1000 sq ft

Sources: American Society of Heating, Refigerating, and Air-Conditioning (ASHRAE)

RECOMMENDED BUILDING TEMPERATURES FOR RELATIVE HUMIDITY

RELATIVE HUMIDITY*	WINTER TEMP†	SUMMER TEMP†
30	68.5–76.0	74.0–80.0
40	68.5–75.5	73.5–79.5
50	68.5–74.5	73.0–79.0
60	68.0–74.0	72.5–78.0

* in percentages
† in °F

FILTER (MERV) RATINGS

	ASHRAE 52.2				ASHRAE 52.1	
MERV	3–10*	1–3*	0.3–1*	Arrestance*	Dust Spot*	Dust Spot‡
1	<20	—	—	<65	<20	
2	<20	—	—	65–70	<20	>10
3	<20	—	—	70–75	<20	
4	<20	—	—	>75	<20	
5	20–35	—	—	80–85	<20	
6	35–50	—	—	>90	<20	3.0–10
7	50–70	—	—	>90	20–25	
8	>70	—	—	>95	25–30	
9	>85	<50	—	>95	40–45	
10	>85	50–65	—	>95	50v55	1.0–3.0
11	>85	65–80	—	>98	60–65	
12	>90	>80	—	>98	70–75	
13	>90	>90	<75	>98	80–90	
14	>90	>90	75–85	>98	90–95	0.3–1.0
15	>90	>90	85–95	>98	>95	
16	>95	>95	>95	>98	>95	
17†	>99	>99	>99	—	>99	
18†	>99	>99	>99	—	>99	0.3–1.0
19†	>99	>99	>99	—	>99	
20	>99	>99	>99	—	>99	

MERV	Application
1–4	Minimum residential, minimum commercial, equipment protection
5–8	Commercial, industrial, superior residential
9–12	Superior commercial, superior industrial, best residential
13–16	Medical facilities, surgical rooms, best commercial, smoke protection
17–20	Clean rooms, hazardous materials

* in microns in %
† future use
‡ in microns

GLOSSARY

A

actuator: A device that accepts signal from a controller, which causes mechanical motion to occur.

acute health effects: Adverse effects on a human or animal in which severe symptoms develop rapidly and often subside after the exposure stops.

adaptive control algorithm: A control algorithm that automatically adjusts its response time based on environmental conditions.

air and water system: Provides both air and water to terminal units serving an occupied space.

air cleaning: A method of reducing air pollution that includes electronic air cleaning, ultraviolet (UV) air cleaning, and the use of filters, which includes carbon filters.

air compressor: A component that takes air from the atmosphere and compresses it to increase its pressure.

air-handler economizer: An HVAC unit that uses outside air for free cooling.

air-handling unit: The part of an HVAC system that operates to distribute air throughout a building.

air-handling unit (AHU) controller: A controller that contains input terminals and output terminals required to operate large central-station air-handling units.

air-line respirator: A type of respirator that uses a small-diameter hose from a compressed air cylinder or air compressor to supply air to the mask.

algorithm: A mathematical equation used by a building automation system controller to determine a desired setpoint.

allergen: A substance that causes an allergic reaction in individuals sensitive to it.

alpha track detector (ATD): A passive monitoring device used for radon testing.

analog capture hood: Causes air to flow over a sensing manifold through a mechanical range selector (user adjustable) and finally through a deflecting-vane anemometer.

analog humidity sensor: An analog input device that measures the amount of moisture in the air and sends a proportional signal to a controller.

analog input device (AI): A sensor that indicates a variable such as temperature, pressure, or humidity and causes a proportional electrical signal change at the building automation system controller.

analog output component: A piece of electrical equipment that receives a continuous signal that varies between two values.

analog pressure sensor: An analog input device that measures the pressure in a duct, pipe, or room and sends a signal to a controller.

anemometer: A device that measures average air velocity or average airflow rate.

application-specific controller (ASC): A controller designed to control only one section or type of HVAC system.

area air-sampling pump: An instrument used to collect air samples from a relatively large area into a special filter container.

area air-sampling pump for mold: An instrument used to collect mold from air samples by vacuuming the mold into a special canister, which is then sent to a laboratory for analysis.

asbestos: A mineral fiber that can pollute air or water and cause cancer or asbestosis when inhaled.

asphyxiate: A substance that causes a lack of oxygen.

asthma: A respiratory disease in which airway obstruction is triggered by various stimuli, such as allergens or a sudden change in air temperature.

auxiliary device: A device used in a pneumatic control system that produces a desired function when actuated by the output signal from a pneumatic controller.

averaging temperature sensor: Used in large duct systems to obtain an accurate temperature reading.

B

bimetallic element: A temperature sensing device that consists of two different metals joined together at one end.

biological contaminants: Living organisms or derivates that can cause harmful health effects when inhaled, swallowed, or otherwise taken into the body.

bleedport: An orifice that allows a small volume of air to be expelled into the atmosphere.

branch line pressure: The pressure in the air line that is piped from the thermostat to the controlled device.

building automation system (BAS): A system that uses microprocessors (computer chips) to control the energy-using components in a building.

building automation system input device: An object such as a sensor that indicates building conditions to a BAS controller.

building automation system output component: A piece of electrical equipment that changes state (ON or OFF) in response to a command from a building automation system controller.

building-related illness (BRI): A diagnosable illness whose cause and symptoms can be directly attributed to a specific pollutant source within a building.

building system commissioning: The systematic process of ensuring and documenting that all building systems perform together properly according to the original design intent and the owner's operational needs.

C

carbon dioxide (CO_2): An odorless, colorless gas that does not support combustion and is normally present in indoor air.

carbon dioxide meter: A meter that measures the amount (in parts per million) of carbon dioxide in an air sample.

carbon filter: An air-cleaning method where an activated carbon-filtering medium is used to remove most odors, gases, smoke, and smog from the air by means of an adsorption process.

carbon monoxide (CO): A colorless, odorless, poisonous gas produced by incomplete fossil fuel combustion.

carbon monoxide meter: A meter that measures the amount (in parts per million) of carbon monoxide in an air sample.

carcinogen: A substance that is capable of causing cancer.

CBR attack: An attack on a building or area that employs airborne chemical, biological, or radiological (CBR) agents.

chemical inventory form: A form used to identify all chemicals found in a building and indicates how the chemicals are used.

chemical smoke test: An IAQ test that shows the airflow inside a space or from one space into another space.

chronic health effects: Health problems in which symptoms occur frequently or develop slowly over a long period of time.

clamp-on ammeter: A test instrument that measures the current in a circuit by measuring the strength of the magnetic field around a single conductor.

cleaning survey: A method by which sources of dirt and debris are identified.

closed-loop control: A type of control system where feedback occurs among the controller, sensor, and controlled component.

commissioning agent (CA): A mechanical contractor who acts as a third-party representative to ensure that the mechanical systems of a building are performing properly.

computer-based MSDS system: A chemical tracking method that uses software to eliminate the difficulty of handling paperwork for MSDS storage, manual updates, and other MSDS maintenance issues.

constructive eviction: A legal issue that occurs when a tenant is unable to use a building space due to special circumstances, such as those related to an unresolved IAQ problem.

controlled device: The object that regulates the flow of fluid or air in a system to provide a heating, air conditioning, or ventilation effect.

control point: The actual value that a building automation system experiences at any given point in time.

control strategy: A BAS software method used to control the energy-using equipment in a building.

cooling coil: A heat exchanger in an HVAC system that allows chilled water to be used for cooling air.

D

damper: A movable piece of metal in a duct used to control the flow of air.

deflecting-vane anemometer: A test instrument that indicates the airflow rate with a pointer or arm that swings on a suspension.

derivative control algorithm: A control algorithm that determines the instantaneous rate of change of a variable.

detector tube: A device that indicates the average contaminant concentrations in a work area or other space for personal exposure monitoring.

differential pressure switch: A digital input device that opens or closes contacts because of the difference between two pressures.

digital (binary) input device: A sensor that produces only an ON or OFF signal.

digital capture hood: A capture hood with a thermoanemometer sensor to provide digital readings of airflow volume.

digital manometer: A pressure and differential pressure-measuring test instrument usually used in portable applications.

digital multimeter (DMM): A test instrument that can measure two or more electrical properties and displays the measured properties as numerical values.

digital output component: A piece of electrical equipment that receives an ON or OFF signal.

direct air measurements: Measurements of air quantities with a test instrument for which no math or conversions must be performed.

direct-coupled actuators: Actuators that are directly attached to the damper or valve without the use of a linkage.

direct digital control (DDC) strategy: A control strategy in which a building automation system performs closed-loop temperature, humidity, or pressure control.

direct digital control (DDC) system: A building automation system in which controllers are wired directly to controlled components to turn them ON or OFF.

direct-return system: A system that is piped so that the terminal unit with the shortest supply pipe also has the shortest return pipe.

disposable dust mask: A type of respiratory protection made of a filter material that is worn over the nose and mouth.

distributed direct digital control (DDC) system: A building automation control system that has multiple central processing units (CPUs) at the controller level.

diverting (bypass) valve: A three-way valve that has one inlet and two outlets.

dry bulb economizer: A type of economizer that operates strictly in connection to the outside-air temperature, with no reference to humidity values.

dual-duct system: A type of air system in which all the air is conditioned in a central fan system and distributed to conditioned spaces through two parallel main ducts.

dual-input (reset) receiver controller: A receiver controller in which the change of one variable, commonly the outside-air temperature, causes the setpoint of the controller to automatically change (reset) to match the changing condition.

duct diameter: The distance across the middle from side-to-side of either a rectangular or round duct.

duct-mounted temperature sensor: Used to sense air temperatures inside ductwork.

dust-holding capacity: A filter's ability to hold dust without seriously reducing the filter's efficiency.

dust spot efficiency: A filter rating that measures a filter's ability to remove large particles from the air that tend to soil building interiors.

DX system: A system in which the refrigerant expands directly inside a coil in the main airstream itself to affect the cooling of the air.

E

electric/pneumatic (EP) switch: A device that enables an electric control system to interface with pneumatic HVAC system components.

electronic air cleaner: A type of air cleaner in which the air is positively charged in an ionizing section and then the air moves to a negatively charged collecting area (i.e., plates) to separate out the dirt.

electronic tachometer: A test instrument that measures the rotational or linear speed of a moving object.

enterprise asset management (EAM): The whole-life, optimal management of the physical assets of an organization to maximize value.

environmental tobacco smoke (ETS): A mixture of smoke from the burning end of a cigarette, pipe, or cigar and smoke exhaled by a smoker.

eye protection: A properly fitting set of goggles or a full-face respirator mask that covers the eyes.

F

fan: A mechanical device that is part of an air-handling unit, which forces conditioned air throughout a building.

feedback: The measurement of the results of a controller action by a sensor or switch.

fiberglass: A material made from small fibers of glass twisted together.

filter: An air-cleaning method in which a mechanical device is used to remove particles from the air.

filter airflow resistance: The pressure drop across a filter at a given velocity.

flow-balancing valve: A special valve that is usually attached to the output side of individual heating and/or cooling coils (terminal units) in the main piping runs in order to control flow.

flow hood: A test instrument that is placed over a diffuser to measure the volume of air moving in cubic feet per minute.

footprint: The amount of building area in contact with the ground.

foot protection: Footwear such as shoes, boots, or covers that protect the foot and seal the bottom of the coverall or bodysuit leg.

force-balance design: A design in which the controller output is determined by the relationship among mechanical pressures.

formaldehyde: A colorless, pungent, and irritating gas used as a preservative in building materials.

formaldehyde passive sampler: An instrument used to collect air samples on a special reactive tape that is sent to a laboratory for testing.

full-face helmet: A type of respiratory protection that covers the entire head and has a filtered air blower to maintain positive pressure in the helmet or hood.

full-face respirator: A type of respirator protection that covers the entire face including the eyes and has filter cartridges that protect against dusts, fumes, and particulates.

full PPE: The level of protection required to perform large-size tasks or to perform a task in a large-size area (over 100 sq ft).

G

gas: Matter that has no definite volume or shape.

global data: Data needed by all controllers in a network.

good IAQ: When the indoor air of a building is free of harmful amounts of chemicals and particles, and the temperature and humidity of the air is comfortable.

H

half-face respirator: A type of respiratory protection that covers the nose and mouth and has filter cartridges that protect against dust, fumes, and particulates.

hand protection: Gloves that protect the skin from contaminants and cleaning solutions.

head protection: A hood or bouffant cap that protects the head from contaminants or mold.

heating coil: Removes moisture from the air as the coil increases air temperature (i.e., sensible heat).

high-limit control: A type of control that ensures that a controlled variable remains below a programmed high-limit value.

high/low signal select: A DDC feature in which a building automation system selects among the highest or lowest values from multiple input signals.

homeland security: A concerted national effort to prevent terrorist attacks within the United States, reduce the United States' vulnerability to terrorism, and minimize the damage and recovery time from attacks that do occur.

housekeeping program: A set of procedures that includes preventing excess dirt from entering an environment, using cleaning products that introduce the least amount of contaminants into the air, and implementing maintenance procedures by trained and monitored personnel.

humidifier fever: A respiratory illness caused by exposure to toxins from microorganisms found in wet or moist areas in humidifiers and air conditioners.

humidistat: A digital input device that indicates that a humidity setpoint has been reached.

HVAC checklist: A document listing any HVAC equipment and maintenance procedures that might have a negative impact on a building's indoor air quality.

HVAC system adjustment: The process of changing the settings of operating devices to obtain the final setpoint of control or air-handling system balancing.

HVAC system balancing: The regulation of the flow of air in a system through the use of acceptable industry testing and balancing procedures.

HVAC system testing: The process of using specific IAQ test instruments to measure variables within an air-handling system in order to evaluate mechanical equipment and system performance.

hydronic manometer: A test instrument that measures water pressures (differential and gauge) to determine the flow rate and to balance a hydronic system.

hygroscopic element: A sensor mechanism that changes its characteristics as the humidity level in the air changes.

hypersensitivity pneumonitis: A disease that irritates a person's lungs after inhaling certain allergens.

I

immersion (well-mounted) temperature sensor: Used to sense water temperature in piping systems.

inclined-tube manometer: A differential pressure-measuring test instrument used mainly in applications where a manometer is permanently attached.

indoor air quality (IAQ): The study of various contaminant levels and air properties of the indoor air of a building.

integrated pest management: A type of pest management where an environmentally sensitive and common-sense approach to pest management is used.

integration control algorithm: A control algorithm that eliminates any offset after a certain length of time.

interface: A device that allows two different types of components or systems to interact with each other.

irritant: A substance that can cause irritation of the skin, eyes, or respiratory system.

K

K factor: A number that represents the actual open area and design characteristics of a specific diffuser.

L

lead: A heavy metal that is a health hazard if breathed or swallowed.

Legionnaires' disease: An infection of the lungs caused by Legionella bacteria.

limited PPE: The level of protection required to perform medium-size tasks or to perform a task in a medium-size area (10 sq ft to 100 sq ft).

limit thermostat: A pneumatic thermostat that causes a

change in a controlled device whenever the set temperature or pressure is reached.

low-limit control: A type of control that ensures that a controlled variable remains above a programmed low-limit value.

M

manometer: An air-handling system test instrument that measures low air pressures, usually in inches of water column (in. WC).

mass psychogenic illness: A condition that occurs when groups of people start feeling sick at the same time, even though there is no physical or environmental reason.

material safety data sheet (MSDS): A form containing data regarding the properties of a particular substance.

method detection limit: The minimum concentration of a contaminant that is reported as zero.

microbial contaminants: Contaminants invisible to the naked eye that are either derived from living organisms or that are living organisms.

minimum-position relay: A relay that prevents outside-air dampers from completely closing.

minimum PPE: The level of protection required to perform small-size IAQ tasks or to perform a task in a small area (under 10 sq ft).

mixed-air system: A unit or section of an HVAC system that brings a specific amount of return air into contact with the outside air entering the HVAC system for space use.

mixing valve: A three-way valve that has two inlets and one outlet.

mixture: Any combination of two or more chemical substances.

mold: A type of fungus that can be found both indoors and outdoors.

multiple chemical sensitivity (MCS): A negative physical reaction to low levels of common chemicals.

multizone system: A type of air system that serves a relatively small number of zones from a single, central air-handling unit.

N

National Environmental Balancing Bureau (NEBB): An organization that sets procedural standards for testing, adjusting, and balancing environmental systems.

negligence: A legal issue that occurs when there is an injury because of a lack of performance by building personnel.

network communication module (NCM): A special controller that coordinates communication from controller to controller on a network and provides a location for operator interface.

neurotoxin: A contaminant that damages the central nervous system.

nitrogen dioxide (NO_2): A toxic gas that is the result of nitric oxide combining with oxygen in the atmosphere.

nitrogen dioxide monitor: A meter that measures the amount (in parts per million) of nitrogen dioxide in an air sample.

nonducted return-air systems: Systems that commonly use hallways or spaces above dropped ceilings as return-air paths or plenums.

nonpoint source: A type of contamination that results from multiple sources or a large-area source.

normally closed (NC) valve: A valve that does not allow fluid to flow when the valve is in its normal position.

normally open (NO) valve: A valve that allows fluid to flow when the valve is in its normal position.

O

off-gassing: The release, over time, of a chemical compound from a material to the air.

offset: The difference between a control point and setpoint.

100% outside-air (OA) system: A system that does not recirculate any return air from building spaces.

one-pipe hydronic system: A type of hydronic piping system that uses a single pipe as both a supply pipe and return pipe.

one-pipe (low-volume) device: A device that uses a small amount of the compressed air supply (restricted main air).

ongoing commissioning: Building commissioning that is centered on analyzing and optimizing heating, ventilation, and air conditioning (HVAC) system operation and control for existing building conditions.

open-loop control: A type of control system where no feedback occurs among the controller, sensor, and controlled components.

operator interface: A device that allows a technician to access and respond to building automation system information.

opposed blade damper: A damper in which adjacent blades move in opposite directions from one another.

ozone: A gas that is a variety of oxygen.

ozone meter: A meter that measures the amount (in parts per million) of ozone in an air sample.

P

paper-based MSDS system: A chemical tracking method that keeps paper MSDS files in three-ring binders at an accessible location.

parallel blade damper: A damper in which adjacent blades are parallel and move in the same direction with one another.

particle: Any very small part of matter, such as a molecule, atom, or electron.

particulate detector and counter: A test instrument used to collect and analyze air samples for the presence and number of particles in the air.

passive dosimeter: A container in which air passes through a chemically treated badge or solution.

pathogen: A disease-causing biological agent.

permanence mitigation: A type of mitigation effort that creates a permanent solution to an indoor air problem compared to efforts that provide temporary solutions (unless the problems are also temporary).

personal air-sampling pump: An instrument carried by a person and used to collect air samples in a special cassette for testing of the air sample by a laboratory.

personal air-sampling pump for asbestos: An instrument used to collect air samples in a special cassette, which is then sent to a laboratory for testing.

personal protective equipment (PPE): Clothing and equipment used to protect personnel from sickness, injury, or death by creating a barrier against IAQ hazards.

pesticides: Substances intended to repel, kill, or control any species that is designated a pest, including weeds, insects, rodents, fungi, bacteria, and other organisms.

picocurie (pCi): A unit for measuring radioactivity and equals the decay of approximately two radioactive atoms per minute.

pitot tube: A pressure-measuring instrument used to measure velocity pressures.

PM work order: A form or ticket that indicates specific preventive maintenance (PM) tasks to perform at regularly scheduled intervals.

pneumatic control system: A control system in which compressed air is used to provide power for the system.

pneumatic/electric (PE) switch: A device that allows an air pressure signal to energize or de-energize an electrical device such as a fan, pump, compressor, or electric heating element.

pneumatic humidistat: A controller that uses compressed air to open or close a device that maintains a certain humidity level inside a duct or area.

pneumatic positioner (pilot positioner): An auxiliary device mounted to a damper or valve actuator that ensures that the damper or actuator moves to a given position.

pneumatic pressurestat: A controller that maintains a constant air pressure in a duct or area.

pneumatic thermostat: A thermostat that uses changes in compressed air pressure to control the temperature in individual rooms inside a commercial building.

point source: A type of contamination that results from a single or small source.

poor IAQ: When the indoor air of a building is filled with a level of chemicals and/or particles that is harmful to occupants, and/or the temperature and the humidity of the air is uncomfortable.

predictive maintenance (PdM): Maintenance that involves evaluating the condition of equipment by performing periodic or continuous equipment monitoring.

pressure regulator: A valve that maintains downstream pressure by restricting or blocking airflow.

preventive maintenance (PM): The periodic inspection and servicing of a building and equipment for the purpose of preventing failures.

proactive maintenance: A maintenance strategy that stabilizes the reliability of equipment using specialized maintenance services.

proportional control algorithm: A control algorithm that positions the controlled component in direct response to the amount of offset in a building automation system.

proportional/integration/derivative (PID) controller: A direct digital control (DDC) system controller that uses proportional, integration, and derivative algorithms.

proportional/integration (PI) control: The combination of proportional and integration control algorithms.

R

radon: A colorless, naturally occurring, radioactive, inert gas formed by the radioactive decay of radium atoms in soil or rocks.

radon detector: A test instrument used to collect and analyze air samples and indicates the amount of radon present on a numeric LED display.

rapidly replenished materials: Materials that are readily available for purchase and whose extraction from the environment has a minimal impact on the environment.

receiver controller: A device that accepts one or more input signals from pneumatic transmitters and produces an output signal based on the setup of the controller.

reheat system: A type of air system that permits zone or space control for areas of different exposure, provides heating or cooling of perimeter areas of unequal exposure, or provides process or comfort control where close temperature control is required.

reliability-centered maintenance (RCM): A maintenance plan used to create a cost-effective maintenance strategy to address the dominant causes of equipment failure.

reset control: A direct digital control (DDC) feature in which a primary setpoint is reset automatically as another value (reset variable) changes.

respirable particles: Airborne particles that can be inhaled into the lungs when breathing.

respirator: A piece of equipment that protects IAQ personnel from inhaling dust, airborne mold, mold spores, and gases.

retro-commissioning: Building commissioning that focuses on analysis of old systems, old buildings, and systems that have been upgraded with new technology.

reverse-return system: A system consisting of pipes in which coil with the shortest supply pipe also has the longest return pipe.

rotating-vane anemometer: A test instrument that measures airflow rate at a specific location and is similar to a small propeller or windmill in design.

round blade damper: A damper with a round blade.

run-to-fail maintenance: Corrective, reactive, and unscheduled maintenance.

S

safety valve: A device designed to automatically open and release pressure from the receiver if and when pressure rises to an unsafe level.

self-contained breathing apparatus (SCBA): A type of respiratory protection that uses air supplied by a compressed air tank on the back of the wearer.

server computer: A large-capacity, hard-drive computer attached to a network.

setback: The unoccupied heating setpoint.

setpoint: 1. The setting of the desired temperature in the room or space. **2.** The desired value to be maintained by a system.

setpoint schedule: A description of the amount of change that occurs when a reset variable resets the primary setpoint.

setup: The unoccupied cooling setpoint.

shelter-in-place: A small, interior room of a building with few or no windows where individuals can take refuge.

sick building syndrome (SBS): A situation in which building occupants experience acute health and comfort effects that appear to be linked to time spent in a building, though no specific illness or cause can be identified.

single-input receiver controller: A receiver controller that is designed to be connected to only one transmitter and to maintain only one temperature, pressure, or humidity setpoint.

single-zone system: A type of air system that serves a single temperature control zone and is the simplest form of air system.

sink effect: The emission of volatile organic compounds (VOCs) from furniture and other material that have absorbed the VOCs from indoor air.

source control: A method of reducing air pollution by identifying strategies to reduce the origin of pollutants.

special clothing: A coverall or bodysuit that is treated to be impervious to specific fibers, gases, and/or chemicals.

stack effect: The flow of air that results from warm air rising, creating a positive pressure area at the top of a building and a negative pressure area at the bottom of a building.

stand-alone control: The ability of a controller to function on its own.

static pressure: The pressure that acts against the sides of a duct and is the same throughout a cross section of the duct.

step-by-step IAQ troubleshooting process: A list of procedures that identifies steps to follow in chronological order until an IAQ problem is found.

switching relay: A device that switches airflow from one circuit to another.

T

thermoanemometer: A permanently installed air-handling system test instrument used to measure air velocity through a duct.

thermostat: A digital input device and controller.

three-pipe hydronic system: A type of hydronic piping system where hot or cold water can be introduced to a terminal unit (e.g., coil).

three-way valve: A valve that has three pipe connections.

total pressure: Static pressure and velocity pressure added together.

traceability: The unbroken chain of paperwork relating a test instrument's accuracy to a known standard.

tracer gas: A gas that is identifiable in an airstream and can be measured with a degree of accuracy.

transducer: An electrical component that changes one type of proportional control signal into another type of signal.

traverse: The process of taking multiple measurements across the area of a duct and averaging the values.

triac: A solid-state switching device used to switch alternating current (AC).

two-pipe (high-volume) device: A device that uses the full volume of compressed air available.

two-pipe hydronic system: A type of hydronic piping system that has a separate supply pipe and return pipe.

U

ultraviolet (UV) air-cleaner system: An air-cleaning system that kills biological contaminants using a specific light wavelength.

unitary controller: A type of controller designed for basic zone control using a standard wall-mounted temperature sensor.

V

valve: A device that controls the flow of fluids in an HVAC system.

variable-air-volume (VAV) air-handling unit controller: A controller that modulates the damper inside a VAV air-handling unit or rpm of the blower motor(s) with signals to electric motor drives to maintain a specific building space temperature.

variable-air-volume (VAV) system: An air-handling unit that provides air at a constant temperature but varies the amount delivered to various loaded zones.

velocity pressure: The pressure caused by airflow pushing against an object in the duct.

velometer: A device that measures the velocity of air flowing out of a register.

ventilation: A method of reducing air pollution by introducing various amounts of outdoor air in order to dilute indoor pollutants.

visual inspection checklist: A checklist that is used to record any repairs needed or possible problems that are identified during a visual inspection.

volatile organic compound (VOC): A carbon-containing compound that easily evaporates at room temperatures.

W

wall-mounted temperature sensor: The most common temperature sensor used in building automation systems and used to sense air temperatures in building spaces.

wood smoke: A complex mixture of gases and fine particles produced when wood and other organic matter burn.

workers' compensation: A legal issue referring to payments that are offered to employees who are temporarily unable to work because of a job-related injury.

INDEX

Using the *Indoor Air Quality Solutions for Stationary Engineers* CD-ROM

Before removing the CD-ROM from the protective sleeve, please note that the book cannot be returned for refund or credit if the CD-ROM sleeve seal is broken.

System Requirements

The *Indoor Air Quality Solutions for Stationary Engineers* CD-ROM is designed to work best on a computer meeting the following hardware/software requirements:

- Intel® Pentium® (or equivalent) processor
- Microsoft® Windows® XP, Windows® 2000, Windows® NT, Windows® Me, Windows 98® SE, or Windows® 95 operating system
- 128 MB of available system RAM (256 MB or more recommended)
- 90 MB of available disk space
- 1024 × 768 16-bit (thousands of colors) color display or better
- Sound output capability and speakers
- CD-ROM drive

Opening Files

Insert the CD-ROM into the computer CD-ROM drive. Within a few seconds, the home screen will be displayed allowing access to all features of the CD-ROM. Information about the usage of the CD-ROM can be accessed by clicking on USING THIS CD-ROM. The Quick Quizzes®, Illustrated Glossary, Flash Cards, Test Tool Procedures, Survey Checklists, Virtual Meters, Equipment Forms, Print Library, Media Clips, and www.ATPeResources.com can be accessed by clicking on the appropriate button on the home screen. Clicking on the American Tech web site button (www.go2atp.com) accesses information on related educational products. Unauthorized reproduction of the material on this CD-ROM is strictly prohibited.

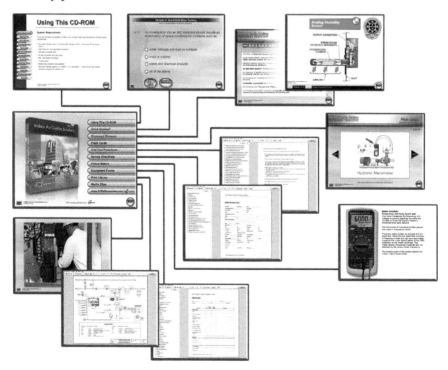